HIGH ENERGY COLLISIONS
OF ELEMENTARY
PARTICLES

HIGH ENERGY COLLISIONS OF ELEMENTARY PARTICLES

BY

R.J.EDEN

Cavendish Laboratory, Cambridge

CAMBRIDGE

AT THE UNIVERSITY PRESS

1967

Published by the Syndics of the Cambridge University Press
Bentley House, 200 Euston Road, London, N.W.1
American Branch: 32 East 57th Street, New York, N.Y. 10022

Library of Congress Catalogue Card Number: 67-24938

Printed in Great Britain
at the University Printing House, Cambridge
(Brooke Crutchley, University Printer)

CONTENTS

Chapter 5. Complex angular momentum and Regge theory

Chapter 6. Asymptotic bounds on the behaviour of cross-sections

Chapter 7. Crossing symmetry and analyticity in the energy

PREFACE

Elementary particles are complicated objects, whose complexity is revealed by their numbers and by the variety of their interactions. This book is about the high energy collisions of hadrons, which are the elementary particles that interact strongly. They include both particles and resonances, between which there does not appear to be any fundamental distinction, although superficial distinctions can be made in terms of their lifetimes. The resonances may decay in times as short as 10^{-22} or 10^{-23} seconds, which are comparable with the duration of the strong interactions in a high energy collision.

The great variety and number of recently discovered particles or resonances has opened up an important new area of research in the study of their strong interactions through high energy collisions at energies ranging from a few Gev to a few tens of Gev. Experiments during the past three years have already indicated the wealth of information that can be obtained, and related theoretical studies have suggested major lines of development that will influence both theory and experiment over the next decade.

This book has been developed from lectures that I have given during 1965–67 to graduate students in the physics department of the University of Maryland, in the Scottish Universities Summer School, and in the Cavendish Laboratory, Cambridge. It is partly a textbook and partly a survey of research on high energy collisions, designed to give both an introduction to the subject and an indication of some of the problems and experiments on which progress can be expected in the next few years. Although it is intended primarily for graduate students and for more senior physicists not familiar with recent developments in the field, parts of the book could also be read by a good final year undergraduate who has attended a course on quantum theory (including scattering theory) and has some knowledge of the physics of elementary particles.

The first chapter is an introductory survey that states in general terms the types of experiment that are made on high energy collisions, and the types of theoretical approach to their interpretation. I have also included in this chapter lists of strongly interacting elementary particles whose quantum numbers are reasonably well established at the present time. The next chapter gives a survey of the main categories of experiment and illustrates them with results from selected experiments. In Chapter 3 a number of standard techniques are described, partly to introduce notation and formulae, and partly in order to make this account of high energy collisions sufficiently self-contained for use by someone new to the field.

Analytic properties of collision amplitudes, regarded as functions of energy and momentum transfer, play a major role in most theoretical studies of strongly interacting elementary particles and provide the guidelines for many experiments. These analytic properties are introduced in Chapter 4 and a number of important results, such as dispersion relations, are described. In Chapter 5, it is shown how analytic properties, based on complex angular momentum, can lead to the Regge theory of high energy collisions. The elementary parts of this theory are introduced in some detail, but the more complicated aspects of the relativistic theory are given only in outline, with illustrative examples and references to original papers for further details. Although some of the special assumptions of Regge theory may not survive future critical experimental and theoretical tests, it is certain that the basic methods of complex angular momentum will remain amongst the most useful techniques for studying high energy collisions. The comparison with experiment of approximations to Regge theory is described in Chapter 9, after three chapters which are concerned with rigorous theoretical results.

In Chapter 6 the main rigorous results on high energy bounds on cross-sections are described. These follow from the axioms of quantum field theory, and they can be obtained also from S-matrix theory with certain assumptions about analytic properties. The next two chapters contain an account of relations, between various experimentally observed quantities, that can be established by theoretical arguments. These include relations at high energy between bounds on the behaviour of a cross-section and bounds on the ratio of the real to imaginary parts of the corresponding scattering amplitude. They include also the Pomeranchuk theorem on the asymptotic equality of particle-target and antiparticle-target cross-sections at high energy.

Chapters 9 and 10 are concerned with phenomenology and the experimental comparison of theoretical models. They include accounts of approximations to Regge theory, and of other models such as the peripheral (absorption) model and the quark model. The successes and defects of these models are illustrated by comparison with experimental results and an indication is given of further critical tests that can be made.

This book is intended partly as a survey of the subject and partly as a guide to the main areas of research on high energy collisions, so I have given a fairly extensive list of references, sometimes to specialised textbooks and review articles, but more frequently to original papers. However, in such an active subject of research, it is difficult to give all the useful references, and a comprehensive list would confuse rather than help the reader new to the field. I hope that the three hundred references will provide him with a foothold in the literature, and I apologise to those authors whose work is omitted from the references or inadequately represented.

I am indebted to many physicists who have increased my knowledge and understanding of the subject, both by discussions and by sending details of their research, often before publication. These include especially Professors G. F. Chew, M. Gell-Mann, N. N. Khuri, and C. N. Yang, and Drs S. Lindenbaum, A. Martin, D. R. O. Morrison, R. J. N. Phillips, L. Van Hove, and A. M. Wetherell, each of whom has influenced my outlook on high energy collisions (sometimes in opposing directions). It is a pleasure to thank my colleagues in the University of Cambridge, and research students in the high energy physics group of the Cavendish Laboratory, for many helpful suggestions and for their assistance in removing errors and obscurities from the manuscript. Valuable help was received from Drs P. V. Landshoff, E. Leader, L. Łukaszuk, K. F. Riley, and J. G. Rushbrooke, and from N. H. Buttimore, S. R. Cosslett, D. R. Dance, N. W. Dean, G. C. Fox, G. D. Kaiser, T. W. Rogers and R. W. Tucker. I am grateful to Mrs H. Maczkiewicz for help with the drawings, Miss Carol McCall and Miss Christine Eden for typing the manuscript, and my wife and family for their help and assistance throughout its writing and preparation. I would also like to express my thanks to the staff of the Cambridge University Press for their help and care in the preparation and printing of this book.

R. J. E.

20 May 1967

CHAPTER 1

INTRODUCTION

1.1 Outline and discussion

Experimental work on high energy collisions of elementary particles in the multi Gev range has developed rapidly in the past few years and it is evident that future experiments will greatly increase the range and variety of processes measured. It is necessary to make use of increasingly sophisticated theoretical methods of studying collision processes both for the planning of high energy experiments and for their detailed analysis. The aim of this book is to describe the main features of theoretical methods used in the study of high energy collisions of elementary particles that have strong interactions.

These theoretical methods are related directly or indirectly to the following main types of experiment:

(1) Total cross-sections $\sigma_t(E)$ for two-body collisions measured as functions of the laboratory energy E at energies from 5 Gev upwards.

(2) The phase of the forward scattering amplitude. The imaginary part of this amplitude is determined by the optical theorem from the total cross-section; its real part and hence its phase is found by Coulomb interference.

(3) Total elastic cross-sections.

(4) The forward peak in differential cross-sections.

(5) Structure of the differential cross-section as a function of momentum transfer. This includes structure related to both the near-forward and near-backward scattering.

(6) Large angle scattering, say $60° < \theta < 120°$ in the centre of mass system.

(7) Two-body reactions $A + B \rightarrow C + D$ including exchange scattering.

(8) Quasi two-body reactions $A + B \rightarrow C + D$ where one or both of C, D, denotes a particle that decays by strong interactions. For each of these, the experiments listed above for elastic scattering are also of great importance.

(9) Polarisation measured as a function of both the angle of scattering and the energy.

(10) Spin correlation measurements, both for two-body reactions and for quasi two-body reactions.

(11) Multiparticle production processes that are not simply related to resonances.

In addition to these experiments involving hadrons (strongly interacting particles), the experiments on electron scattering by protons and neutrons are important in the study of strong interactions through their measurement of electromagnetic form factors.

In most of the above experiments the measurements depend on energy as well as on one or more angles (or momentum transfers), and it is evident that their value will be enhanced by a suitable choice of variables in situations where a multi-dimensional plot is not feasible. This choice should depend at least as much on theoretical motivation as on practical convenience.

Theoretical methods for studying high energy collisions can be divided into the following categories:

(1) Standard techniques such as those of relativistic kinematics or of helicity state analysis.

(2) Methods that make use of rigorous consequences of quantum field theory. For example the Froissart bound on total cross-sections can be derived by these methods, and one can also establish relations between the rates of change at high energy of certain total cross-sections and the phase of a forward scattering amplitude.

(3) Methods that have been developed from assumptions not yet fully proved from fundamental axioms but which involve techniques whose importance will surely remain even if there is some change in their motivation. These include dispersion theory and the Mandelstam representation, and Regge theory using complex angular momentum.

(4) Theoretical models which can be expected to apply to a limited range of experimental results and to evolve as one begins to understand more clearly the essential approximations involved. These include the peripheral model and simple approximations to Regge theory.

(5) The use of symmetries. These include crossing symmetry, the use of spin and isospin groups and of SU 3.

(6) Special approximations, for example those based on perturbation theory, perhaps with the addition of final state interactions.

A book that attempted to cover all these categories in detail would be of such size as to discourage the most determined reader (not to mention the author). I have therefore selected theoretical methods with two criteria in mind. First that they should be sufficiently representative that the book provides a survey of the whole field of collisions of strongly interacting elementary particles. Secondly I have concentrated more on general methods and their objectives, rather than on special techniques or the detailed methods of calculation from particular models. As techniques are both important and unavoidable, they occur in a number of examples to illustrate the general theory and references for further reading are given with each example.

A general survey and classification of experiments is given in Chapter 2. Experimental results are used to illustrate the main characteristics of each class of experiment, and in each class the relevant theoretical topics are noted. Later, in the context of these theoretical methods, there is further discussion about what form of experimental comparison would be valuable. This discussion is not exhaustive in any instance but it is hoped that in conjunction with the theory described and references given, it may help towards more complete and comprehensive experiments than have so far appeared. It is becoming increasingly clear, for example, that future theoretical studies will require single-run experiments covering multidimensional plots, like differential cross-sections for resonance production, for a wide range of energies and momentum transfers.

In Chapter 3 a number of standard techniques are described, partly to introduce notation and partly in order to make this account of high energy collisions sufficiently self-contained for use by someone new to the field. These techniques are described briefly and include relativistic kinematics, phase space analysis, the optical theorem, partial wave and helicity analysis. In Chapter 4 an account is given of dispersion relations for scattering amplitudes. These are not derived from fundamental axioms but their relation to analytic properties of the amplitudes is discussed. Particular dispersion relations are given with energy as the independent variable, and the role of momentum transfer in determining subtractions is described. This chapter includes a description of the Mandelstam representation which is a double dispersion relation with two independent variables (invariant energy and momentum transfer). Although it has not been proved from fundamental axioms, it provides a useful illustration of analytic properties and crossing symmetry relations, many of which would remain correct even if some aspect of the Mandelstam representation (such as the subtraction assumption) turns out to be incorrect. The same comment applies to the Regge theory using complex angular momentum which is described in Chapter 5. Although some of the special assumptions may not survive critical analysis, it is certain that the methods of complex angular momentum will remain amongst the useful techniques for parametrising and studying high energy collisions.

In Chapter 6 the main rigorous results on high energy bounds are described. These results follow from the analytic properties of scattering amplitudes that have been derived from the axioms of quantum field theory. Crossing symmetry and analyticity in the energy for fixed momentum transfer have also been established for scattering amplitudes from the basic axioms. It is shown in Chapter 7 how they lead to relations at high energy between bounds on the rate of growth of an amplitude and bounds on its phase. One consequence of this symmetry and analyticity is the Pomeranchuk theorem on the asymptotic equality of total cross-sections for particle-particle and antiparticle-particle scattering. Conditions for this are described in Chapter 8 and some consequences of other symmetries are discussed.

Regge theory is an example of a very general model that satisfies explicitly many of the unitary and analytic properties required of a scattering amplitude. Symmetries can also be built into the theory. It may be that the implicit requirements of unitarity and analyticity lead to complications of Regge theory such as inconvenient branch cuts in the complex angular momentum plane. However, even if this does transpire, it seems very likely that simple approximations to the full theory will be useful for many years in the analysis of high energy experiments. Some of these approximations are presented in Chapter 9 to illustrate both their relation to the general theory and their experimental consequences. These are principally related to collisions near the forward direction or near the backward direction.

Sum rules are described within the context of Regge theory although in principle they have a more general basis. A number of more special models are described in Chapter 10 including the peripheral model and the quark model.

Electron scattering, to second order in the fine structure constant, involves strong interactions of elementary particles through a single vertex. Electromagnetic interactions have been described extensively in the literature (see Hofstadter (1963), Drell (1966) and Wilson (1967)), and they will be mentioned only for illustration or comparison. Similarly unitary symmetry SU 3 has been fully described elsewhere (Gell-Mann and Ne'eman, 1964; Carruthers, 1966; Lipkin, 1965) and no detailed account will be given of its consequences at high energy.

However it is shown in Chapter 8 how internal symmetry groups of elementary particles combined with analytic properties of collision amplitudes lead to relations between cross-sections at high energy. These symmetry groups are illustrated by the isospin group and the extension of the results to $SU3$ is indicated by some examples.

1.2   Particles and resonances

The following lists of particles and resonances are presented to illustrate the general scope of experiments on high energy collisions. These experiments include for example, quasi two-body reactions of the type

$$A + B \to C + D, \tag{1.2.1}$$

where B is a stable particle (neutron or proton), A is stable or decays only under weak interactions; 'particles' C and D may be any pair of particles or resonances. The study of high energy collisions includes the energy dependence, angular distribution and polarisation in all

Table 1.2.1. *Mesons (masses are given in Mev).*

$Q\overline{Q}$ model	J^P	C	Nonet			
			$Y = 0, T = 1$	$Y = 0, T = 0$		$Y = \pm 1,$ $T = \frac{1}{2}$
$^1S_0$	O^-	$+$	π 140	η 549	η' 958 $(E\ 1424)$	K 494
$^3S_1$	1^-	$-$	ρ 760	ω 783	ϕ 1019	K^* 892
$^1P_1$	1^+	$-$	$(B\ 1208)$	—	—	—
$^3P_0$	O^+	$+$	1003 (?)	—	1068 (?)	—
$^3P_1$	1^+	$+$	A_1 1079 (?)	D 1285	$(E\ 1424)$	K^* 1320
$^3P_2$	2^+	$+$	$(A_2\ 1306)$	f 1254	f' 1514	K^* 1411
$^1D_2$	2^-	$+$	—	—	—	—
$^3D_1$	1^-	$-$	—	—	—	—
$^3D_2$	2^-	$-$	$(B\ 1208)$	—	—	—
Others	?	?	δ 963	—	—	—
	V	$-$	ρ 1637	—	—	—
	A	$+$	π 1640	—	—	—
	A	$-$	—	—	—	K^* 1789
	—	—	S 1929	—	—	—
	—	—	T 2195	—	—	—
	—	—	U 2382	—	—	—

(...) denotes spin assignment not yet confirmed.
(?) denotes existence not confirmed.

The following have been omitted from Table 1.2.1: σ (410), S (720), K (720), H (975), $K_1 K_1$ (1440), $\rho\rho$ (1410), K^* (1090), K^* (1215), $K^*_{\frac{3}{2}}$ (1175), K^* (1270), and $R_1 R_2 R_3$ (*circa* 1700). For some of these the assignment to a place in the table is uncertain; for some their existence is not yet confirmed.

such reactions. It includes also decay correlations and multiple production involving both particles and resonances. The resonances that were observed up to 1966 have been reviewed by Goldhaber (1966, meson resonances), Murphy (1966, $S = O$ baryon resonances) and Ferro-Luzzi (1966, $S \neq O$ baryon resonances). I am indebted to Dr K. F. Riley and Mr S. R. Cosslett for consolidating the experimental details in the tables. These are drawn on the basis of states that could occur in the quark model, so it may transpire that some of the gaps in the tables will not be filled; it is also possible that new resonances may be observed that have the same quantum numbers but different masses compared with those shown in the tables. Further details are given in the references above, and in the annual reviews of particles and resonances by Rosenfeld (1966, et seq.). Some of the resonances listed in Table (1.2.2) can have spins and parities assigned on the basis of the Regge model (see Chapter 9), but unless this has been confirmed experimentally, they are listed amongst 'others' at the foot of the table.

Table 1.2.2. *Baryons* (*masses are given in Mev*).

J^P	Singlet Y_0 $I=0,$ $Y=0$	Octet N $I=\frac{1}{2},$ $Y=1$	Octet $I=0,$ $Y=0$	Octet $I=1,$ $Y=0$	Octet $I=\frac{1}{2},$ $Y=-1$	Decuplet Y_1 $I=\frac{3}{2},$ $Y=1$	Decuplet $I=1,$ $Y=0$	Decuplet $I=\frac{1}{2},$ $Y=-1$	Decuplet $I=0,$ $Y=-2$
$\frac{1}{2}^-$	1405	—	—	—	—	—	—	—	—
$\frac{1}{2}^+$	—	938	1115	1193	1318	—	—	—	—
		(1450)	—	—	—	—	—	—	—
$\frac{3}{2}^+$	—	—	—	—	—	1238	1385	(1530)	(1675)
$\frac{3}{2}^-$	1520	1518	(1698)	1660	(1816)	—	—	—	—
$\frac{5}{2}^-$	—	—	—	[1775]	—	—	[1775]	—	—
$\frac{5}{2}^+$	—	1688	1815	(1900)	—	—	—	—	—
$\frac{7}{2}^+$	—	—	—	[2030]	—	1920	[2030]	—	—
$\frac{7}{2}^-$	—	2190	2100	—	—	—	—	—	—
$\frac{9}{2}^-$	—	—	—	—	—	—	—	—	—
$\frac{9}{2}^+$	—	—	—	—	—	—	—	—	—
$\frac{11}{2}^+$	—	—	—	—	—	(2420)	—	—	—
Others	—	1700(?)	2340	—	[1933]	2360	—	[1933]	—
	—	1680(?)	—	—	—	2850	—	—	—
	—	2650	—	—	—	3230	—	—	—
	—	3030	—	—	—	—	—	—	—
	—	3350	—	—	—	—	—	—	—

(...) denotes resonance parameters not yet confirmed.
 (?) denotes existence not confirmed.
[...] denotes SU3 multiplet uncertain (both possibilities are shown).

CHAPTER 2

EXPERIMENTAL SURVEY

In this chapter we will survey the main types of experiments on the collisions of strongly interacting particles at high energy. A few of these experiments, particularly those on total cross-sections, have been carried out with great precision, but with most of them the experimental uncertainties are larger than one would like for use in a detailed theoretical study. However, many important characteristics are revealed by existing experimental results and these form the subject of the survey in this chapter. It is hoped that their description here, and the related theory in later chapters, will help to show which characteristics should receive the most detailed experimental and theoretical study. Some of the most important experiments, for example those involving polarisation and spin parameters, are of considerable technical difficulty, but now that their significance is recognised it is certain that many new experiments over the next few years will achieve results that would have been declared impossible only two or three years ago.

The diagrams in this chapter showing experimental results are intended to illustrate their main features, especially those of significance for the theoretical study of the strong interactions of elementary particles. For detailed experimental results, references are given to original papers but the reader should also refer to the current proceedings of the international conferences on high energy physics. He can expect to find results of considerable importance to our understanding of high energy collisions in the 1968 and 1970 Conference Reports. However there are other aspects of asymptotic behaviour of collision cross-sections that will have to await the development of the 200 or the 300 Gev accelerator.

Many important experiments are only briefly summarised in this chapter. Further details and references to the original papers may be found in the review articles by Czyzewski (1966), Lindenbaum (1965), Morrison (1966), Wetherell (1966) and Van Hove (1966). The last two appear in the Berkeley Conference Proceedings where there are also a number of reviews of experiments at lower energies, in the resonance range, that were mentioned in § 1.2.

Kinematic notation and units

In this chapter we make some use of the invariants s, t and u that are associated with a two-body collision. The reader who is not acquainted with relativistic kinematics will find these variables defined in more detail in § 3.1. For this chapter it is sufficient to know that s is the centre of mass energy squared, t is minus the momentum transfer squared and u is minus the exchange momentum transfer squared. In subsequent chapters, particularly those concerned with Regge theory, we will sometimes use t for the energy variable and s for the momentum transfer variable. These variables are relativistic invariants and have the dimensions of energy squared. They will be given in terms of $(\text{Gev})^2$ or of $(\text{Mev})^2$. We will take units so that the velocity of light is equal to one, and so that $h/2\pi = 1$. This means that we will have the same value for the invariant t in units of $(\text{Gev})^2$ as in units of $(\text{Gev}/\text{c})^2$.

2.1 Total cross-sections and forward scattering amplitudes

(*a*) *Total cross-sections*

The high energy behaviour of total cross-sections is indicated in Fig. 2.1.1. On the scale of this diagram the errors are small enough that the main features should not change as experiments improve. At a more detailed level, the most accurate measurements of total cross-sections for the range 8 to 29 Gev/c are those of Lindenbaum and collaborators (see Foley *et al.* 1967 and Lindenbaum, 1967).

The following important features of the total cross-sections shown in Fig. 2.1.1 should be noticed:

  (i) Each total cross-section decreases, apparently towards a constant value in each case.

 (ii) The total cross-sections for particle-target and for antiparticle-target collisions are tending towards equality at high energy.

(iii) There is an indication that, at asymptotic values of the energy above the range of existing experiments, the baryon-baryon total cross-sections may all be about 36 mb and the meson-baryon total cross-sections may all be in the range 18–24 mb.

These observations point to many open questions. The question whether cross-sections tend to a constant limit cannot be decisively answered by present or future experiments on existing accelerators. However, it would be very interesting to know whether the asymptotic decrease is really as smooth as it looks in Fig. 2.1.1. A variation from the smooth decrease in the slope would indicate a significant partial cancellation between contributing terms in Regge theory and probably would be equally significant in any other theory.

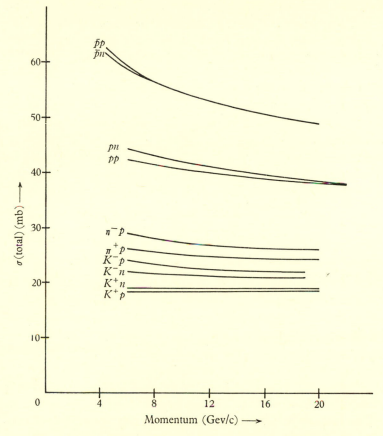

Fig. 2.1.1. Total cross-sections at high energy (from Lindenbaum, 1965).

A related question is whether the difference between two cross-sections, (Fig. 2.1.2) for example,

$$\Delta\sigma_1 = \sigma(\text{total}, \pi^-p) - \sigma(\text{total}, \pi^+p) \qquad (2.1.1)$$

decreases smoothly to zero, or whether it changes sign or oscillates at higher energies. If it does not change sign, does its derivative change sign and does this difference $\Delta\sigma_1$ show a peak at any high energy value? The processes $(\pi^\pm, p)$ are taken only for illustration, these questions are of equal importance for the difference of any particle-target and anti-particle-target cross-sections.

It was first suggested by Pomeranchuk (1956) that total cross-sections should tend asymptotically to constant values, and in Regge theory this is attributed to the exchange of a 'non-particle', called the Pomeranchuk particle or Pomeron. Pomeranchuk also suggested that charge-exchange cross-sections should tend to zero and that particle-target and anti-particle-target cross-sections become asymptotically equal (Pomeranchuk, 1958). We will see in Chapter 8 that there is a

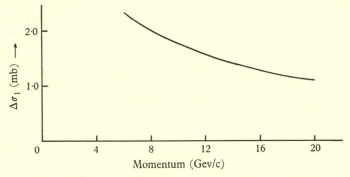

Fig. 2.1.2. The difference $\Delta\sigma_1$ between π^-p and π^+p total cross-sections (sketched from Galbraith *et al.* 1965),

$$\Delta\sigma_1 = \sigma_t(\pi^-p) - \sigma_t(\pi^+p).$$

theoretical difference in the assumptions required for the last hypothesis depending on whether particle and anti-particle belong to the same isospin multiplet. Thus the experimental study of

$$\Delta\sigma_2 = \sigma(\text{total}, \bar{p}p) - \sigma(\text{total}, pp), \tag{2.1.2}$$

which also appears to tend to zero may involve some different theoretical foundations from those involved in (2.1.1). At a more phenomenological level, differences of total cross-sections provide important information for determining Regge trajectories (Chapters 5 and 9).

It is believed that isospin amplitudes become equal for collisions between particles in two multiplets at high energy. The pp and np total cross-sections certainly tend to equality; again one should ask whether their difference changes sign. There are similar predictions that correspond to SU3 invariance but their precise comparison with theory awaits a really convincing method for taking into account the effects of broken symmetry. The same remark applies to tests of the Johnson–Treiman relation (1965), which predicts a relation between $\Delta\sigma_1$ (2.1.1) and the difference between the K^+p and K^-p total cross-sections.

(b) *The forward scattering amplitude*

The imaginary part of the forward scattering amplitude F can be found from a measurement of total cross-sections using the optical theorem. Using a relativistic normalisation for F (see §§ 3.2, 3.3),

$$\text{Im}\, F(E, 0) = 2m_2(E^2 - m_1^2)^{\frac{1}{2}} \sigma(\text{total}), \tag{2.1.3}$$

where E denotes the laboratory (lab.) energy of m_1.

The real part of the forward amplitude, $\text{Re}\, F(E, 0)$, can be measured from the effects of Coulomb interference in the near-forward direction. For 10 Gev. $\pi^+ p$ scattering, this interference is strong for

$$O < -t < 0\cdot01 \quad (\text{Gev/c})^2, \tag{2.1.4}$$

where t denotes the invariant momentum transfer squared,

$$t = (q_i - q_f)^2, \tag{2.1.5}$$

and q_i, q_f denote the initial and final four momenta of the pion.

For πp scattering, the amplitude neglecting the Coulomb interaction can be written (see § 3.6),

$$F(E, t) = (8\pi s^{\frac{1}{2}}) [f + i\boldsymbol{\sigma} . \mathbf{n} g], \tag{2.1.6}$$

where $\mathbf{n}$ is a vector perpendicular to the plane of scattering, and s is the total energy squared in the c.m. system. The differential cross-section for $\pi^- p$ scattering near $\theta = 0$, including Coulomb interference, has the form (see §§ 3.2, 3.6),

$$\frac{d\sigma}{d\Omega} = \left| -\frac{G(t)}{|t|} \exp(2i\delta) + \text{Re} f(s, t) + i \, \text{Im} f(s, t) \right|^2, \tag{2.1.7}$$

where δ denotes the relative phase shift introduced by the Coulomb interaction (Bethe, 1958; Solov'ev, 1966), and $G(t)$ is the product of the nucleon and pion form factors. The spin-flip term $g(s, t)$ is zero in the forward direction ($t = 0$), and one can approximate G using electron scattering measurements.

Using the optical theorem and measured values of the total cross-section, the experimental results on the differential cross-section can be used to determine the ratio of the real part to the imaginary part of the forward scattering amplitude,

$$\alpha(\pi^-, p) = \frac{\text{Re}\, F(E, 0)}{\text{Im}\, F(E, 0)} = \frac{\text{Re} f(s, 0)}{\text{Im} f(s, 0)}. \tag{2.1.8}$$

The first measurements of the important quantities $\alpha(\pi^-, p)$ and $\alpha(\pi^+, p)$ were made by Foley et al. (1965a). More recently the same group have completed an extensive experiment to determine more accurately the real parts of the forward amplitude for these and other processes (Foley et al. 1967 and Lindenbaum, 1967). The author is indebted to Dr Lindenbaum and collaborators for sending advance information about their results for inclusion in this book; some of these results are given in Figs. 2.1.3, 4 and 5.

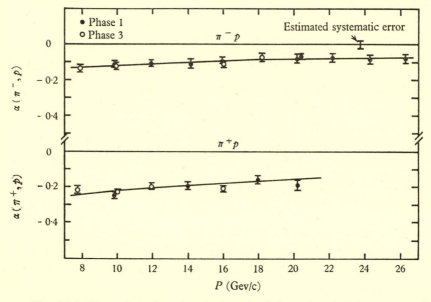

Fig. 2.1.3. The ratios, $\alpha(\pi^-, p)$ and $\alpha(\pi^+, p)$, of the real to imaginary parts of the forward scattering amplitudes for π^-p scattering and π^+p scattering, plotted as functions on the incident momenta, (from Foley et al. 1967, see also Lindenbaum, 1967). The experimental points are obtained using the Solov'ev (1966) form for δ, and the continuous lines are calculated from dispersion relations.

The experimental results for the ratios α, for a given process, will depend on the value of the relative phase δ in (2.1.7). The calculation of δ by Bethe (1958) is not fully relativistic, and there is some ambiguity in formulating the relativistic calculation of Solov'ev (1966); this leaves a small residual uncertainty in the measured values of α. The results of Foley et al., that are shown in Fig. 2.1.3 for $\alpha(\pi^-, p)$ and $\alpha(\pi^+, p)$, are based on the values of δ obtained by Solov'ev. The continuous lines in this figure are obtained from a calculation of Re $F(E, 0)$, using a dispersion relation and the measured total cross-sections (see §4.10 for a discussion of this method).

Experimental results for the sum $\alpha(\pi^-, p) + \alpha(\pi^+, p)$ are shown in Fig. 2.1.4, using both the values of δ given by Bethe and the values given by Solov'ev. The continuous line shows the result of the calculation using a dispersion relation. The advantage of using the sum of these two ratios is that the systematic errors tend to cancel, as also do the effects of the uncertainty in δ (Lindenbaum, private communication).

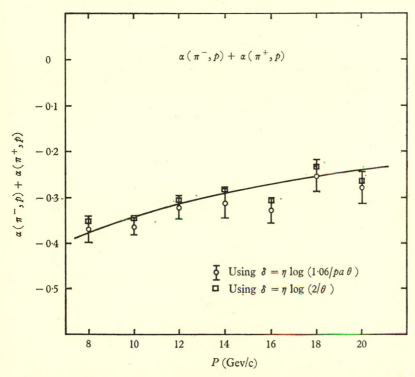

Fig. 2.1.4. The sum of the ratios $\alpha(\pi^-, p) + \alpha(\pi^+, p)$, plotted as a function of the incident momentum (Foley *et al.* 1967; Lindenbaum, 1967). The small circles denote the results obtained from the form of δ given by Bethe (1958), and the small squares denote results obtained using δ given by Solov'ev (1966).

We will see in § 8.2 that there are strong theoretical arguments that both $\alpha(\pi^-, p)$ and $\alpha(\pi^+, p)$ should tend to zero as the energy tends to infinity. The new experimental results shown in Fig. 2.1.3 and Fig. 2.1.4 are consistent with this theoretical prediction.

The theoretical situation is different for proton-proton scattering. There are no fundamental theoretical reasons why the ratio $\alpha(p, p)$ should tend to zero as the energy tends to infinity, although such a result is desirable from the viewpoint of Regge theory (see §5.3). However, if $\alpha(p, p)$ does *not* tend to zero, one can deduce from general theoretical assumptions (see §8.2), that the ratio $\alpha(\overline{p}, p)$ for anti-proton-proton scattering must be such that

$$\frac{\alpha(\overline{p}, p)}{\alpha(p, p)} \to -1 \quad \text{as} \quad E \to \infty. \tag{2.1.9}$$

The results of Foley *et al.* (1967) for the ratio $\alpha(p, p)$, shown in Fig. 2.1.5 together with some earlier results, are inclined to favour the conclusion that it tends to zero as E tends to infinity. If this is the case,

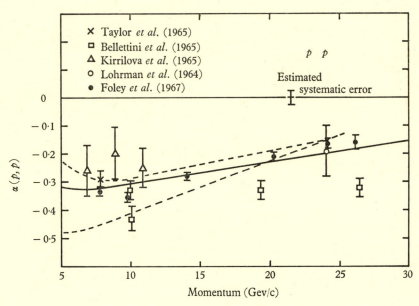

Fig. 2.1.5. The ratio $\alpha(p, p) = (\text{Re } F)/(\text{Im } F)$, for proton-proton scattering, from Lindenbaum (1967). The solid curve corresponds to the dispersion theory calculation of Soding (1966), and the broken lines denote the bounds given by the calculation of Levintov (1966).

then the result (2.1.9) need not hold. The continuous line in Fig. 2.1.5 shows the result of a dispersion relation calculation, and the broken lines represent upper and lower bounds that are also based on dispersion relations (see also Carter, 1966). The experimental results do not rule out the possibility that $\alpha(p, p)$ has an asymptotic value in the region of -0.2, so we will include this possibility in our later discussion. Further details and references relating to these experimental results are given by Lindenbaum (1967).

2.2   The elastic forward peak and nearby structure

(a) *The forward peak*

Within the range

$$0\cdot02 < -t < 0\cdot5\,(\mathrm{Gev/c})^2, \qquad (2.2.1)$$

the differential cross-sections for any elastic process show a forward peak which has very little structure for energies above 10 Gev. The peak can be summarised by the formula (Foley *et al.* 1965*b*; Lindenbaum, 1965; Wetherell, 1966),

$$\frac{d\sigma}{dt} = \left[\frac{d\sigma}{dt}\right]_{t=0} \exp bt. \qquad (2.2.2)$$

The coefficient b is measured in units $(\mathrm{Gev/c})^{-2}$. Its value for all elastic processes is in the range 7 to 12 (possibly up to 16). Its dependence on energy can be obtained directly from measurements of $(d\sigma/dt)$. For fixed energy $(d\sigma/dt)$ is strongly peaked in the forward direction as shown in Fig. 2.3.1. If $b(E)$ increases with energy the peak will shrink.

The simplest one-pole Regge model would predict that $b(E) \sim \log E$ and increases with E. This behaviour is not required for energies up to 30 Gev for a Regge model using several poles (see Chapter 9), and with more complicated Regge models it is not required at all.

The observed behaviour of $b(E)$ depends on the process being studied. For pp, pn and K^+p, one finds that b increases with E. For π^+p, π^-p, K^-p, no variation of b with E is observed. For $\bar{p}p$ one finds that b decreases with E. The energies at which these observations are made range from 5 to 27 Gev/c for pp; 8 to 24 Gev/c for π^-p, but only 4 to 5·5 Gev/c for K^-p. Only the pp, pn and π^-p data can be considered to be at reasonably high energies from the viewpoint of a theory concerned with asymptotic behaviour.

(b) *Structure at medium momentum transfer*

For intermediate values of the momentum transfer variable in the range

$$0\cdot5 < -t < 1\cdot5\,(\mathrm{Gev/c})^2, \qquad (2.2.3)$$

some structure is observed when energies are not too large. For example π^+p and π^-p both show a dip at $t \sim -0\cdot7\,(\mathrm{Gev/c})^2$, and a bump at $t \sim -1\cdot2\,(\mathrm{Gev/c})^2$. This structure is observed below 4 Gev energy but it decreases rapidly with energy and is very small at 8 Gev.

We will see in Chapter 9 that, on the basis of Regge theory, it is probably easier to observe structure in exchange reactions. However, there may be some circumstances where structure is present in an elastic cross-section but has been 'subtracted away' in an exchange reaction. Such structure might arise from an interference involving an exchange 'particle' that is allowed in elastic scattering but disallowed for a particular exchange process. The enhancement of structure in a suitably chosen exchange scattering process is confirmed experimentally (see § 2.4).

2.3 Large angle scattering

It is probable that phenomena observed for large angle scattering at existing energies, for example, angles in the range

$$50° < \theta < 130° \quad \text{at } 10 \, \text{Gev}, \tag{2.3.1}$$

will hold for an increasingly wide range of angles as the energy increases. In that case these experiments are concerned with the asymptotic behaviour as $E \to \infty$ at any fixed angle. The energy at which asymptotic behaviour can be assumed to dominate will certainly depend on the angle. The range (2.3.1) seems characteristic down to about 8 Gev energy.

A typical fixed energy differential cross-section is shown in Fig. 2.3.1. The striking feature is the rapid decrease with angle through five orders of magnitude for pion-nucleon scattering at energy 8 Gev. This rapid decrease is a feature of all high energy two-body reactions. For πp and pp scattering the dependence at large angles is approximately

$$f(t) \exp \left[-C \sqrt{|t|} \right], \tag{2.3.2}$$

where $f(t)$ is a polynomial.

A number of curve fitting formulae have been suggested that give improvements on the order of magnitude form (2.3.2). A suggestion by Orear (1964) for pp scattering has been successful in giving a close approximation to the data. This is

$$\frac{d\sigma}{d\Omega} = \frac{A}{s} \exp \left[-ap \sin \theta \right], \tag{2.3.3}$$

where $A = 595 \, (\text{Gev})^2 \, \text{mb/sr}$, $a^{-1} = 0.158 \, \text{Gev/c}$, $s = $ (total centre of mass energy)^2, and $p \sin \theta$ is the transverse momentum transfer for the proton.

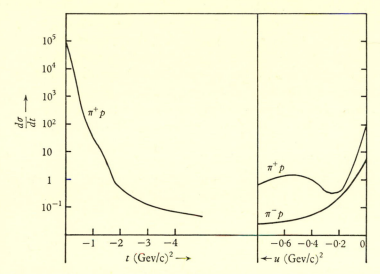

Fig. 2.3.1. The differential cross-sections for π^+p and π^-p scattering at 8 (Gev/c) measured in $\mu b/(Gev/c)^2$. The cross-sections for π^+p and π^-p are similar in the forward direction and out to large angles, π^-p being slightly larger than π^+p. There is a significant difference near the backward direction ($u = 0$), which is shown on an enlarged scale. References to the experimental work are given by Wetherell (1966).

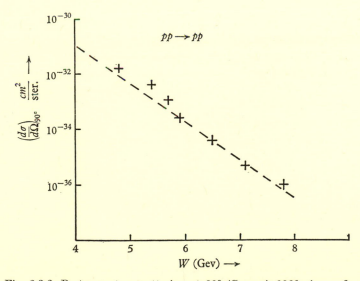

Fig. 2.3.2. Proton-proton scattering at 90° (Cocconi, 1966 gives references to the experiments). The broken line follows the Orear formula (2.3.3), ($W \approx 2p$, $\theta = 90°$), which is approximated by

$$\frac{d\sigma}{d\Omega} = 10^{-(25\cdot3)} \exp\left[-W/(0\cdot31 \text{ Gev})\right].$$

According to (2.3.3) a plot of the pp data:

$$\log\left[s\frac{d\sigma}{d\Omega}\right] \quad \text{against} \quad [p\sin\theta] \tag{2.3.4}$$

should be a straight line with slope $-a$. This is shown for $\theta = 90°$ in Fig. 2.3.2 with an indication of the errors. The experimental points range through energies from 8 to 25 Gev.

There is also quite good agreement between the Orear formula (2.3.3) and data for pn and $\overline{p}p$ scattering (Białas & Czyzewski, 1966 and see § 10.4). The formula should not be regarded as sacrosanct and it is less satisfactory for πN scattering, however it is an interesting first approximation in an experimental area where no satisfactory theory exists.

2.4   Exchange scattering and near-backward scattering

These two categories are taken together because scattering at or near the backward direction for the elastic scattering process

$$a^+ + b^- \rightarrow a^+ + b^- \tag{2.4.1}$$

is equivalent to forward scattering for the exchange process

$$a^+ + b^- \rightarrow b^- + a^+. \tag{2.4.2}$$

In some instances this is simply charge exchange but in others it may involve baryon exchange and strangeness exchange (see below and Fig. 2.4.2 in particular).

(a) Charge exchange scattering

Differential cross-sections that have been measured at high energy include those for the following exchange processes:

$$p + n \rightarrow n + p, \tag{2.4.3}$$

$$p + \overline{p} \rightarrow n + \overline{n}, \tag{2.4.4}$$

$$\pi^- + p \rightarrow \pi^0 + n, \tag{2.4.5}$$

$$K^- + p \rightarrow \overline{K}^0 + n. \tag{2.4.6}$$

The first of these (2.4.3) taken in the forward direction corresponds to backward pn elastic scattering. The experimental features are reviewed by Wetherell (1966), who gives further references.

The total charge exchange cross-sections can be defined by integrating the differential cross-sections for charge exchange (except for

(2.4.3) since this would give the elastic cross-section). We denote them by

$$\sigma_t(\bar{p}p \to n\bar{n}); \quad \sigma_t(\pi^-p \to \pi^0 n); \quad \sigma_t(K^-p \to \bar{K}^0 n). \qquad (2.4.7)$$

These cross-sections are observed to decrease rapidly with energy, in contrast to the total elastic cross-sections which are either constant or slowly decreasing at high energy. For example,

$$\sigma_t(\bar{p}p \to n\bar{n}) \sim 5\,\text{mb} \quad \text{at } 2\,\text{Gev.}$$
$$\sim 0{\cdot}5\,\text{mb} \quad \text{at } 5\,\text{Gev.}$$
$$\sim 0{\cdot}2\,\text{mb} \quad \text{at } 10\,\text{Gev.} \qquad (2.4.8)$$

This is typical of the rate of decrease observed for charge exchange cross-sections.

If isospin invariance is assumed, the forward charge exchange cross-sections can be related to cross-sections for elastic scattering. For example, the imaginary part of the charge exchange amplitude for (2.4.6) can be found from the optical theorem in terms of $\sigma(\text{total})$ for the scattering of K^-p and K^-n. Then (see §3.3), a measurement of the quantity

$$\left[\frac{d\sigma(K^-p \to \bar{K}^0 n)}{dt}\right]_{\theta=0} - \frac{1}{16\pi}[\sigma_t(K^-p) - \sigma_t(K^-n)]^2, \qquad (2.4.9)$$

where θ denotes the scattering angle, will give an experimental value at $\theta = 0$, for

$$\frac{[\text{Re}\,F(K^-p \to \bar{K}^0 n)]^2}{64\pi M^2(E^2 - m^2)}. \qquad (2.4.10)$$

Experimental values reported by Astbury *et al.* (see Wetherell, 1966) indicate that the forward exchange amplitude is almost real near 10 Gev, i.e. the first of the two terms in (2.4.9) is much larger than the second.

The forward exchange differential cross-sections all decrease with energy. For example,

$$\left[\frac{d\sigma(\pi^-p \to \pi^0 n)}{dt}\right]_{t=0} \sim A\left(\frac{E}{E_0}\right)^{-\frac{1}{2}} \qquad (2.4.11)$$

with $E_0 \sim 1$ Gev, and $E > 8$ Gev.

The non-forward exchange differential cross-sections also decrease with energy at fixed momentum transfer. They show a forward peak. For pn (2.4.3) the peak is unusually narrow in the forward direction ($|t| < 0{\cdot}2\,(\text{Gev/c})^2$), but has typical exponential decrease at larger t values out to $-1{\cdot}0\,(\text{Gev/c})^2$, (see (2.7.10)).

The π^-p charge exchange cross-section shows significant structure outside the forward peak. This is illustrated in Fig. 2.4.1 for two values of the energy of the incident pion (4 and 18 Gev). The structure shows a dip at $t = -0.6\,(\text{Gev/c})^2$, which remains equally significant at the higher energy. This is in contrast to the structure in near forward πp elastic cross-sections, which disappears rapidly with increasing energy. This difference of behaviour, as well as the existence of the dip in the

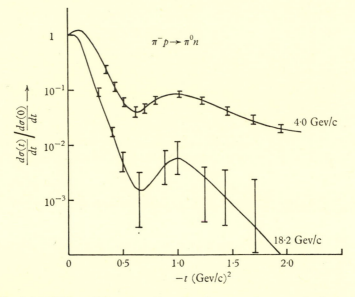

Fig. 2.4.1. Differential cross-sections for charge exchange scattering, $\pi^-p \to \pi^0 n$ at 4 Gev/c and 18·2 Gev/c (Borgeaud *et al.* 1966). Not all experimental points are shown.

cross-section, has an interesting interpretation in Regge theory (Chapter 9). This suggests strongly that experiments on two-body reactions can be usefully classified according to the exchanged quantum numbers, and, more particularly, according to the particles (or Regge poles) that can carry these quantum numbers. On this basis charge exchange is no longer a special category (see § 2.7), but should be considered with other two-body reactions.

(b) Near-backward scattering

Backward scattering for the process,

$$\pi^- + p \to \pi^- + p,$$

$$(2.4.12)$$

shown in Fig. 2.4.2 (a) corresponds to forward scattering for the exchange process which we write as

$$\pi^- + p \rightarrow p + \pi^-. \tag{2.4.13}$$

This process is shown in Fig. 2.4.2 (b). It can proceed by the exchange of the quantum numbers of B^{++} shown in Fig. 2.4.2 (c), namely those in the process

$$\pi^+ + p \rightarrow \pi^+ + p. \tag{2.4.14}$$

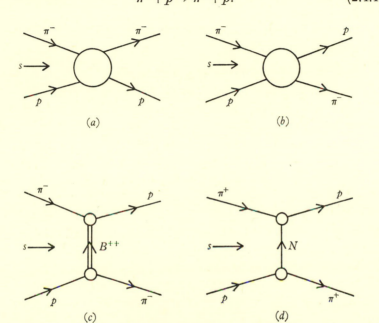

Fig. 2.4.2. Diagrams relevant for backward pion-proton scattering: (a) π^-p scattering denoted by $\pi^-p \rightarrow \pi^-p$; (b) the exchange scattering process denoted $\pi^-p \rightarrow p\pi^-$; (c) a diagram showing the exchange process that may dominate π^-p backward scattering; (d) a diagram that may dominate π^+p backward scattering but is not allowed for π^-p backward scattering.

The particle B^{++} could be the resonance N^{*++} (1238) for example. This may be contrasted with backward scattering for the process (2.4.14) itself which can proceed as shown in Fig. 2.4.2 (d) by exchange of a neutron in addition to the possibility of exchange of the resonance N^{*0}(1238). The fact that neutron exchange is allowed for one process but not the other has important experimental consequences. We will consider their theoretical interpretation in more detail in Chapter 9.

Experimental results on near-backward scattering for π^-p and π^+p show a striking difference both in their magnitude and in structure.

For π^+p there is a significant dip near an *exchange* momentum transfer squared, $u \approx -0\cdot2\,(\text{Gev/c})^2$, and there is a strong narrow backward peak. In contrast π^-p has no dip and a less strong, less narrow backward peak (see Fig. 2.3.1).

We will consider other backward scattering processes in §2.7, but will then describe them as forward exchange reactions and classify them according to the exchanged quantum numbers as noted above.

2.5 Energy dependence for backward scattering and for fixed momentum transfer

(a) Backward scattering for medium energies

At low energies, differential cross-sections may be dominated by an individual resonance near to its mean energy value. The corresponding angular distribution at this value is then dominated by the resonance, and has the form (for angular momentum l),

$$\frac{d\sigma}{d\Omega} \sim |c_l P_l(\cos\theta)|^2 \tag{2.5.1}$$

$$\sim |c_l(-1)^l|^2 \quad \text{for} \quad \theta \sim \pi. \tag{2.5.2}$$

The coefficient $c_l(E)$ will be larger at the resonance than the individual contributions from other partial waves. When there are no resonances, the backward scattering cross-section is observed to decrease with energy much faster than E^{-1}, which is the rate of decrease of the contribution of an individual partial wave (in an order of magnitude estimate). Experiment therefore indicates that non-resonant partial waves tend to cancel in the backward direction. If we write them as $B(E)$, and let c_l denote a partial wave with a resonance at E_0, we will have

$$\frac{d\sigma}{d\Omega} \sim |B(E) + c_l(E)(-1)^l|^2 \quad \text{at} \quad \theta = \pi. \tag{2.5.3}$$

If the cancellation of non-resonant partial waves is such that

$$|B(E_0)| \sim |c_l(E_0)|, \tag{2.5.4.}$$

there will be strong interference near the resonance value (recalling that the phase of $c_l(E)$ changes rapidly near the resonance value).

This interference phenomenon is observed at medium energies for backward πp scattering, and a number of resonances give strong peaks or dips depending on whether the interference is constructive or destructive. The observed structure decreases rapidly at angles away from the backward direction. The $\pi^- p$ backward scattering is shown as a function of energy in Fig. 2.5.1 (Kormanyos *et al.* reviewed by Murphy, 1966). Other experiments to detect baryon resonances are reviewed by Murphy and by Ferro-Luzzi (1966) but, since they are not directly high energy phenomena, we will not consider them here. A list of observed baryon resonances and of boson resonances is given in § 1.2.

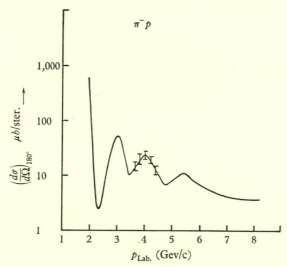

Fig. 2.5.1. Experimental results for backward scattering for $\pi^- p \rightarrow \pi^- p$ as a function of the energy (Kormanyos *et al.* 1966). Only a few of the experimental points are shown, to indicate the order of magnitude of their uncertainty. There may be more structure than is indicated here (see Barger & Cline, 1966).

(b) High energies—fixed momentum transfer

At higher energies and fixed t, according to Regge theory (see Chapters 5 and 9) a differential cross-section may have the characteristic form,

$$\frac{d\sigma}{dt} \sim \left| \sum_n a_n(t) \left(\frac{E}{E_0} \right)^{\alpha_n(t)-1} \right|^2, \tag{2.5.5}$$

where the sum over n involves only a small number of terms that dominate the cross-section. The number of these terms may be severely restricted for certain exchanged quantum numbers. Near the backward direction at fixed u, where u denotes minus the exchange momentum transfer squared, Regge theory indicates that

$$\frac{d\sigma}{du} \sim \left| \sum_m b_m(u) \left(\frac{E}{E_0} \right)^{\alpha_m(u)-1} \right|^2 \quad as \ E \rightarrow \infty. \tag{2.5.6}$$

The point to note here is that from a theoretical viewpoint it would be very useful to have good measurements for fixed t, and varying E, near the foward direction; and for fixed u and varying E near the backward direction. For fixed t, as $E \to \infty$ we always tend to the forward direction. Similarly, for fixed u as $E \to \infty$, we have $\theta \to \pi$ (see Chapters 3 and 4).

For various exchange reactions there is already some data available on (2.5.5) and (2.5.6), (reviewed by Morrison, 1966). When this data is more extensive, one would hope to evaluate the parameters α, β, a and b in the above equations at different values of t and u.

2.6   Polarisation

In pion-nucleon elastic scattering the polarisation $P(t)$ of the recoil nucleon is given by

$$P(t)\frac{d\sigma}{d\Omega} = 2\,\mathrm{Im}\,[f^*g], \qquad (2.6.1)$$

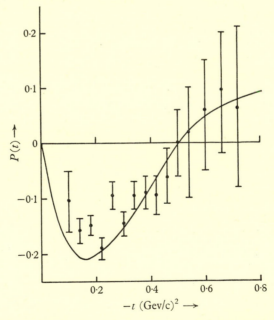

Fig. 2.6.1. Polarisation $P(t)$ for $\pi^- p \to \pi^- p$ at 6 Gev/c, (Borghini *et al.* reported by Van Hove, 1966).

where f and g denote the non-spin-flip and the spin-flip amplitudes respectively. In Fig. 2.6.1 the polarisation data is shown for $\pi^- p$ at 6 Gev/c (see the review by Van Hove, 1966). The continuous line denotes the polarisation given by a Regge model.

In the charge exchange reaction

$$\pi^- + p \to \pi^0 + n, \tag{2.6.2}$$

it is expected from a Regge model that the polarisation $P(t)$, should decrease with energy (see Chapter 9). The energy dependence of $P(t)$ will be of considerable importance for any model of high energy collisions. For the charge exchange process (2.6.2), $P(t)$ is observed to have a maximum at $t \sim -0.15$ (Gev/c)^2, and it takes values (Wetherell, 1966),

$$P(t) \sim 15 \pm 3 \% \quad \text{at } 6 \, \text{Gev/c}. \tag{2.6.3}$$

Its value at $11 \, \text{Gev/c}$ is less well measured but there is no decisive evidence that $P(t)$ decreases with energy.

Other spin correlation parameters can in principle be measured, and probably will be measured over the next few years. They are of great value in testing certain aspects of Regge theory and may be crucial in assessing the validity or accuracy of simplifying assumptions (see Chapter 9, particularly § 9.5).

2.7   Inelastic two-body reactions

The number of inelastic two-body reactions observed is already extensive. In order to test theoretical models for high energy collisions one requires both the energy dependence and the momentum transfer dependence of all available quasi two-body processes. It can be expected that these experiments will increase rapidly in detail and content over the next few years. Instead of giving a detailed account of observed properties, I will list the main types of observation and indicate some of the relevant reactions and some of the observed data. A detailed report on these reactions has been given by Morrison (1966).

The reactions discussed in this section include many that involve resonances rather than particles. These are described as quasi two-body reactions, for example,

$$\pi^+ + p \to \rho^0 + N^{*++} \to (\pi^+\pi^-) + (p\pi^+). \tag{2.7.1}$$

These will be discussed from the viewpoint of experiments that have selected out the quasi two-body process by a suitable choice of the energy of the decay products,

$$E(\pi^+\pi^-) \approx E(\rho^0) \quad \text{and} \quad E(p\pi^+) \approx E(N^{*++}). \tag{2.7.2}$$

This invariant separation of a resonance from the background will be discussed in Chapter 3.

In addition to quasi two-body reactions there are simple inelastic two-body reactions like

$$\pi^+ + p \to K^+ + \Sigma^+.$$

All these processes give information on strong interactions additional to that from elastic reactions like charge exchange discussed in § 2.4 and simple elastic scattering. To some extent we will use the results for elastic scattering and reactions as a basis for comparison. Experimentally there is a difference in the techniques employed, but the results for quasi two-body reactions show many of the same features that were noted previously for elastic scattering.

(a) *Elastic and inelastic reaction cross-sections*

We begin by comparing reaction cross-sections

$$\sigma(\text{total}, a+b \to c+d), \tag{2.7.3}$$

with total elastic cross-sections, and with total cross-sections, namely

$$\sigma(\text{total}, a+b \to a+b) \quad \text{and} \quad \sigma(\text{total}, a+b). \tag{2.7.4}$$

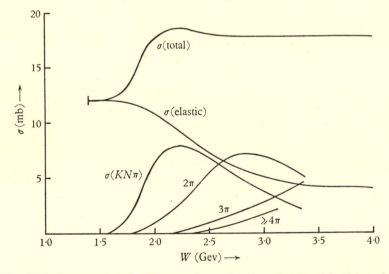

Fig. 2.7.1. Cross-sections for K^+p reactions, showing $\sigma(\text{total})$, $\sigma(\text{elastic})$, and σ for production of 1, 2, 3 and 4 or more pions. Experimental references are given by Morrison (1966) and by Ferro-Luzzi (1966).

The comparison is shown in Fig. 2.7.1. for K^+p reactions (CERN–Brussels collaboration reported by Morrison, 1966). The features that stand out are: (i) the constancy of $\sigma(\text{total})$ for large E and the near constancy of $\sigma(\text{elastic})$ at about 20 % of $\sigma(\text{total})$; (ii) the typical behaviour of an inelastic process is to increase above threshold and then decrease. The latter decrease is not always to zero, as we see from Fig. 2.7.2. This figure shows the variation with E of σ (total reaction) for several reactions of the type

$$p + p \rightarrow p + N^{*+}, \tag{2.7.5}$$

where N^* denotes a resonance.

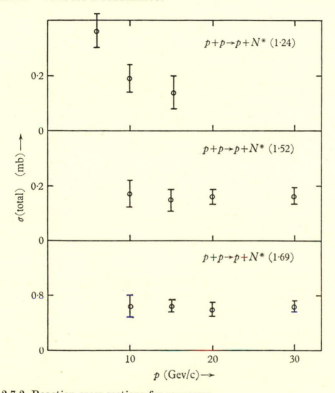

Fig. 2.7.2. Reaction cross-sections for processes,

$$p + p \rightarrow p + N^{*+},$$

for N^{*+} equal to the resonances at 1·24, 1·52 and 1·69 Gev (Anderson *et al.* 1966).

There are two kinds of behaviour observed for these total reaction cross-sections. When N^* is the 1·24 Gev resonance the reaction cross-section decreases as one would expect for a production process; note that for this resonance $(I, J^P) = (\frac{3}{2}, \frac{3}{2}^+)$. When N^* denotes the 1·52 or the 1·69 or the 2·19 resonance the reaction cross-section appears to be nearly constant at high energy. The latter resonances have $I = \frac{1}{2}$, and $J^P = \frac{3}{2}^-, \frac{5}{2}^+$ and $\frac{7}{2}^-$ respectively. It appears that the associated reactions are quasi elastic in the sense that isospin is conserved, and the total reaction cross-sections behave like the total elastic cross-sections. They are discussed in § 8.4, (this type of behaviour is sometimes called diffraction dissociation). Quasi two-body processes like (2.7.5) might be expected to be quasi elastic if they can proceed without exchange of any object except one having the quantum numbers of the vacuum. It is of considerable interest to know if this rule holds for all two-body reactions. Other reaction cross-sections have been observed, for example

$$K^+ + p \to K + N^*. \tag{2.7.6}$$

However, the observations for this reaction have not been at sufficiently high energies to give information on the asymptotic cross-sections.

(b) Differential cross-sections for quasi two-body reactions

It is found in general that these cross-sections show a forward peak rather like that for elastic scattering, for small $|t|$,

$$\frac{d\sigma(a + b \to c + d)}{dt} = C \exp{(bt)}. \tag{2.7.7}$$

The slope b has a wider range of values than for elastic scattering. It is about $5\,(\mathrm{Gev/c})^{-2}$ for the reactions,

$$K^- p \to \bar{K}^0 n; \quad \pi^+ p \to N^* \eta; \quad pp \to pN^*(1\cdot52). \tag{2.7.8}$$

It is found that $b \approx 17\,(\mathrm{Gev/c})^{-2}$ for the reaction

$$pp \to pN^*(1\cdot24). \tag{2.7.9}$$

For small values of $|t| < 0\cdot2\,(\mathrm{Gev/c})^2$, the slope for the peak of pn charge exchange is unusually large,

$$pn \to np, \quad b \approx 40\,(\mathrm{Gev/c})^{-2} \quad \text{for} \quad |t| < 0\cdot2\,(\mathrm{Gev/c})^2. \tag{2.7.10}$$

however, the slope becomes much less when $|t| > 0\cdot2\,(\mathrm{Gev/c})^2$. Except for the last process the slopes remain nearly constant (with t) out to $t \sim -0\cdot5\,(\mathrm{Gev/c})^2$, at energies more than $10\,\mathrm{Gev}$.

The variation of slope with energy is important for testing any detailed high energy model. There do not appear to be any simple systematic characteristics (Morrison, 1966).

(c) Structure in $\dfrac{d\sigma}{dt}$ for quasi two-body reactions

The reactions

$$\pi^+ + p \to \eta^0 + N^{*++}, \tag{2.7.11}$$

$$\pi^+ + p \to \omega^0 + N^{*++}, \tag{2.7.12}$$

both show maxima outside the forward peak near

$$t = -0.2\,(\text{Gev}/c)^2. \tag{2.7.13}$$

These resemble the exchange reaction ($\pi^- p \to \pi^0 n$) which has a similar peak at $t = -0.1\,(\text{Gev}/c)^2$.

In the charge exchange reaction (see Fig. 2.4.1)

$$\pi^- + p \to \pi^0 + n, \tag{2.7.14}$$

there is evidence for a second peak ($t \approx -1.2\,(\text{Gev}/c)^2$) in addition to the peak at $t = -0.1\,(\text{Gev}/c)^2$, (Sonderegger *et al.* 1966). The variation observed in the slope (2.7.7) of the forward peak may be attributable in part to the existence of structure. One should certainly, in any theoretical analysis, allow for the possibility of interference producing a greater than normal slope.

(d) Angular correlations

The simplest possibility in high energy reactions is that they are dominated by single particle exchange. This is certainly not valid in general, but it can be tested by observing angular correlations. In the

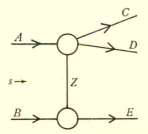

Fig. 2.7.3. One particle exchange. If this process dominates the reaction then angular correlations will be particularly simple.

process indicated in Fig. 2.7.3, if this diagram dominates the reaction, and if the exchanged particle Z has no spin, then there can be no dependence of the cross-section on the azimuthal angle in the centre of mass system of particles C and D (Treiman & Yang, 1962). The azimuthal dependence for arbitrary spin has been given by Gottfried & Jackson (1964).

More generally, angular correlation measurements will give information about the interference between two or more exchanged particles. The reaction,

$$\pi^+ + p \to \rho + N^*, \tag{2.7.15}$$

was measured by Aachen, Berlin, CERN (1966), and shows similarity in angular distribution with the reaction

$$K^+ + p \to K^{*0} + N^{*++}, \qquad (2.7.16)$$

measured by CERN–Brussels. The significance of this similarity has been discussed by Białas & Kotanski (1966).

(e) Baryon exchange reactions

It seems probable that properties of high energy reactions in forward and near-forward directions are determined largely by the quantum numbers that are exchanged between the colliding particles (by suitable re-labelling this includes near-backward scattering). We list here some of the reactions that are presently being studied that involve baryon exchange. We noted in § 2.4 that there are special properties observed for the reaction, ($\pi^+ p$ backward scattering),

$$\pi^+ + p \to p + \pi^+. \qquad (2.7.17)$$

In this case neutron exchange is allowed (see Fig. 2.7.4 (a)) and experiments show a dip in the differential cross-section when

$$t \approx -0{\cdot}2 \,(\text{Gev/c})^2.$$

This is conventionally described as $\pi^+ p$ backward scattering with a dip at $u \approx -0{\cdot}2 \,(\text{Gev/c})^2$, and a backward peak (for 8 Gev incident energy).

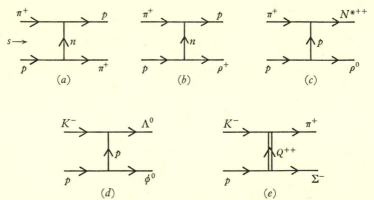

Fig. 2.7.4. Exchange diagrams that may dominate the corresponding reactions near the backward direction.

Evidence for diffraction peaks is observed for the reactions

$$\pi^+ + p \to p + \rho^+, \qquad (2.7.18)$$

$$\pi^+ + p \to N^{*++} + \rho^0. \qquad (2.7.19)$$

In the reaction (2.7.18) neutron exchange is allowed and in (2.7.19) proton exchange is allowed. The relevant diagrams are shown in Fig. (2.7.4(b), (c)).

Reactions involving hyperons are discussed in the review by Morrison (1966). These include

$$K^- + p \rightarrow \pi^- + \Sigma^+, \tag{2.7.20}$$

$$\rightarrow \rho^- + \Sigma^+, \tag{2.7.21}$$

$$\rightarrow \pi^0 + \Lambda^0, \tag{2.7.22}$$

$$\rightarrow \omega^0 + \Lambda^0, \tag{2.7.23}$$

$$\rightarrow \phi^0 + \Lambda^0. \tag{2.7.24}$$

For the forward direction each of these can proceed by meson exchange and there is a large forward peak. A backward peak is observed except for the reaction (2.7.24). The exchange relevant to the backward direction for (2.7.24) is shown in Fig. 2.7.4(d); the absence of a backward peak could be explained by a small (ϕpp) coupling strength.

In the reaction

$$K^- + p \rightarrow \pi^+ + \Sigma^-, \tag{2.7.25}$$

there is no forward peak. The exchange diagram relevant to the forward directions is shown in Fig. 2.7.4(e). A meson with double charge and strangeness plus one would have to be exchanged. No such particle has been observed, so the charge would have to be carried by two mesons (or a hyperon plus antibaryon). Evidently this produces a much weaker effect than a single particle and accounts for the absence of a forward peak.

2.8 Multiple production of particles

There are quite extensive measurements on multiple production of particles, which have been reviewed by Czyzewski (1966). The wealth of parameters in production processes make it difficult to give systematic features that are likely to be relevant to a future theory. For low multiplicity, say three final particles, it is probable that Regge theory will have some validity, but this is less likely for high multiplicity.

High multiplicity at present accelerator energies is taken (arbitrarily) to mean events with five or more pions in the final state. The observed events must have no more than one neutral particle so they are only a fraction of the actual events. For 6-prong events, $(n = 5)$, at 8 Gev energy in the reaction

$$\pi^+ + p \to n\pi + p, \qquad (2.8.1)$$

the observed cross-section is $1\cdot65 \pm 0\cdot15$ mb. For 8-prong events it is $0\cdot21 \pm 0\cdot03$ mb. If one makes the assumption that the various isospin states of the $n\pi$ system that are allowed by charge and isospin conservation, have about equal probabilities, then the multiple production cross-sections can be estimated. At 8 Gev, multipion production $(n \geqslant 5)$ is responsible for about half the inelastic cross-section giving

$$\sigma(\text{multipion production}) \approx 8 \text{ mb}. \qquad (2.8.2)$$

In order to study multiple production one should remove two-body and quasi-two-body events. This is partly achieved by taking the multiplicity to be high. The practical procedure is to assume that 'central' collisions correspond to large angle scattering, and 'peripheral' collisions to small angle scattering. The central collisions are assumed to correspond to multi-pion production. The two may be distinguished experimentally by taking an angular distribution for central collisions to be given by $(d\sigma/d\Omega)$ for $\theta > \frac{1}{2}\pi$, and to be symmetric about $\theta = \frac{1}{2}\pi$, where θ denotes the angular variation of baryons in the centre of mass system. This is rather a crude procedure and can certainly be refined as our understanding of multiplicity develops. It gives a lower figure than that in (2.8.2) at 10 Gev, being about 5 mb. Experimentally the resulting total cross-section for central collisions decreases with energy; approximately

$$\sigma(\text{total central}) \sim CW^{-1}, \qquad (2.8.3)$$

where W is the centre of mass energy.

The dependence of multiplicity on transverse momentum has also been studied systematically. The results are not very striking. At 8 Gev incident energy the multiplicity decreases slightly with transverse momentum. A more striking result is found in a plot of average transverse against average longitudinal momentum of a pion for a given multiplicity. The Dubna–Bucharest results reported by Van Hove (1966) are shown in Fig. 2.8.1 for 8 Gev pion-nucleon reactions,

$$\pi^- + p \to \pi^+ + \pi^- + \pi^- + n\pi^0 + p. \qquad (2.8.4)$$

The dip for small longitudinal momentum is probably a geometric effect that depends on the shape of the pion energy spectrum,

(Bardardin–Otwinowska *et al.* 1966). For further details of multi-particle production the reader is referred to the review by Czyzewski (1966).

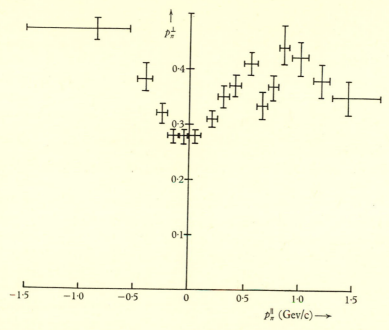

Fig. 2.8.1. The dependence on the average transverse momentum $p_\pi^\perp$ of the π^-, for different intervals of the longitudinal momentum in the centre of mass scattering for the reaction $\pi^- p \rightarrow p\pi^+\pi^-\pi^- + (n\pi^0)$. Reference Belyakov (1966).

CHAPTER 3

THEORETICAL BACKGROUND

3.1 Relativistic kinematics

An energy-momentum four-vector will be denoted p or $(p_0, \mathbf{p})$, and we will use the metric giving

$$p^2 = p_0^2 - \mathbf{p}^2. \tag{3.1.1}$$

Units will normally be chosen so that $\hbar = c = 1$.

We will consider the kinematics for the reaction shown in Fig. 3.1.1a,

$$1 + 2 \to 3 + 4. \tag{3.1.2}$$

The four-momenta of particles 1 and 2 will be denoted by p_1, p_2 and those of the final states 3, 4 will be denoted by $-p_3$, $-p_4$ respectively. Then conservation of total energy and momentum gives

$$p_1 + p_2 + p_3 + p_4 = 0. \tag{3.1.3}$$

The particles will be on the mass shell in their initial and final states,

$$p_r^2 = m_r^2 \quad (r = 1, 2, 3, 4). \tag{3.1.4}$$

The reaction in the centre of mass system is shown in Fig. 3.1.1b. In this system the invariant energy variable s becomes the square of the energy of particles 1 and 2 (or of 3 and 4). In general it is given by (Møller, 1932, 1945),

$$s = (p_1 + p_2)^2 = (p_{10} + p_{20})^2 - (\mathbf{p}_1 + \mathbf{p}_2)^2. \tag{3.1.5}$$

The 'invariant momentum transfer' t is defined by

$$t = (p_1 + p_4)^2 = (p_2 + p_3)^2, \tag{3.1.6}$$

and the invariant exchange momentum transfer u is,

$$u = (p_1 + p_3)^2 = (p_2 + p_4)^2. \tag{3.1.7}$$

The invariants s, t and u are not independent on account of the conservation law (3.1.3) and the mass shell conditions (3.1.4) which give,

$$s + t + u = \sum_1^4 m_r^2. \tag{3.1.8}$$

For equal mass particles, elastically scattered, these invariants take a particularly simple form in terms of the centre of mass three-momentum $\mathbf{k}$ and the centre of mass scattering angle θ. Write k^2 for $\mathbf{k}^2$, then

$$s = 4(m^2 + k^2), \tag{3.1.9}$$

$$t = -2k^2(1 - \cos\theta), \tag{3.1.10}$$

$$u = -2k^2(1 + \cos\theta). \tag{3.1.11}$$

Also in this special case the laboratory energy E, with one initial particle at rest, satisfies

$$s = 2m(E + m). \tag{3.1.12}$$

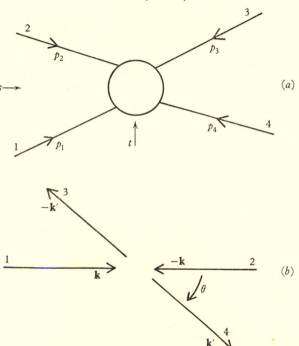

Fig. 3.1.1. (a) Notation for the four-momenta in the reaction $1 + 2 \rightarrow 3 + 4$. The variable s denotes the square of the centre of mass energy, and t denotes minus the square of the momentum transfer; (b) the reaction $1 + 2 \rightarrow 3 + 4$ in the centre of mass system, $\mathbf{k}$ is usually used for centre of mass three-momentum but sometimes $\mathbf{q}$ is used.

The detailed treatment and use of kinematics for general masses is described by Baldin, Goldanski & Rosental (1961) and by Källen (1964). A few formulae will be noted here for subsequent use. Writing E for the energy of particle 1 in the laboratory system with particle 2 at rest,

$$E = E_1(\text{lab.}) = \frac{1}{2m_2}(s - m_1^2 - m_2^2). \tag{3.1.13}$$

Using (3.1.4) this gives for the momentum,

$$|\mathbf{p}_1(\text{lab.})| = |\mathbf{p}| = \frac{1}{2m_2}\lambda(s, m_1^2, m_2^2). \qquad (3.1.14)$$

The function λ occurs frequently in relativistic kinematics. It is defined by
$$\lambda(x, y, z) = [x^2 + y^2 + z^2 - 2xy - 2yz - 2zx]^{\frac{1}{2}}. \qquad (3.1.15)$$

Note that our definition of λ includes the square root and so differs from that of Källen. The velocity v of particle 1 in the laboratory system is required for defining cross-sections; it is

$$v = v_1(\text{lab.}) = \frac{\lambda(s, m_1^2, m_2^2)}{s - m_1^2 - m_2^2}. \qquad (3.1.16)$$

The centre of mass energy of particle 1 is,

$$p_1^0(\text{c.m.}) = (s + m_1^2 - m_2^2)/(2\sqrt{s}). \qquad (3.1.17a)$$

The centre of mass three-momentum $\mathbf{k}$ (or $\mathbf{q}$) has magnitude

$$|\mathbf{k}| = \frac{\lambda(s, m_1^2, m_2^2)}{2\sqrt{s}}. \qquad (3.1.17b)$$

The centre of mass scattering angle θ for the deflection from 1 to 4 is given by, $(\theta = \theta_{14})$,

$$\cos\theta = -\frac{(\mathbf{p}_1 \cdot \mathbf{p}_4)}{|\mathbf{p}_1|\,|\mathbf{p}_4|} = \frac{t - m_1^2 - m_4^2 - 2p_1^0 p_4^0}{2|\mathbf{p}_1|\,|\mathbf{p}_4|}, \qquad (3.1.18)$$

$$= \frac{s^2 + s(2t - \Sigma m_r^2) - (m_1^2 - m_2^2)(m_3^2 - m_4^2)}{\lambda(s, m_1^2, m_2^2)\,\lambda(s, m_3^2, m_4^2)}, \qquad (3.1.19)$$

where we have used (3.1.17a) in the last step (note that p_4^0 is negative). The corresponding laboratory angle ϕ_{14} for $1 \to 4$ satisfies,

$$\cos\phi_{14} = \frac{(s - m_1^2 - m_2^2)(s - m_1^2 - m_3^2) + t(s + m_2^2 - m_1^2) - 2m_2^2(m_1^2 + m_4^2)}{\lambda(s, m_1^2, m_2^2)\,\lambda(u, m_2^2, m_4^2)}. \qquad (3.1.20)$$

In using these formulae the relation (3.1.8) between s, t and u should be noted.

We will use pion-nucleon scattering as an example on several occasions. For this we have

$$m_1 = m_4 = m; \quad m_2 = m_3 = M, \qquad (3.1.21)$$

where m denotes m_π and M denotes m_N (sometimes μ and m are used instead of m and M). Then

$$\lambda(s, m_1^2, m_2^2) = [s^2 + m^4 + M^4 - 2sm^2 - 2sM^2 - 2m^2M^2]^{\frac{1}{2}}. \qquad (3.1.22)$$

The centre of mass momentum $\mathbf{k}$ (or $\mathbf{q}$) has magnitude

$$|\mathbf{k}| = (4s)^{-\frac{1}{2}}\lambda(s, m^2, M^2), \tag{3.1.23}$$

giving,

$$k^2 = \frac{(s - m^2 - M^2)^2 - 4m^2M^2}{4s}. \tag{3.1.24}$$

The velocity of the pion in the laboratory system is

$$v = v_1(\text{lab.}) = \frac{|\mathbf{p}_1(\text{lab.})|}{E_1(\text{lab.})} = \frac{\lambda(s, m^2, M^2)}{s - m^2 - M^2}, \tag{3.1.25}$$

giving

$$v_1^2 = 1 - \left(\frac{2mM}{s - m^2 - M^2}\right)^2. \tag{3.1.26}$$

The laboratory momentum of the pion satisfies

$$|\mathbf{p}_1(\text{lab.})| = \frac{ks^{\frac{1}{2}}}{m}. \tag{3.1.27}$$

Further results on pion-nucleon kinematics will be given in §4.6.

3.2   The S-matrix, scattering amplitudes and cross-sections

The S-matrix is important both for its direct relation to experimental data and for its role in the theoretical study of strong interactions. It was invented by Wheeler (1937) for use in nuclear reaction theory. Heisenberg (1943, 1944) proposed its use as the basis for a new formulation of the theory of elementary particles (see also Møller, 1945, 1946). Although this proposal is not yet fully realised, much progress has been made in this direction. The study of the analytic properties of the S-matrix is one of the central features of the theory; these are described in *The Analytic S-Matrix* by Eden, Landshoff, Olive & Polkinghorne (1966), and in a book of the same title by Chew (1966). The first of these books is primarily concerned with establishing the analytic properties, the second concentrates more on the general framework of the theory. Some technical aspects that are useful for applications are described by Jacob & Wick (1959) and Trueman & Wick (1964). We will introduce the S-matrix in this section, and discuss its analytic properties and some applications in Chapter 4. Technical aspects of the theory will be introduced partly in Chapter 5 and later as they are required for applications and comparison with experiment.

The S-matrix describes the transformations from initial states to final states of a system of colliding particles. Denote the state of n noninteracting (i.e. widely separated) incoming particles with four-momenta $p_1, p_2, ..., p_n$, by

$$|p_1, p_2, ..., p_n, \text{in}\rangle = |\alpha, \text{in}\rangle. \qquad (3.2.1)$$

The set of all such states is a complete set. From this initial state, the particles come together, interact and separate to form a final outgoing state,

$$|p_1', p_2', ..., p_n', \text{out}\rangle = |\beta, \text{out}\rangle. \qquad (3.2.2)$$

The S-matrix is defined as the operator that transforms the outgoing state to the incoming state

$$|\alpha, \text{in}\rangle = S |\alpha, \text{out}\rangle. \qquad (3.2.3)$$

The probability for a transition, $\alpha \to \beta$, is therefore proportional to the modulus squared of the matrix element $S_{\beta\alpha}$,

$$S_{\beta\alpha} = \langle \beta, \text{out}|\alpha, \text{in}\rangle = \langle \beta, \text{out}| S |\alpha, \text{out}\rangle. \qquad (3.2.4)$$

Since both the set of incoming states, and the set of outgoing states forms a complete set, the S-matrix is unitary

$$S^\dagger S = S S^\dagger = 1. \qquad (3.2.5)$$

Hence, instead of (3.2.3), we could have used

$$\langle \beta, \text{in}| S = \langle \beta, \text{out}| \qquad (3.2.6)$$

giving $\qquad S_{\beta\alpha} = \langle \beta, \text{out}|\alpha, \text{in}\rangle = \langle \beta, \text{in}| S |\alpha, \text{in}\rangle. \qquad (3.2.7)$

Each initial or final particle will be on the mass shell,

$$p_r^2 = m_r^2 + \mathbf{p}_r^2. \qquad (3.2.8)$$

Each S-matrix element must conserve total energy and momentum, so there will be a factor

$$\delta(\Sigma p_r' - \Sigma p_r) = \prod_{i=1}^{4} \delta(\Sigma p_{ri}' - \Sigma p_{ri}). \qquad (3.2.9)$$

A δ-function that contains a vector argument is to be interpreted as a product of δ-functions for each component.

The scattering amplitude, or more generally the reaction amplitude, $T_{\beta\alpha}$, is defined by

$$S_{\beta\alpha} = \delta_{\beta\alpha} + i(2\pi)^4 \delta(\Sigma p_r' - \Sigma p_r) T_{\beta\alpha}. \qquad (3.2.10)$$

The amplitude $T_{\beta\alpha}$ for spinless particles is an invariant if a covariant normalisation is chosen. We take,

$$\langle p'|p\rangle = 2p_0(2\pi)^3 \delta(\mathbf{p}' - \mathbf{p}), \qquad (3.2.11)$$

which gives

$$(2\pi)^{-3} \int |p'\rangle \, d^4 p' \, \theta(p_0') \, \delta(p'^2 - m^2) \langle p'|p\rangle = |p\rangle. \qquad (3.2.12)$$

The total cross-section σ_t for collision of two particles 1 and 2 involves summation over all allowed final states,

$$\sigma_t = \frac{1}{2\lambda(s, m_1^2, m_2^2)\,(2\pi)^{3n-4}} \sum_n \int dq_1 \ldots \int dq_n$$

$$\times \prod_1^n \delta(q_r^2 - m_r^2) \, \theta(q_{r0}) \, \delta(p_1 + p_2 - \Sigma q_r) \sum_{\text{spin}} |\langle q_1, \ldots, q_n, \text{in} | T | p_1, p_2, \text{in}\rangle|^2.$$

$$(3.2.13)$$

For elastic scattering of spinless particles of equal mass this becomes

$$\sigma(\text{elastic}) = \frac{1}{64\pi^2 s} \int d\phi \, d(\cos\theta) \, |F(s,t)|^2, \qquad (3.2.14)$$

$$= \frac{1}{16\pi} \int_{4m^2 - s}^{0} dt \, \frac{|(F(s,t)|^2}{s(s - 4m^2)}, \qquad (3.2.15)$$

where t is minus the invariant momentum transfer squared (see 3.1.10) and $F(s,t)$ is the scattering amplitude (equal to a matrix element of T). The differential cross-section with respect to t, for equal mass elastic scattering, is defined by

$$\frac{d\sigma}{dt} = \frac{|F(s,t)|^2}{16\pi s(s - 4m^2)}. \qquad (3.2.16)$$

In the general case, one can obtain the partial or differential cross-section with respect to a particular invariant by inserting a suitable δ-function in (3.2.13). To obtain the differential cross-section for two of the final state particles to have an invariant energy s_{ij}, with

$$s_{ij} = (q_i + q_j)^2, \qquad (3.2.17)$$

one inserts the δ-function

$$\delta(s_{ij} - (q_i + q_j)^2), \qquad (3.2.18)$$

into (3.2.13), giving

$$\frac{d\sigma}{dt_{ij}} = \frac{1}{2\lambda(2\pi)^{3n-4}} \sum_n \int dq_1 \ldots \int dq_n \, \delta(s_{ij} - (q_i + q_j)^2)$$

$$\times \prod_1^n \delta(q_r^2 - m_r^2) \, \theta(q_{r0}) \, \delta(p_1 + p_2 - \Sigma q_r) \, \Sigma |\langle \ldots | T | \ldots \rangle|^2. \qquad (3.2.19)$$

Whether one requires any summation over allowed states n depends on the experiment concerned. The matrix element of the scattering amplitude T in (3.2.13) is a function only of the invariants. For a two-body reaction it can be written $F(s, t)$ and taken outside the integral, which then gives for unpolarised particles,

$$\frac{d\sigma}{dt} = \frac{1}{16\pi[\lambda(s, m_1^2, m_2^2)]^2} \sum_{\text{spin}} |F(s, t)|^2. \tag{3.2.20}$$

Using (3.1.19) we obtain the differential cross-section in the centre of mass system,

$$\frac{d\sigma}{d\Omega} = \frac{\lambda(s, m_1^2, m_2^2)\,\lambda(s, m_3^2, m_4^2)}{4\pi s} \frac{d\sigma}{dt}$$

$$= \frac{\lambda(s, m_3^2, m_4^2)}{64\pi^2 s\lambda(s, m_1^2, m_2^2)} \sum_{\text{spin}} |F(s, t)|^2. \tag{3.2.21}$$

For $m_1 = m_4$ and $m_2 = m_3$ this becomes

$$\frac{d\sigma}{d\Omega} = \frac{1}{64\pi^2 s} \sum_{\text{spin}} |F(s, t)|^2. \tag{3.2.22}$$

The scattering amplitude f is often defined so that

$$\frac{d\sigma}{d\Omega} = \sum_{\text{spin}} |f(s, t)|^2. \tag{3.2.23}$$

This is therefore related to our amplitude $F(s, t)$, when (3.2.22) holds, by

$$f(s, t) = \frac{F(s, t)}{8\pi s^{\frac{1}{2}}}. \tag{3.2.24}$$

For a production process, by using a δ-function

$$\delta(t_{ij} - (p_i + q_j)^2), \tag{3.2.25}$$

one obtains the differential cross-section with respect to the invariant momentum transfer squared,

$$t_{ij} = (p_i + q_j)^2, \tag{3.2.26}$$

where $-q_j$ denotes the outgoing four-momentum of particle j in the reaction

$$1 + 2 \to 1' + 2' + \ldots + n'. \tag{3.2.27}$$

By expressing t_{ij} in terms of a scattering angle θ_{ij} one can obtain the differential cross-section with respect to this angle.

When some of the particles have non-zero spin, experimental quantities may involve several invariant amplitudes. For example, in pion-nucleon elastic scattering there are two independent invariant amplitudes (Chew, Goldberger, Low & Nambu, 1957). Then the scattering amplitude has the form,

$$F = 2m\bar{u}(p_2)\left[A(s,t) - \tfrac{1}{2}iB(s,t)\,\gamma_\mu(q_1+q_2)^\mu\right]u(p_1), \qquad (3.2.28)$$

where u, $\bar{u}$ denote Dirac spinors, and q_1, q_2 are the pion four-momenta. Taking account also of charge and assuming charge independence A and B can each be expressed, for pion-nucleon scattering, in terms of isospin amplitudes in the usual way (for example see Källen 1964). We will consider the effects of spin and isospin on scattering amplitudes in §§ 3.6, 4.9, 5.5, and later in contexts when they are relevant for experimental comparison as in Chapter 9.

3.3   Unitarity and the optical theorem

When the S-matrix is expressed in terms of the T-matrix by means of (3.2.10), the unitarity condition (3.2.5) gives,

$$i(T^\dagger_{\beta\alpha} - T_{\beta\alpha}) = (2\pi)^4 \sum_\gamma \int d\gamma\, \delta(\Sigma p_r - \Sigma q_r)\, T^\dagger_{\beta\gamma} T_{\gamma\alpha}, \qquad (3.3.1)$$

where γ denotes the states $q_1 \ldots q_n$ that are allowed by energy-momentum conservation as indicated by the δ-function in (3.3.1).

In the particular case when $\beta = \alpha$, we obtain $T_{\alpha\alpha}$, the forward scattering amplitude. For two-body scattering, (3.3.1) becomes

$$\mathrm{Im} \langle p_1, p_2, \mathrm{in}| \, T \, |p_1, p_2, \mathrm{in}\rangle = \tfrac{1}{2}(2\pi)^4 \sum_\gamma \int d\gamma\, \delta\!\left(\sum_1^2 p_r - \sum_1^n q_r\right) |T_{\gamma\alpha}|^2. \tag{3.3.2}$$

The sum and integral over the states γ is a shorthand for the sum and integral in equation (3.2.13) for the total cross-section σ_t for two-body scattering. The left-hand side of (3.3.2) involves only the scattering amplitude F in the forward direction, $\cos\theta = 1$. Hence the unitarity condition (3.3.2), on comparison with (3.2.13), gives

$$\mathrm{Im}\, F(s,0) = \lambda(s, m_1^2, m_2^2)\, \sigma_{\mathrm{total}}, \qquad (3.3.3)$$

where $0 = t(s, \cos\theta = 1)$ is the value of the invariant momentum transfer squared in the forward direction. The invariant λ can be evaluated with particle 2 at rest, giving

$$\mathrm{Im}\, F(s,0) = 2m_2(\mathbf{p}_1^2)^{\frac{1}{2}}\, \sigma_{\mathrm{total}}. \qquad (3.3.4)$$

For scattering of equal mass particles,

$$\mathrm{Im}\, F(s,0) = 2ks^{\frac{1}{2}}\sigma_{\mathrm{total}}. \qquad (3.3.5)$$

From (3.2.16) the elastic differential cross-section at $t = 0$, involves $|F|^2$, and gives

$$\frac{d\sigma(t=0)}{dt} = \frac{[\mathrm{Re}\,F(s,0)]^2 + [\mathrm{Im}\,F(s,0)]^2}{16\pi s(s-4m^2)}. \qquad (3.3.6)$$

Hence
$$[\mathrm{Re}\,F(s,0)]^2 = 64\pi sk^2\frac{d\sigma(0)}{dt} - 4sk^2(\sigma_{\text{total}})^2. \qquad (3.3.7)$$

Note that $d\sigma(0)/dt$ denotes $d\sigma(t)/dt$ evaluated at $t = 0$. An analogous relation holds for the general mass case with modification only of the phase space factors.

The relation (3.3.7) gives in principle a method for evaluating the real part of the forward amplitude. In practice the method can be used most accurately for charged particles and it is then necessary to allow for Coulomb effects before extrapolating $d\sigma/dt$ to the forward direction.

For pion-nucleon scattering the scattering amplitude with neglect of Coulomb effects can be written,

$$\left(\frac{1}{8\pi s^{\frac{1}{2}}}\right) F(s,t) = f(s,t) + i\boldsymbol{\sigma}\cdot\mathbf{n}\,g(s,t), \qquad (3.3.8)$$

where
$$\mathbf{n} = \frac{[\mathbf{q}_1 \wedge \mathbf{q}_2]}{|\mathbf{q}_1|\,|\mathbf{q}_2|}, \qquad (3.3.9)$$

is a vector perpendicular to the initial and final three-momenta of the pions, $\mathbf{q}_1$ and $\mathbf{q}_2$. The amplitudes f and g refer to non-spin-flip and spin-flip respectively (see §3.6).

The differential cross-section near $t = 0$ including Coulomb interference (Lindenbaum, 1965) for $\pi^+ p$ scattering, is

$$\left(\frac{q^2}{\pi}\right)\frac{d\sigma}{dt} = \left|-\frac{G}{t}\exp(2i\delta) + f_R + if_I\right|^2, \qquad (3.3.10)$$

where $f_R = \mathrm{Re}\,f$ and $f_I = \mathrm{Im}\,f$; and δ is the relative phase shift introduced between the nuclear and Coulomb phase shifts by the long range Coulomb interaction (Bethe, 1958; Solov'ev, 1966), and G is the product of nucleon and pion form factors (see §4.3). Approximating the latter two to be equal, (note that our normalisation differs from that of Lindenbaum),

$$G \approx \left(\frac{2e^2\pi}{q}\right)\left[G_E^2(t) - \frac{t}{4M^2}G_M^2(t)\right]\Big/[1-t/4M^2], \qquad (3.3.11)$$

where
$$G_M \approx \mu G_E \approx \mu[1 + (2\cdot77)\,t]; \quad -t < 0\cdot1\,(\text{Gev}/\text{c})^2. \qquad (3.3.12)$$

It is reasonable to assume for small t, ($|t| < 0 \cdot 1 \, (\mathrm{Gev}/\mathrm{c})^2$), that

$$f_I = \mathrm{Im} f \approx \exp\left[(a+bt)/2\right], \tag{3.3.13}$$

where a is determined from total cross-section measurements using the optical theorem.

Define the real to imaginary ratio α by

$$\alpha(s) = \frac{\mathrm{Re} f(s,0)}{\mathrm{Im} f(s,0)} = \frac{f_R(s,0)}{f_I(s,0)}. \tag{3.3.14}$$

Then,

$$\left(\frac{q^2}{\pi}\right)\frac{d\sigma}{dt} = \frac{G^2}{|t|^2} - \frac{2G}{|t|}(\mathrm{Im} f)(\alpha\cos 2\delta + \sin 2\delta) + (1+\alpha^2)(\mathrm{Im} f)^2. \tag{3.3.15}$$

For $\pi^- p$ scattering, the signs of G and of $\sin\delta$ in this expression must be changed.

By plotting $(d\sigma/dt)$ against t, at small momentum transfer, $|t| < 0 \cdot 01 \, (\mathrm{Gev}/\mathrm{c})^2$, the ratio $\alpha(s)$ in (3.3.14) can be determined giving results like those indicated in Fig. 2.1.3.

For one uncharged particle there is no Coulomb interference and, in principle, one can extrapolate to $t = 0$ to give $\alpha(s)$ directly, However, in practice, the measurements are less exact since for example in pn scattering, the target neutron will be in a deuterium target and one has to remove the effects of the target proton.

3.4 Phase space methods and Dalitz diagrams

Final state interactions can be observed by measuring the differential cross-section with respect to the invariant energy for two (or more) final state particles in a production reaction. For example, consider the reaction shown in Fig. 3.4.1a,

$$1 + 2 \rightarrow 3 + 4 + 5, \tag{3.4.1}$$

as a function of the invariant energy s_{34},

$$s_{34} = (p_3 + p_4)^2. \tag{3.4.2}$$

The corresponding differential cross-section is given by (3.2.19), which becomes in this example,

$$\begin{aligned}
P(s, s_{34}) &= \frac{d\sigma}{ds_{34}}, \\
&= \frac{1}{2\lambda(s, m_1^2, m_2^2)(2\pi)^5} \int dp_3 \, dp_4 \, dp_5 \, \delta\left(\sum_1^5 p_r\right) \\
&\quad \times \delta(s_{34} - (p_3 + p_4)^2) \prod_3^5 \delta(p_r^2 - m_r^2)\, \theta(p_{r0}) \, \Sigma \, |T|^2,
\end{aligned} \tag{3.4.3}$$

where $s = s_{12}$ is the invariant energy of the initial particles 1 and 2. The production amplitude T will be a function of five independent invariant energies (or momentum transfers) chosen from

$$s_{ij} = (p_i + p_j)^2 \quad (i, j = 1, 2, ..., 5). \tag{3.4.4}$$

If the individual particles have spin, then T will also involve scalars formed from the products of spin and momentum, such as $\gamma_\mu p_i^\mu$. For simplicity we will assume spinless particles. Then

$$T = T(s_{12}, s_{13}, s_{14}, s_{15}, s_{34}). \tag{3.4.5}$$

If particles 3 and 4 are produced partly through the intermediate production of a particle B of mass m_{34}, which then decays as indicated in Fig. 3.4.1(b), the amplitude T can be written

$$T = T_0 + T(s_{12}, s_{15}, m_{34}) \left[\frac{G}{s_{34} - m_{34}^2 + i\Gamma m_{34}} \right], \tag{3.4.6}$$

where T_0, G and Γ only vary slowly with s_{34}. An example, where this separation is certainly useful, is

$$\pi^- + p \to \Sigma^- + K^+ \atop \llcorner \! \to n + \pi^-. \tag{3.4.7}$$

Since Σ^- travels an appreciable distance before decaying, the process in Fig. 3.4.1(b) can be observed experimentally. However, even with decay via strong interactions and a large background term T_0 in (3.4.6), the resonant form of T can often be observed as a peak in $P(s, s_{34})$ near $s_{34} = m_{34}^2$. An example is given by

$$\pi^- + p \to \rho^0 + n \atop \llcorner \! \to \pi^+ + \pi^-. \tag{3.4.8}$$

To check whether there is a resonant term in (3.4.6), it is necessary to find the contribution from the background term T_0 which, by definition, can be regarded as nearly constant with respect to variations of s_{34} near m_{34}^2. The dependence of T_0 (for fixed s) on the other independent variables is neglected. With $T = T_0 = $ constant in (3.4.3), we obtain the background differential cross-section $P_0(s, s_{34})$. This quantity is called the phase space factor or contribution. Resonances are observed as departures from the phase space contribution. The phase space integral $P_0(s, s_{34})$ has a typical form as shown in Fig. 3.4.2(a). The detailed shape (and the cut-off value) depends on the value of the fixed energy s, as well as the masses of the particles.

More detailed information about a resonance may be obtained from a two-dimensional plot of events in a reaction

$$1 + 2 \rightarrow 3 + 4 + 5. \tag{3.4.9}$$

Each event is plotted as a point on a diagram in which the coordinates are the invariant energies s_{35} and s_{45} (or one could use the laboratory energies of particles 4 and 3 respectively). This type of diagram was

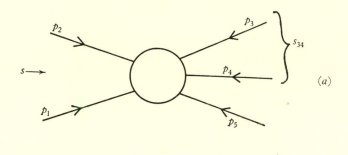

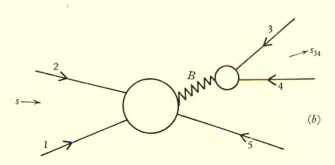

Fig. 3.4.1. (a) The production reaction $1 + 2 \rightarrow 3 + 4 + 5$; (b) a final state interaction via a resonance B.

first proposed by Dalitz (1953) and a typical example is shown in Fig. 3.4.2 (b). Events outside the closed curve are forbidden by energy-momentum conservation. The density of points in any neighbourhood is proportional to

$$P(s, s_{35}, s_{45}) = \frac{d^2\sigma}{ds_{35}\, ds_{45}}. \tag{3.4.10}$$

This quantity can be obtained from (3.4.3) by the substitution of the appropriate δ-functions,

$$\delta(s_{35} - (p_3 + p_5)^2)\, \delta(s_{45} - (p_4 + p_5)^2). \tag{3.4.11}$$

The resulting background contribution that is obtained by taking

$T = T_0 =$ constant, gives within the closed curve shown in Fig. 3.4.2(b),

$$P_0(s, s_{35}, s_{45}) = \frac{C}{\lambda(s, m_1^2, m_2^2)} \int d\Omega_3 \, d\phi_4 \, |T_0|^2$$

$$= \text{constant for fixed } s. \tag{3.4.12}$$

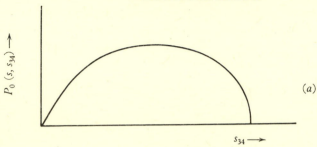

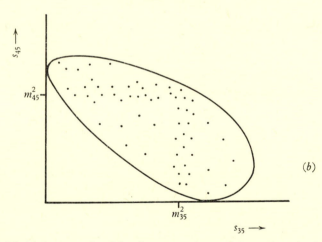

Fig. 3.4.2. (a) An example of the phase space integral (3.4.3) for a production process $1+2 \rightarrow 3+4+5$; (b) An example of a Dalitz diagram for a production process. Each dot indicates an observed event and the density of dots is proportional to $P(s, s_{35}, s_{45})$. A uniform density would indicate there was no resonance. In this figure, a resonance is suggested in the 3, 5 system at m_{35}^2, and one in the 4, 5 system at m_{45}^2.

The departure from a constant distribution of events shown in Fig. 3.4.2(b) indicates a resonance at m_{35}^2 in the 3, 5 system and one at m_{45}^2 in the 4, 5 system. This situation occurs for example in the Dalitz diagram for the reaction

$$K^- + p \rightarrow \pi^+ + \pi^- + \Lambda^0. \tag{3.4.13}$$

The increases in the density of events correspond to the resonances Y_1^{*+} and Y^{*-} at 1385 Mev.

When the final state particles are identical, differerent variables can be used that take better account of the symmetry (Dalitz, 1953). Further refinements of great importance allow for the variation with angle of the matrix element T when the particle B (Fig. 3.4.1(b)) is either pseudo-scalar or has non-zero spin, (Stevenson *et al.* 1962). An account of these further details is given by Källen (1964).

It is becoming of increasing importance in high energy collisions not only to know that a resonance exists, but also to know its differential cross-section with respect to invariant momentum transfer and the energy dependence of this differential cross-section. In terms of Fig. 3.4.1(b) one requires $d\sigma/dt$ for production of the particle B. This is therefore evaluated with s_{34} near the resonance value m_{34}^2,

$$P(s, t, m_{34}) = \frac{d^2\sigma}{ds_{34}\,dt} \quad \text{with} \quad s_{34} \approx m_{34}^2. \qquad (3.4.14)$$

where
$$t = (p_1 + p_5)^2. \qquad (3.4.15)$$

The production amplitude T will certainly depend on the value of t, which is itself directly related to the scattering angle for production of the unstable particle or resonance B. Experimental information on this dependence is likely to be as important as in the differential cross-sections for two-body reactions. Processes in which all final state particles are associated, as a particle plus a resonance, or as two resonances, are described as quasi two-body reactions. Some of their experimental properties were described in § 2.7.

3.5 Partial waves, inelastic effects, Levinson's theorem

In this section we consider partial waves for the elastic scattering process
$$1 + 2 \to 3 + 4, \qquad (3.5.1)$$

in which $m_1 = m_4$ and $m_2 = m_3$ (making u the exchange momentum variable). In the centre of mass system, let $\mathbf{q}$, $\mathbf{q}'$ denote the initial and final three-momenta with
$$k = |\mathbf{q}| = |\mathbf{q}'|, \qquad (3.5.2)$$

and let W denote the total energy and θ the scattering angle. Then

$$W = \sqrt{s} = [m_1^2 + m_2^2 + 2k^2 + 2(k^2 + m_1^2)^{\frac{1}{2}}(k^2 + m_2^2)^{\frac{1}{2}}]^{\frac{1}{2}}, \quad (3.5.3a)$$

$$4k^2 = s - 2(m_1^2 + m_2^2) + \frac{(m_1^2 - m_2^2)^2}{s}. \qquad (3.5.3b)$$

From (3.1.19), after substituting $t = \Sigma m_r^2 - s - u$, we obtain

$$-\cos\theta = \frac{s^2 + s(2u - \Sigma m_r^2) - (m_1^2 - m_2^2)^2}{[\lambda(s, m_1^2, m_2^2)]^2}, \tag{3.5.4}$$

where we recall (3.1.15),

$$[\lambda(s, m_1^2, m_2^2)]^2 = s^2 + m_1^4 + m_2^4 - 2s(m_1^2 + m_2^2) - 2m_1^2 m_2^2. \tag{3.5.5}$$

The scattering process (3.5.1) is physical when

$$s > (m_1 + m_2)^2 \quad \text{and} \quad |\cos\theta| \leqslant 1. \tag{3.5.6}$$

We will consider the boundaries of the physical region in more detail in §4.6. The boundaries of this region are given by: $\cos\theta = -1$ for backward scattering, for which (3.5.4) gives the boundary curve shown in Fig. 4.6.1, having the equation

$$su = (m_1^2 - m_2^2)^2; \tag{3.5.7}$$

and $\cos\theta = +1$ for forward scattering (for which $t = 0$), which gives

$$s + u - 2(m_1^2 + m_2^2) = 0. \tag{3.5.8}$$

The partial wave expansion of the scattering amplitude is

$$\frac{1}{8\pi W} F(s, t) = \frac{1}{k} \sum_{l=0}^{\infty} (2l+1) f_l(s) P_l(\cos\theta). \tag{3.5.9}$$

The partial wave amplitude is given by

$$f_l(s) = \frac{1}{2} \int_{-1}^{1} d(\cos\theta) \frac{kF(s,t)}{8\pi W} P_l(\cos\theta). \tag{3.5.10}$$

If we consider values of W for which only elastic scattering is possible, the unitarity conditions (3.3.1) takes the simple form,

$$\frac{ik}{8\pi W} \{F(s,t) - F^*(s,t)\} = -\int_{-1}^{1} d(\cos\theta) \left(\frac{k}{8\pi W}\right)^2 F^*(s,t') F(s,t''), \tag{3.5.11}$$

where $t' = (p_1 - x)^2$ and $t'' = (p_4 - x)^2$ with x the four-momentum of the intermediate state. Substituting from (3.5.9), and using the orthogonality of Legendre polynomials one finds that

$$i[f_l^*(W) - f_l(W)] = 2|f_l(W)|^2. \tag{3.5.12}$$

Hence
$$f_l(W) = \frac{\exp(2i\delta_l) - 1}{2i}, \tag{3.5.13}$$

where the phase shift $\delta_l(W)$ is real for W in the elastic region.

More generally, if inelastic scattering can occur, for example through production processes, there will be additional terms on the right of the unitarity condition (3.5.11) that contain production amplitudes as in (3.3.1). Then the partial wave S-matrix will take the form,

$$(S_l) = 1 + 2i \begin{pmatrix} f_l(W), & g_l^{23}(W, \alpha), & \dots, & g_l^{2n}(W, \alpha') \\ g_l^{32}(W, \alpha''), & - & \dots, & - \\ - & & \dots, & - \\ g_l^{n2}(W, \alpha'''), & - & \dots, & - \end{pmatrix}, \quad (3.5.14)$$

where α denotes the extra invariants for production processes, $2 \to r$ particles.

The unitarity condition that includes inelastic effects is,

$$i(f_l^* - f_l) = 2f_l^* f_l + 2 \sum_3^n \int d\alpha (g_l^{2r})^* (g_l^{2r}). \quad (3.5.15)$$

Writing $2\mathscr{P}_l$ for the last term in (3.5.15), we have

$$i(f_l^* - f_l) = 2f_l^* f_l + 2\mathscr{P}_l. \quad (3.5.16)$$

The probability for elastic scattering must not exceed one (whether or not inelastic processes are possible),

$$|S_l(\text{elastic})| = |1 + 2if_l(W)| \leqslant 1. \quad (3.5.17)$$

Hence,
$$0 \leqslant |f_l|^2 \leqslant \operatorname{Im} f_l \leqslant 1, \quad (3.5.18)$$

and
$$0 \leqslant \operatorname{Im} f_l - |f_l|^2 \leqslant \tfrac{1}{4}, \quad (3.5.19)$$

giving
$$0 \leqslant \mathscr{P}_l \leqslant \tfrac{1}{4}. \quad (3.5.20)$$

It is convenient for some purposes to write the partial wave amplitude as

$$f_l(W) = \frac{\eta_l \exp 2i\delta_l - 1}{2i}, \quad (3.5.21)$$

where δ_l is real and η_l takes account of inelastic effects. We can choose η_l to be real and

$$0 \leqslant \eta_l \leqslant 1. \quad (3.5.22)$$

The inelasticity coefficient η_l is related to $\mathscr{P}_l$ by

$$\mathscr{P}_l = \tfrac{1}{4}(1 - \eta_l^2).$$

In the elastic region, $\eta_l = 1$.

When there are inelastic processes allowed by energy and other conservation laws, we can define three associated partial wave cross-sections, σ^l(elastic), σ^l(inelastic) and σ^l(total). The latter is related to the total collision cross-section by,

$$\sigma(\text{total}) = \sum_{l=0}^{\infty} \sigma^l(\text{total}), \tag{3.5.23}$$

and there are similar relations for σ^l(elastic) and σ^l(inelastic). From the optical theorem (3.3.3),

$$\sigma(\text{total}) = \frac{\text{Im}\, F(s, 0)}{\lambda(s, m_1^2, m_2^2)}, \tag{3.5.24}$$

$$= \frac{8\pi W}{k\lambda(s, m_1^2, m_2^2)} \sum_l (2l+1)\, \text{Im}\, f_l(W)\, \mathscr{P}_l(1). \tag{3.5.25}$$

We therefore have, for example, in the equal mass case, $(\lambda = 2Wk)$,

$$\sigma^l(\text{total}) = \frac{(2l+1)\,\pi}{k^2}\, (2 - 2\eta_l \cos 2\delta_l), \tag{3.5.26}$$

$$\sigma^l(\text{elastic}) = \frac{(2l+1)\,\pi}{k^2}\, (1 - 2\eta_l \cos 2\delta_l + \eta_l^2), \tag{3.5.27}$$

$$\sigma^l(\text{inelastic}) = \frac{(2l+1)\,\pi}{k^2}\, (1 - \eta_l^2). \tag{3.5.28}$$

The inelastic scattering for any partial wave is a maximum when $\eta_l = 0$, for which we obtain (using equal masses),

$$\sigma^l(\text{elastic}) = \sigma^l(\text{inelastic}) = \frac{2l+1}{k^2}. \tag{3.5.29}$$

However this does not give the maximum *proportion* of inelastic scattering in the partial wave. In order to obtain this we must maximise the R function,

$$R_l(W) = \frac{\sigma^l(\text{total})}{\sigma^l(\text{elastic})} = \frac{2 - 2\eta_l \cos 2\delta_l}{1 - 2\eta_l \cos 2\delta_l + \eta_l^2}. \tag{3.5.30}$$

With no inelastic scattering, $\eta_l = 1$ and $R_l = 1$. If $\eta_l = 0$, then f_l is pure imaginary and $R_l = 2$, which makes σ^l(inelastic) have its largest value but not its largest share of σ^l(total). The latter is obtained when $R_l \to \infty$.

For some methods using partial wave dispersion relations, it is necessary to use an assumed form of $R_l(W)$ for large values of W (see §4.7). Therefore one would like to know the limiting value of $R_l(W)$ as $W \to \infty$. This is an open question at the present time. Its allowed range of values is

$$1 \leqslant R_l(W) \leqslant \infty. \tag{3.5.31}$$

The lower limit is achieved if the scattering becomes purely elastic, which seems a very unlikely possibility since at high energy one would expect many inelastic channels to be open. The value $R_l = 2$ corresponds to pure absorption with $\eta_l = 0$. However, if we have $\delta_l \to 0$, or $n\pi$, (where n is an integer) as $W \to \infty$, then

$$R_l \sim \frac{2}{(1-\eta_l)}. \tag{3.5.32}$$

Since $|1-\eta_l| \leqslant 1$, this can take any limiting value in the range

$$2 \leqslant R_l \leqslant \infty. \tag{3.5.33}$$

We see that the behaviour of the real part of the phase, namely $\delta_l(W)$ plays an important role in determining the possible limits for $R_l(W)$ as $W \to \infty$. If δ_l does not tend to an integer multiple of π, then R_l must remain bounded. For non-relativistic elastic scattering (for which $\eta_l = 1$) Levinson's theorem states that $\delta_l \to n\pi$, where n is an integer (which may be negative). We will outline a proof of Levinson's theorem.

Levinson's theorem for non-relativistic potential scattering

Reference Levinson (1949) (see also Vaughn, Aaron & Amado, 1961, whose derivation is partly used here). The theorem states that the phase shift $\delta_l(E)$ for elastic potential scattering changes by an integer multiple of π as the energy E varies from 0 to ∞,

$$\delta_l(0) - \delta_l(\infty) = (n_b - n_p)\pi, \tag{3.5.34}$$

where n_b denotes the number of bound states of angular momentum l and n_p denotes the number of CDD poles (reference Castillejo, Dalitz & Dyson, 1956 and see below). Assuming that the phase shift $\delta_l(E)$ is bounded as $E \to \infty$, we can define (Omnes, 1958, 1961), a function $D(E)$ by

$$D(E) = \exp\left[-\frac{(E-E_0)}{\pi}\int_0^\infty dE' \frac{\delta(E')}{(E'-E)(E'-E_0)}\right], \tag{3.5.35}$$

where we have omitted the suffix l for simplicity. Then $D(E)$ is real for $E < 0$, and for $E > 0$ it has the phase $-\delta(E)$. Hence the function $N(E)$ will be real for $E > 0$, where N is defined by

$$N = N_l(E) = \frac{1}{k}f_l(E)D_l(E), \quad k^2 = E, \tag{3.5.36}$$

for each partial wave. Now,

$$f_l = \frac{1}{2i}[\exp(2i\delta_l) - 1], \tag{3.5.37}$$

and for $E > 0$, δ_l is real so that,

$$\text{Im}\,(f_l)^{-1} = 1. \tag{3.5.38}$$

Hence, writing $(E + i0)$ for the limit of $(E + i\epsilon)$ as $\epsilon \to 0$,

$$\frac{1}{2ik}[D_l(E + i0) - D_l(E - i0)] = N_l(E)\,\text{Im}\,[f_l]^{-1} = N_l(E). \tag{3.5.39}$$

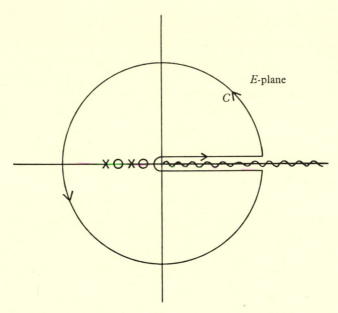

Fig. 3.5.1. The complex energy-plane (the E-plane), showing the contour C used in equations (3.5.43 and 44). Zeros of $D(E)$ are shown by a small circle and poles (CDD poles) by a small cross.

The partial wave S-matrix element is related to D_l by,

$$S_l(E) = \exp(2i\delta_l) = \frac{D_l^*(E)}{D_l(E)} = \frac{D_l(E - i0)}{D_l(E + i0)}, \tag{3.5.40}$$

so the phase shift δ_l is obtained from

$$\delta_l(E) = \frac{1}{2i}\log S_l(E) = \frac{1}{2i}[\log D_l(E - i0) - \log D_l(E + i0)]. \tag{3.5.41}$$

Consider now the integral, round the closed contour C shown in Fig. 3.5.1, of the function of E,

$$\frac{1}{D(E)}\left(\frac{dD}{dE}\right). \tag{3.5.42}$$

By Cauchy's theorem

$$\int_C \frac{dE}{D}\left(\frac{dD}{dE}\right) = 2\pi i(n_b - n_p), \tag{3.5.43}$$

where n_b denotes the number of zeros of $D = D_l(E)$ in the complex E-plane cut along the real axis, and n_p denotes the number of poles. It can be shown that the zeros of $D_l(E)$ in the E-plane occur only for E real and negative, and that they determine all the bound state poles of the amplitude f_l (subject to conditions on the potential, e.g. a Yukawa potential). The poles of $D_l(E)$ are CDD poles (see Castillejo *et al.* 1956), and in this context they can be regarded as a definition of what is meant by CDD poles.

From (3.5.41) and Fig. 3.5.1, neglecting the contribution from the curved part of C (which is justified by our assumption that $\delta_l \to$ constant at infinity),

$$\int_C \frac{dE}{D}\left(\frac{dD}{dE}\right) = \int_0^\infty dE \, \frac{d}{dE}\left[\log D(E+i0) - \log D(E-i0)\right]$$

$$= \int_0^\infty dE \, \frac{d}{dE}\left[-2i\,\delta(E)\right]. \tag{3.5.44}$$

From (3.5.43) and (3.5.44), Levinson's result follows:

$$\delta_l(0) - \delta_l(\infty) = (n_b - n_p)\,\pi. \tag{3.5.45}$$

It is usual to define $\delta_l(0)$ to be zero, so Levinson's theorem states that $\delta_l(E) \to n\pi$ as $E \to \infty$, where n is an integer or zero.

We will return to further discussion of the 'N over D' method in §4.7 when we consider partial wave dispersion relations. Levinson's theorem can be extended to include inelastic potential scattering, but it has not been proved in the physical case where relativistic effects include the production of particles in coupled channels. If a relativistic extension can be established it is likely to apply to the eigenphases and will probably be only approximate since one would need in practice to limit the number of channels considered.

3.6   Spin and isospin; helicity amplitudes

When the colliding particles have non-zero spin, rotational invariance shows that the total angular momentum is conserved. For each angular momentum J, there are several scattering amplitudes corresponding to different orientations of the spin. When the particles are charged each of the spin amplitudes is replaced by several amplitudes corresponding to the different charge states or isospin states. For illustration we will consider pion-nucleon scattering and will show how the various ways of writing the full scattering amplitude are related.

(a) Invariant amplitudes and isospin amplitudes

Write p_i, q_i for the nucleon and pion four-momenta in the initial $(i = 1)$ and final $(i = 2)$ states. The S-matrix element (3.2.10) between final (β) and initial (α) states becomes

$$S_{\beta\alpha} = \delta_{\beta\alpha} + i(2\pi)^4 \delta^{(4)}(p_2 + q_2 - p_1 - q_1) \, 2m\bar{u}_\beta(p_2) \, T_{\beta\alpha} u_\alpha(p_1), \quad (3.6.1)$$

where u_α, u_β are the Dirac spinors for the initial and final states of the nucleon. The differential cross-section (3.2.22) becomes

$$\frac{d\sigma}{d\Omega}(\pi_1 p_1 \to \pi_2 p_2) = \frac{m^2}{16\pi^2 s} \sum_{\text{(spin)}} |\bar{u}_\beta T_{\beta\alpha} u_\alpha|^2. \quad (3.6.2)$$

The scattering amplitude $T_{\beta\alpha}$ can be expressed in terms of invariant amplitudes A and B, (Chew, Goldberger, Low & Nambu 1957),

$$T_{\alpha\beta} = A(s,t) - \tfrac{1}{2}iB(s,t)\gamma_\mu(q_1 + q_2)^\mu. \quad (3.6.3)$$

For different charge states, one can introduce extra indices on A and B, or express them in terms of amplitudes for particular isospin states. For example

$$A(\pi^+ p \to \pi^+ p) = A(\tfrac{3}{2}),$$

$$A(\pi^- p \to \pi^- p) = \tfrac{1}{3}[A(\tfrac{3}{2}) + 2A(\tfrac{1}{2})], \quad (3.6.4)$$

$$A(\pi^- p \to \pi^0 n) = \frac{\sqrt{2}}{3}[A(\tfrac{3}{2}) - A(\tfrac{1}{2})],$$

with similar equations for B. The charge or isospin indices can be introduced similarly in any of the following equations, with the usual Clebsch–Gordan relations. They will not be shown explicitly, in order to avoid an unduly complicated notation, unless they are of significant importance for the discussion.

The invariant amplitudes $A(s,t)$ and $B(s,t)$ are useful in that their analytic properties are simple. For example, they each satisfy the Mandelstam representation (see § 4.4) if it holds for the scattering of spinless particles. However, two other kinds of amplitude are extensively used, particularly for experimental comparison. These are the spin-flip and the non-spin-flip amplitudes, and the helicity amplitudes. We consider the latter first.

(b) Helicity amplitudes

A helicity eigenstate for a particle a of spin λ, is a state for which the spin has a definite eigenvalue λ_a in the direction of the three-momentum of particle a in the centre of mass system of the colliding particles. A helicity amplitude is a scattering amplitude taken between helicity eigenstates. A detailed account of them is given by Jacob & Wick (1959).

For pion-nucleon scattering there are two 'conventional' amplitudes f_1 and f_2. It is convenient first to express the scattering amplitude in terms of the 2×2 Pauli scattering matrix M, for which

$$\frac{d\sigma}{d\Omega} = \Sigma \, |\langle\beta|\, M \,|\alpha\rangle|^2, \tag{3.6.5}$$

$$\langle\beta|\, M \,|\alpha\rangle = \frac{m}{4\pi s^{\frac{1}{2}}} \overline{u}_\beta \, T_{\beta\alpha} \, u_\alpha. \tag{3.6.6}$$

In terms of the pion-nucleon conventional amplitudes f_1, f_2,

$$M = f_1(s,t) + (\boldsymbol{\sigma}.\hat{\mathbf{q}}_2)\,(\boldsymbol{\sigma}.\hat{\mathbf{q}}_1) f_2(s,t), \tag{3.6.7}$$

where $\hat{\mathbf{q}}_1$ and $\hat{\mathbf{q}}_2$ are unit vectors along the initial and final pion three-momenta in the centre of mass system (note that $\mathbf{q}_1 = -\mathbf{p}_1, \mathbf{q}_2 = -\mathbf{p}_2$). There are two helicity amplitudes to consider,

$$F_{++} = F_{--} = (f_1 + f_2) \cos{(\tfrac{1}{2}\theta)}, \tag{3.6.8}$$

$$F_{+-} = -F_{-+} = (f_1 - f_2) \sin{(\tfrac{1}{2}\theta)}, \tag{3.6.9}$$

where θ is the scattering angle in the s channel centre of mass system. These are the helicity non-flip amplitude F_{++} and the helicity flip amplitude F_{+-}. The left-hand equalities in (3.6.8, 9) follow from parity and time reversal invariance (which were also used in asserting that there are only two invariant amplitudes A and B in (3.6.3)).

The differential cross-section for scattering with no change in helicity is,

$$\frac{d\sigma(++)}{d\Omega} = |F_{++}|^2 = [|f_1|^2 + |f_2|^2 + (f_1^* f_2 + f_2^* f_1)] \cos^2 \tfrac{1}{2}\theta. \tag{3.6.10}$$

With helicity flip,

$$\frac{d\sigma(+-)}{d\Omega} = |F_{+-}|^2 = [|f_1|^2 + |f_2|^2 - (f_1^* f_2 + f_2^* f_1)] \sin^2 \tfrac{1}{2}\theta. \tag{3.6.11}$$

The polarisation of the recoil nucleon (for non-polarised initial state) is given by

$$P(\theta)\frac{d\sigma}{d\Omega} = \sin\theta \, \mathrm{Im} \, [f_1^* f_2].$$
(3.6.12)

It should be noted that the advantages of helicity amplitudes over the invariant amplitudes become really significant only when more complicated spins are involved than in the pion-nucleon system.

From (3.6.3) and (3.6.6, 7) one can relate the amplitudes, f_1 and f_2, to the invariant amplitudes,

$$f_1 = \frac{(E+m)}{8\pi W}[A + (W-m)B],$$
(3.6.13)

$$f_2 = \frac{(E-m)}{8\pi W}[-A + (W+m)B],$$
(3.6.14)

where E is the nucleon energy, and W the total energy in the centre of mass system,

$$E^2 = m^2 + p^2, \quad W = s^{\frac{1}{2}}.$$
(3.6.15)

In terms of the non-spin-flip and the spin-flip amplitudes f and g, (Ashkin, 1959; Donnachie, 1966),

$$M = f(s,t) + i\boldsymbol{\sigma} \cdot \left[\frac{(\hat{\mathbf{q}}_2 \wedge \hat{\mathbf{q}}_1)}{|\hat{\mathbf{q}}_2 \wedge \hat{\mathbf{q}}_1|}\right] g(s,t),$$
(3.6.16)

giving

$$f = f_1 + f_2 \cos\theta,$$
(3.6.17)

$$g = f_2 \sin\theta.$$
(3.6.18)

The total cross-section by the optical theorem is

$$\sigma(\mathrm{total}) = \frac{4\pi}{|\mathbf{q}_1|} \, \mathrm{Im} f(s, t = 0).$$
(3.6.19)

The differential cross-section is

$$\frac{d\sigma}{d\Omega} = |f(s,t)|^2 + |g(s,t)|^2.$$
(3.6.20)

(c) *Polarisation*

The polarisation of the recoil nucleon is $P(\theta)$, where

$$P(\theta)\frac{d\sigma}{d\Omega} = 2 \, \mathrm{Im} \, [f^* g].$$
(3.6.21)

Written as a vector, the polarisation is

$$\mathbf{P} = \frac{2 \, \mathrm{Im} \, (f^* g)}{|f|^2 + |g|^2} \frac{\mathbf{q}_1 \wedge \mathbf{q}_2}{|\mathbf{q}_1 \wedge \mathbf{q}_2|}.$$
(3.6.22)

The Wolfenstein parameters R and A, for double polarisation, are given by, (Wolfenstein, 1954; Phillips & Rarita, 1965),

$$R\frac{d\sigma}{d\Omega} = [|f|^2 - |g|^2]\cos(\theta - \psi) - 2\,\mathrm{Re}\,(fg^*)\sin(\theta - \psi), \quad (3.6.23)$$

$$A\frac{d\sigma}{d\Omega} = [|f|^2 - |g|^2\sin(\theta - \psi) + 2\,\mathrm{Re}(fg^*)\cos(\theta - \psi)], \quad (3.6.24)$$

where ψ is the laboratory angle between the recoil nucleon and the incident pion.

(d) Helicity partial waves

We will not be concerned, in our account of high energy scattering, with details of phase shift analysis. There is an account of pion-nucleon phase shift studies by Donnachie (1966) and Donnachie & Hamilton (1965) who give earlier references (see also Hamilton, 1964; Hamilton & Woolcock, 1963). At the present time there is no very reliable way of extending phase shift analysis to energies above 2 Gev, although it is possible that methods for this analysis may be developed when we have a better understanding of collisions at higher energies. It is undoubtedly an area of importance to the study of strong interactions and it has been omitted from the subject matter of this book with some reluctance. The results of phase shift analysis in identifying resonances and their properties are of considerable importance to Regge theory (Chapters 5 and 9), and they will be quoted as required. We will later make use of partial wave analysis in the general theory of high energy scattering, and will briefly outline the analogous development for helicity amplitudes.

The helicity partial wave expansion for the reaction

$$a + b \to c + d \quad (3.6.25)$$

with helicities $\lambda_a, \lambda_b, \lambda_c, \lambda_d$, has been given by Jacob & Wick (1959). It is

$$(c,d|\,F\,|a,b) = q^{-1}\sum_{J}(J + \tfrac{1}{2})\,(\lambda_c\lambda_d|f^J(W)|\lambda_a\lambda_b)\exp[i(\lambda - \mu)\,\phi]\,d^J_{\lambda\mu}(\theta), \quad (3.6.26)$$

where W is the centre of mass energy, and q the momentum, and

$$\lambda = \lambda_a - \lambda_b; \quad \mu = \lambda_c - \lambda_d. \quad (3.6.27)$$

Jacob & Wick (1959) also give general intensity and polarisation formulae, and recurrence relations for the d functions $d^J_{\lambda\mu}(\theta)$. For spinless particles,

$$d^l_{00}(\theta) = P_l(\cos\theta). \qquad (3.6.28)$$

For general spin, the $d^J_{\lambda\mu}(\theta)$ are best evaluated and studied in terms of Jacobi polynomials $P^{(b,c)}_A(\cos\theta)$, (Szego, 1948; (Wang, 1966),

$$d^J_{\lambda\mu}(\theta) = \pm \left[\frac{(J+M)!\,(J-M)!}{(J+N)!\,(J-N)!}\right]^{\frac{1}{2}}$$
$$\times (\cos\tfrac{1}{2}\theta)^{|\lambda+\mu|}\,(\sin\tfrac{1}{2}\theta)^{|\lambda-\mu|}\,P^{(|\lambda-\mu|,\,|\lambda+\mu|)}_{(J-M)}(\cos\theta), \quad (3.6.29)$$

where $\qquad M = \max.\,(|\lambda|,|\mu|), \quad N = \min.\,(|\lambda|,|\mu|).$

A number of useful properties and relations between d functions are given by Andrews & Gunson (1964).

For pion-nucleon scattering,

$$\lambda_a = \pm\tfrac{1}{2}; \quad \lambda_c = \pm\tfrac{1}{2}; \quad \lambda_b = \lambda_d = 0. \qquad (3.6.30)$$

Using (3.6.8, 9), for πN scattering, F_{++} and F_{+-} are given by,

$$F_{++}_{+-} = \sum_J (2J+1) f^J_{++}_{+-}(W)\exp\left[i(\tfrac{1}{2}\pm\tfrac{1}{2})\,\phi\right]d^J_{\frac{1}{2},\pm\frac{1}{2}}(\theta). \qquad (3.6.31)$$

The sum over J is for $J = \tfrac{1}{2}, \tfrac{3}{2}, \ldots.$

The corresponding expansions for f_1 and f_2 are (Jacob, 1963),

$$f_1 = \sum_J [f^J_{++}\{P'_{J+\frac{1}{2}} - P'_{J-\frac{1}{2}}\} + f^J_{-+}\{P'_{J+\frac{1}{2}} + P'_{J-\frac{1}{2}}\}], \qquad (3.6.32)$$

$$f_2 = \sum_J [f^J_{++}\{P'_{J+\frac{1}{2}} - P'_{J-\frac{1}{2}}\} - f^J_{-+}\{P'_{J+\frac{1}{2}} + P'_{J-\frac{1}{2}}\}], \qquad (3.6.33)$$

where a prime, P'_l denotes $(d/dx)\,P_l$.

For each value of total angular momentum J, there are two values of orbital angular momentum $J+\tfrac{1}{2}$, $J-\tfrac{1}{2}$, denoted by $(l+1)_-$ and l_+ respectively, where $l = J - \tfrac{1}{2}$. The parity for given l' is $-(-1)^{l'}$. Hence

$$J, l_+ \text{ has parity } (-1)^{J+\frac{1}{2}}, \qquad (3.6.34)$$

$$J, (l+1)_- \text{ has parity } -(-1)^{J+\frac{1}{2}}. \qquad (3.6.35)$$

For some applications one requires partial wave amplitudes of definite parity. These can be defined from the partial wave helicity states

$$2^{-\frac{1}{2}}[|J,m,\lambda\rangle \pm |J,m,-\lambda\rangle], \qquad (3.6.36)$$

which are eigenstates of parity. For the sum of these states in (3.6.36), the corresponding scattering amplitude is

$$f_{(J-\frac{1}{2})+} = f_{l+} = f^J_{++} + f^J_{-+}. \qquad (3.6.37)$$

For the difference in (3.6.36) one has,

$$f_{(J+\frac{1}{2})-} = f_{(l+1)-} = f^J_{++} - f^J_{-+}. \tag{3.6.38}$$

The partial wave expression for $f_{l\pm}$ is

$$f_{l\pm} = \frac{1}{2} \int_{-1}^{1} dx [f_1 P_l(x) + f_2 P_{l\pm 1}(x)], \tag{3.6.39}$$

giving a partial wave expansion for f_1 and for f_2,

$$f_1(s,t) = \sum_l [f_{l+} P'_{l+1} - f_{l-} P'_{l-1}], \tag{3.6.40}$$

$$f_2(s,t) = \sum_l [f_{l-} - f_{l+}] P'_l. \tag{3.6.41}$$

The phase shift expansion is obtained by writing,

$$f_{l\pm} = \frac{\exp(i\delta_{l\pm}) \sin \delta_{l\pm}}{q}. \tag{3.6.42}$$

When Coulomb corrections are incorporated these expansions form the basis of the phase shift analysis mentioned earlier.

CHAPTER 4

ANALYTIC PROPERTIES
OF COLLISION AMPLITUDES

If collision amplitudes are derived from potential theory or from quantum field theory, it is found that they are analytic functions of certain physical parameters. In the relativistic case these parameters are the invariants s, t, ... associated with energy and momentum transfer. In S-matrix theory they are assumed to be analytic in the neighbourhood of physical values of the variables and, using unitarity and crossing symmetry, larger analyticity domains are derived. These analytic properties form the basis for a large part of the theory of strong interactions of elementary particles and for the interpretation of experiments.

In this chapter we will begin by describing the various approaches to the study of analytic properties of amplitudes. Next we will describe their physical interpretation and will give the main results, some of which can be proved from basic assumptions and some of which are still heuristic. We will indicate how some of these results can be derived in perturbation theory, and state what has been proved rigorously. A number of technical complications will be discussed that arise when particles have charge and spin, and we will describe the comparison of dispersion relations with experiment.

The strictly S-matrix approach to analyticity will be described very briefly and only an indication will be given of procedures that have been suggested for obtaining self-consistent equations relating masses and coupling constants of elementary particles. Both of these topics are considered in *The Analytic S Matrix* by Chew (1966). An account of the derivation and use of analytic properties, both from perturbation theory and from the S-matrix approach, is given in *The Analytic S-Matrix* by Eden, Landshoff, Olive & Polkinghorne (1966).

4.1 Assumptions and methods

The analytic behaviour of collision amplitudes as functions of energy and momentum transfer plays a central role both in establishing rigorous results at high energies and in providing the basis for approximation procedures. One of the most successful aspects of analytic behaviour has been the use of dispersion relations, first proposed in the context of quantum field theory by Gell-Mann, Goldberger & Thirring (1954) and Goldberger (1955). In order to establish a dispersion relation, it is necessary to show that there are no singularities of the amplitude in some simple domain of the complex energy (or momentum) variable. In the simplest form, this domain is the complex plane cut along part of the real axis. It is also necessary to establish the behaviour of the amplitude as the complex variable tends to infinity.

Some of the singularities of collision amplitudes are directly related to special characteristics of experimental behaviour. Branch points in the energy variable occur at 'normal thresholds' at which the energy becomes large enough so that a new physical process (for example the production of new particles) is allowed by conservation of total energy and momentum (Eden, 1952). Poles in the energy variable may correspond to bound states or to resonances (Dirac, 1958). Thus the experimental study of resonances gives important information on analytic properties that can rarely be proved on purely theoretical grounds.

Objectives

The objectives of the study of analytic properties of scattering amplitudes or collision amplitudes are:

(1) To derive approximate relations between different experimental quantities, or to summarise several experimental results by establishing some simple analytic properties. For example, scattering lengths may be related to bound state energies, or phase shifts to the location of a resonance pole in the amplitude. These examples illustrate the simplest aspects of the dominance of nearby singularities.

(2) To set up a consistent procedure of approximation, either in terms of nearby singularities, and/or by taking into account high energy behaviour—i.e. the effect of distant singularities.

(3) To study self-consistency conditions within the above approximation procedure (2). This probably the most ambitious aspect of analyticity studies. Often called the 'bootstrap' method, the most optimistic objective is to show that masses, coupling constants, and perhaps symmetries, for resonances and particles are defined either uniquely, or to within a few parameters, by self-consistency of analyticity and unitarity for collision amplitudes.

(4) To prove rigorous results on asymptotic behaviour at high energies based on analytic properties established from the axioms of quantum field theory. These results usually take the form of high energy bounds or inequalities, and asymptotic equalities. Within this approach one can also include rigorous consequences of assumed heuristic analytic properties; these may be particularly important in providing an experimental test of the assumed properties. Chapters 6, 7 and 8 will be concerned with rigorous methods and results.

(5) The development of models for scattering or collision processes that are consistent with analytic properties that have been rigorously established, and which are also consistent with some parts of the unitarity and crossing conditions. An example of this is given by Regge theory. An important aspect of this approach is the study of the simplifying assumptions, including both those made for the practical use of the model and those inherent in the foundations of the model itself. We consider the heuristic basis for Regge theory in Chapter 5 and its applications in Chapter 9.

Methods

The methods used for studying analytic properties of collision amplitudes fall into the following main categories: (*a*) Axiomatic quantum field theory, (*b*) Axiomatic *S*-matrix theory, (*c*) Perturbation solutions of quantum field theory, (*d*) Heuristic methods guessed from a simplified model, e.g. from potential theory. We will briefly consider the special features of these methods:

(*a*) Axiomatic quantum field theory

This approach is based on the fundamental assumptions:

(i)   There exists a field that interpolates the asymptotic physical states,

(ii)   Lorentz invariance,

(iii) There exists a unique vacuum state relative to which physical states have positive energy,

(iv)   Physical states form a complete set,

(v)   Local causality, the field operators at two space-like points commute with each other,

(vi)   Expectation values of operators correspond to tempered distributions in coordinate space,

(vii) Unitarity.

It is known that for some purposes not all these assumptions are necessary. For example a class of dispersion relations has been proved without the use of (vi) (Jaffe, 1966). Also there are a variety of ways of formulating, satisfying or increasing these assumptions (see for example, Wightman, 1960; Lehmann, Symanzik & Zimmerman, 1955; Källén & Wightman, 1958; Goldberger, 1960). We will be concerned with the results established from these axioms in § 4.8 where further references are given.

A major difficulty in the approach to analyticity from axiomatic quantum field theory is that analytic continuation is often blocked by the existence of a branch cut. Future research will undoubtedly lead to methods that permit more easy continuation round a singularity or through a branch cut on to another Riemann sheet. Both these difficulties are reduced if it is assumed that the perturbation series for an amplitude provides the correct analyticity properties. More generally, the methods of axiomatic S-matrix theory give some information about other Riemann sheets.

(b) Axiomatic S-matrix theory

This aims to develop a theory based only on states describing particles on the mass shell. These states are sufficient to define all matrix elements of S that correspond to collision amplitudes of stable particles. The possibility of this approach and its importance was first noted by Heisenberg (1943, 1944) and more recently was greatly stimulated by the work of Chew (1962) (see also Chew, 1966). The basic assumptions include: the superposition principle; completeness of physical (mass shell) states; cluster-decomposition of matrix elements; analyticity near the physical scattering regions and the existence of paths of continuation related to crossing; unitarity.

At present the assumptions of the axiomatic S-matrix approach appear to be more extensive than those of quantum field theory, though explicit reference is not made to local fields. It is possible that eventually the two approaches will be found to be equivalent. However, at present, the S-matrix approach makes specific assumptions about domains of analyticity and is therefore less suitable for some purposes than quantum field theory which has *established* domains of analyticity from its basic axioms. Our starting point for discussing the rigorous high energy behaviour of scattering amplitudes will be these known domains of analyticity. It is possible that they can eventually be derived from axioms that do not involve quantum field theory but this has not yet been achieved. One of the important contributions of S-matrix theory is the proof that certain singularities must exist if the basic axioms of unitarity and analyticity are to be satisfied. These are of considerable significance in setting limits to the type of domain of analyticity that can be established.

Accounts of axiomatic S-matrix theory are given by Eden, Landshoff, Olive & Polkinghorne (1966, hereafter denoted by ELOP (1966)), Landshoff (1965) and by Chew (1966). Papers of particular importance in its development include those of Stapp (1962 a, b), Gunson (1963) and Olive (1964).

(c) *Perturbation theory*

The perturbation series for the S-matrix provides a formal solution to the equations of quantum field theory. If the series converges, then it may be possible to establish certain analytic properties of the S-matrix by studying the analytic properties of each term in the series. This method can be developed rigorously for potential scattering (Blankenbecler, Goldberger, Khuri & Treiman, 1960) but has not been made rigorous in quantum field theory. It does however provide a convenient model for testing the plausibility of assumptions about analyticity.

In practice the use of perturbation theory has provided the basis for some heuristic proposals of simple analytic properties and has been useful in disproving others. There are three modes of use. One, which has proved very valuable, is to obtain the analytic properties of the first few terms of the series, and then to assume that some of these properties hold for all terms of the series and/or for its sum. For example, Mandelstam's study of fourth-order terms in a scattering process led to his important proposal of a double dispersion relation. The second use of perturbation theory is to establish domains of holomorphy for every term (i.e. for a general term) of the perturbation series. Taking care to avoid obviously divergent points, one can then deduce with reasonable confidence that the sum of the series has as holomorphy domain any such domain that is common to every term of the series. The third use of perturbation theory is the location of singularities. One must be careful to avoid possible cancellation but there is a reasonable chance that a singularity of an individual term will also be a singularity of the sum.

Methods for obtaining the singularities of perturbation theory are described by ELOP (1966) and by Bjorken & Drell (1965). They are so important to the study and use of analyticity that a brief account will also be given in this chapter in §4.5.

(d) Heuristic methods

Some of these have been mentioned above in discussing perturbation theory. In a different class, there is the development of Regge theory for high energy scattering. The original work of Regge (1959, 1960) was concerned solely with potential scattering by a Yukawa potential, for which asymptotic behaviour for large momentum transfer was shown to be related to singularities of partial wave amplitudes for (complex) angular momentum. The significance for high energy scattering comes when Regge's method is combined with the crossing symmetry of a relativistic theory. Its applicability to a relativistic theory can be made plausible but has not been established rigorously. We will discuss Regge theory and the use of complex angular momentum in Chapter 5.

In the next section we will introduce the basic mathematical ideas in analyticity and dispersion theory and will illustrate them by simple examples. In the remainder of the chapter we will describe more complicated developments of analytic properties and applications. The aim is not so much to give an exhaustive account, but to describe the main framework of ideas and results that can be used in the study of strong interactions and high energy collisions.

4.2   Introduction to analytic properties

(a) Dispersion relations, crossing symmetry

Let $f(w)$ be a function of the complex variable $w = u + iv$, that has no singularities inside or on a closed curve C in the w plane. Then Cauchy's theorem shows that if z is inside C,

$$f(z) = \frac{1}{2\pi i} \int_C \frac{f(w)\,dw}{w-z}, \tag{4.2.1}$$

where the integral is taken round the contour C. In particular if (i) the function $f(w)$ has no singularities in the complex w-plane except when w is real and greater than a and (ii) $f(w)$ tends to zero as w tends to infinity, then C can be taken to be the limit for infinite radius of the contour shown in Fig. 4.2.1. From condition (ii) the contribution from the circular part of C tends to zero as the radius tends to infinity. Hence

$$f(z) = \frac{1}{2\pi i} \int_a^\infty du \frac{f(u+i\epsilon)-f(u-i\epsilon)}{u-z} \quad \text{with } \epsilon \to +0. \tag{4.2.2}$$

If also, (iii) $f(u)$ is real for $u \leqslant a$, then Schwarz' reflection principle shows that

$$f(z^*) = f^*(z). \tag{4.2.3}$$

Then (4.2.2) can be written

$$f(z) = \frac{1}{\pi} \int_a^\infty du \frac{f_1(u)}{u - z}, \tag{4.2.4}$$

where $f_1(u) = \mathrm{Im} f(u)$ is assumed to be finite.

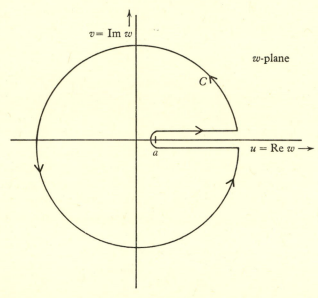

Fig. 4.2.1. The complex w-plane for the function $f(w)$ used in equation (4.2.1).

The formula (4.2.4) is called a Hilbert transform (see for example Titchmarsh, 1939). When $f(z)$ is a scattering amplitude that satisfies the above conditions, then (4.2.4) is called a Dispersion Relation. In a simple case z could denote the energy, or the invariant energy s, or the square of the invariant momentum transfer t, defined in § 3.1.

It may happen that condition (ii) on the asymptotic behaviour is not satisfied but

$$\left| \frac{f(z)}{z} \right| \to 0 \quad \text{as} \quad |z| \to \infty. \tag{4.2.5}$$

Then the function

$$\frac{f(z) - f(z_0)}{z - z_0} \tag{4.2.6}$$

will satisfy the necessary conditions, and the corresponding dispersion relation gives

$$f(z) = f(z_0) + \frac{(z - z_0)}{\pi} \int_a^\infty \frac{du \, f_1(u)}{(u - z_0)(u - z)}. \tag{4.2.7}$$

This is described as a dispersion relation for $f(z)$ with one subtraction. The subtraction point is z_0, and $f(z_0)$ is usually an unknown parameter that has to be determined from experiment.

One of the basic problems in collision theory is to prove that a scattering amplitude satisfies a dispersion relation, and if s is the integration variable, to ascertain for which values of the 'momentum transfer' t the dispersion relation remains valid. For production amplitudes the analytic properties are usually too complicated for a simple dispersion relation to be valid.

It is generally believed that the validity of a dispersion relation for a scattering amplitude is a consequence of causality. Certainly, in its usual formulations, quantum field theory satisfies causality in the sense of commutation of operators at space-like points. It is much more difficult to establish a direct relation between causality and analyticity in S-matrix theory (Eden & Landshoff, 1965; Chandler & Stapp, 1966; Omnes, 1966). However, one can see that there is reason to expect such a relation by considering classical dispersion relations (reviewed by Hamilton, 1959) or from the following simple model.

Let $A(z,t)$ be a wave packet moving in the z direction with speed v and Fourier component $a(w)$,

$$A(z,t) = (2\pi)^{-\frac{1}{2}} \int_{-\infty}^{\infty} dw\, a(w) \exp\left[iw\left(\frac{z}{v} - t\right) \right]. \qquad (4.2.8)$$

If this wave packet is scattered at the origin $z = 0$, the scattered wave may be written as,

$$B(r,t) = r^{-1}(2\pi)^{-\frac{1}{2}} \int_{-\infty}^{\infty} dw\, f(w)\, a(w) \exp\left[iw\left(\frac{z}{v} - t\right) \right], \qquad (4.2.9)$$

where r is the distance from the origin. From the inverse of (4.2.8)

$$a(w) = (2\pi)^{-\frac{1}{2}} \int_{-\infty}^{\infty} dt\, A(0,t) \exp(iwt), \qquad (4.2.10)$$

we see that if the incident wave does not reach the origin before $t = 0$, that is if

$$A(0,t) = 0 \quad \text{for } t < 0, \qquad (4.2.11)$$

then $a(w)$ is regular in the upper half complex w-plane, $\mathrm{Im}\,(w) > 0$.

Causality in this model, requires that there is no scattered wave at a distance r, until a time r/v after the incident wave reaches the scatterer at $z = 0$. Thus we must require that

$$B(r,t) = 0 \quad \text{for } (vt - r) < 0. \qquad (4.2.12)$$

From this condition and the inverse of (4.2.9) analogous to (4.2.10), we see that $f(w)\,a(w)$ is analytic in $\mathrm{Im}\,(w) > 0$. Hence the scattering amplitude $f(w)$ is analytic in $\mathrm{Im}\,(w) > 0$, except possibly at zeros of $a(w)$.

The difficulty in making the above discussion rigorous arises from condition (4.2.11). This assumes precise localization in time, which is not compatible with the S-matrix requirement that the incident energy is known. The difficulty is discussed further by ELOP (1966) and by Stapp (1966).

We can use the analyticity of $f(w)$ to derive a dispersion relation. Since $B(0, t)$ is zero for $t < 0$, we can write

$$r\,B(r, t)|_{r=0} = b(t)\tfrac{1}{2}\{1 + \epsilon(t)\}, \tag{4.2.13}$$

where $\epsilon(t) = 1$ for $t > 0$ and -1 for $t < 0$. We can define

$$b(-|t|) = b(|t|), \tag{4.2.14}$$

since $b(t)$ is always multiplied by zero when $t < 0$. Then the inverse of (4.2.9) analogous to (4.2.10) can be written

$$f(w)\,a(w) = (2\pi)^{-\frac{1}{2}} \int_{-\infty}^{\infty} dt\,b(t)\exp(iwt)\tfrac{1}{2}\{1 + \epsilon(t)\}$$

$$= (2\pi)^{-\frac{1}{2}} \int_{0}^{\infty} dt\,b(t)\cos wt + i(2\pi)^{-\frac{1}{2}} \int_{0}^{\infty} dt\,b(t)\sin wt \tag{4.2.15}$$

$$= f_2(w)\,a(w) + if_1(w)\,a(w). \tag{4.2.16}$$

The equality of (4.2.15) and (4.2.16) defines f_1 and f_2. If the incident wave has equal positive and negative frequency components, $a(-w) = a(w)$; then we can deduce

$$f_2(-w) = f_2(w), \tag{4.2.17}$$

$$f_1(-w) = -f_1(w). \tag{4.2.18}$$

This is the crossing condition in our model, analogous to the crossing condition in quantum field theory that we will use later. For real w, it gives

$$f(-w + i0) = f^*(w + i0). \tag{4.2.19}$$

Assume that $f(w) \to 0$ as $w \to \infty$, in $\mathrm{Im}\,(w) \geqslant 0$ (we relax this assumption later in (4.2.22)). Assume also that $a(w)$ is real when w is pure imaginary, then $f(w)$ is real when w is pure imaginary. Define $z = w^2$ and write

$$F(z) = f(w). \tag{4.2.20}$$

Then $F(z)$ is analytic in the z-plane except for a branch cut from $z = 0$ to $z = \infty$, and satisfies a dispersion relation,

$$F(z) = \frac{1}{\pi} \int_0^\infty \frac{dx \, \mathrm{Im} \, F(x)}{x - z}. \tag{4.2.21}$$

If $F(z)$ does not tend to zero as $z \to \infty$, but F/z does do so, then we must introduce a subtraction and write

$$F(z) - F(0) = \frac{z}{\pi} \int_0^\infty \frac{dx \, \mathrm{Im} \, F(x)}{x(x - z)}. \tag{4.2.22}$$

The imaginary part $\mathrm{Im} \, F(x)$ is related to the cross-section. Taking the real part of this equation, we get

$$\mathrm{Re} \, F(x) - \mathrm{Re} \, F(0) = \frac{x}{\pi} P \int_0^\infty \frac{dx' \, \mathrm{Im} \, F(x')}{x'(x' - x)}, \tag{4.2.23}$$

where x is real. This is the analogue of the Kramers–Krönig dispersion relation for the scattering of light by free electrons (Toll, 1952, 1956; and Gell-Mann, Goldberger & Thirring, 1954).

(b) Potential scattering

The analytic properties of the amplitude for scattering by a superposition of Yukawa potentials have been considered by Blankenbecler, Goldberger, Khuri & Treiman (1960). The amplitude $F(s, t)$ is a function of the energy s and momentum transfer squared $(-t)$,

$$s = k^2, \tag{4.2.24}$$

(we will later re-define s when we consider relativistic scattering, see (4.2.34)),

$$t = -(\mathbf{k}_i - \mathbf{k}_f)^2 = -2k^2(1 - \cos\theta). \tag{4.2.25}$$

Consider a single Yukawa potential of range μ,

$$V(r) = g \frac{\exp(-\mu r)}{r}. \tag{4.2.26}$$

The Born approximation for the scattering amplitude is

$$F_B(t) = -\frac{m}{2\pi\hbar^2} \int d^3r \, V(r) \exp\{i(\mathbf{k}_i - \mathbf{k}_f).r\}$$

$$= -\frac{2mg}{\hbar^2} \left(\frac{1}{\mu^2 - t} \right). \tag{4.2.27}$$

This simple pole in the Born amplitude at $t = \mu^2$ is also present in the full amplitude.

The full scattering amplitude $F(s,t)$ has been studied by two main methods. One is by means of the perturbation series or Fredholm series. This method was used by Blankenbecler *et al.* (1960), and earlier by Khuri (1957). The second method makes use of the partial wave expansion and its generalisation by analytic continuation in the angular momentum l. This method was first used in the present context by Regge (1959) and will be discussed in Chapter 5. We will summarise here only the results that relate to dispersion relations.

The physical scattering amplitude $F(s,t)$ has t real and negative, and s real and positive. It can be shown that this physical amplitude is the boundary value of a function $F(z,t)$ of the complex variable z, that is analytic in the entire z-plane cut from $z=0$ to ∞. The physical amplitude is obtained by letting z tend to real values on the branch cut, from the upper half plane. Thus

$$F(s,t)_{\text{physical}} = \lim_{\epsilon \to 0} F(s+i\epsilon, t), \qquad (4.2.28)$$

for s real and positive.

In practice the variable s is often used to denote complex as well as real values, and we will follow this practice except where special emphasis is desired. For fixed real values of $t < 4\mu^2$, where μ^{-1} is the range of the Yukawa potential $F(s,t)$ is analytic in the cut s-plane so it satisfies a dispersion relation. If $F(s,t)$ tends to zero as $|s| \to \infty$,

$$F(s,t) = \frac{1}{\pi} \int_0^\infty \frac{dx \, F_1(x,t)}{x-s}, \qquad (4.2.29)$$

where

$$F_1(x,t) = \operatorname{Im} F(x,t) = \frac{1}{2i} \lim_{\epsilon \to 0} [F(x+i\epsilon,t) - F(x-i\epsilon,t)]. \quad (4.2.30)$$

This 'fixed transfer' dispersion relation will require subtractions if $F(s,t)$ does not tend to zero as $|s| \to \infty$. In general the behaviour as $s \to \infty$ will depend on the value of t. If, with $\alpha(t)$ real,

$$|F(s,t)| \sim s^{\alpha(t)} \quad \text{as } s \to \infty, \qquad (4.2.31)$$

it will be necessary to introduce N' subtractions, where N' is the smallest integer greater than $\alpha(t)$. Instead of (4.2.29) the fixed transfer dispersion relation will have the form,

$$F(s,t) = \sum_0^{N'-1} a_n s^n + \frac{s^{N'}}{\pi} \int_0^\infty \frac{dx \, F_1(x,t)}{x^{N'}(x-s)}. \qquad (4.2.32)$$

For fixed values of the energy s, the imaginary part of the amplitude

$F_1(s, t)$ is itself an analytic function of t. We will make use of this result when we discuss double dispersion relations and the Mandelstam representation in §4.4.

(c) Forward dispersion relations

In the forward direction, non-relativistic potential scattering gives a dispersion relation, ($s = $ energy)

$$F(s, 0) = \sum_0^{N-1} a_n s^n + \frac{s^N}{\pi} \int_0^\infty \frac{dx \operatorname{Im} F(x, 0)}{x^N(x - s)}. \qquad (4.2.33)$$

For a relativistic theory, there is a branch cut along the left of the real axis in addition to the right-hand axis. For equal mass particles, instead of (4.2.24), we take s to be the square of the centre of mass energy,

$$s = 4(m^2 + k^2), \qquad (4.2.34)$$

and the branch cuts lie along $(-\infty, 0)$ and $(4m^2, \infty)$. The dispersion relation takes the form,

$$F(s, 0) = \sum_0^{N-1} a_n s^n + \frac{s^N}{\pi} \int_{-\infty}^0 \frac{dx \operatorname{Im} F(x, 0)}{x^N(x - s)} + \frac{s^N}{\pi} \int_{4m^2}^\infty \frac{dx \operatorname{Im} F(x, 0)}{x^N(x - s)}. \qquad (4.2.35)$$

For $x > 4m^2$, we can use the optical theorem (3.3.5),

$$\operatorname{Im} F(x, 0) = 2kx^{\frac{1}{2}} \sigma(\text{total}), \qquad (4.2.36)$$

giving the integrand of the last integral of (4.2.35) in terms of experimental quantities. We will see in §4.4 that crossing symmetry relates the value of $\operatorname{Im} F(x, 0)$ for $x > 4m^2$, to its value for $x < 0$. For equal mass uncharged bosons we find (4.4.20) for $x < 0$, which gives

$$\operatorname{Im} F(-x, 0) = -\operatorname{Im} F(4m^2 + x, 0). \qquad (4.2.37)$$

This gives

$$F(s, 0) = \sum_0^{N-1} a_n s^n + \frac{s^N}{\pi} \int_{4m^2}^\infty dx\, x^{\frac{1}{2}}(x - 4m^2)^{\frac{1}{2}} \sigma_t(x)$$
$$\times \left[\frac{(-1)^N}{(x - 4m^2)^N (x + s - 4m^2)} + \frac{1}{x^N(x - s)} \right]. \qquad (4.2.38)$$

If we assume that the total cross-section $\sigma_t(x)$ is constant asymptotically as $x \to \infty$, only two subtractions are required. Using the symmetry relation (4.2.37) only one non-zero subtraction constant is required. From this forward dispersion relation, using experimental results for σ_t and an assumption for σ_t above the experimentally available energies, it is possible to calculate $\operatorname{Re} F(s, 0)$ and compare with results from Coulomb interference experiments (see §4.10).

4.3 Vertex parts and form factors

In electron-proton scattering we can work to lowest order in the fine structure constant. There is therefore only a single photon exchanged between the electron and the proton and, from Lorentz invariance, the scattering amplitude A can depend only on the square of the four-momentum of the exchanged photon,

$$t = q_0^2 - \mathbf{q}^2. \tag{4.3.1}$$

Neglecting spin, we have

$$A(t) = \frac{e^2}{t}\{1 + tF(t)\}. \tag{4.3.2}$$

The vertex part $F(t)$ describes the modification to the Coulomb interaction due to the strong interactions of the proton as represented in Fig. 4.3.1 (a).

It can be shown that $F(t)$ is an analytic function, regular in the complex t-plane cut along the real axis from $4m^2$ to ∞, where m is the pion mass. It satisfies a dispersion relation (neglecting possible subtractions),

$$F(t) = \frac{1}{\pi}\int_{4m^2}^{\infty} \frac{\operatorname{Im} F(p^2)\, d(p^2)}{(p^2 - t)}. \tag{4.3.3}$$

The modification to the Coulomb interaction includes effects arising from the diagram (b) in Fig. 4.3.1. The integral in (4.3.3) is over positive values of p^2 so that $\operatorname{Im} F(p^2)$ will include contributions from real intermediate states of two pions when $p^2 > 4m^2$. These are the intermediate states of lightest mass and determine the start of the branch cut of $F(t)$.

The experimental values of t for electron-proton scattering must have $t < 0$, since the initial and final states are physical. On the other hand if $t > 4M^2$, where M is the proton mass, the amplitude (4.3.2) represents the creation of a proton antiproton pair by electron positron annihilation. These two different physical processes correspond to the same function $A(t)$, but to different parts of the t-plane, as indicated in Fig. 4.3.1 (c). Since there is a branch cut on the positive real axis one has to decide whether the physical amplitude for the process

$$e^- + e^+ \to p + \overline{p}, \tag{4.3.4}$$

is the boundary value of $A(t)$ above or below the cut. The usual convention, which results for example from the Feynman rules for pertur-

bation theory, is that the physical amplitude for (4.3.4) corresponds
to the limit on to the cut from the upper half plane,

$$A(t) = \lim_{\epsilon \to +0} A(t + i\epsilon),$$ (4.3.5)

for $t > 4M^2$.

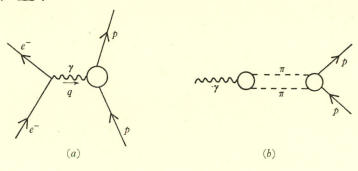

(a)   (b)

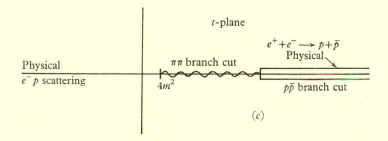

(c)

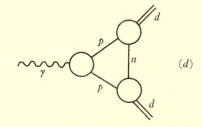

(d)

Fig. 4.3.1. (a) The single photon exchange diagram for electron-proton scattering;
(b) the photon-proton vertex showing an intermediate two-pion state; (c) the complex
t-plane for the vertex function, showing branch cuts, and physical values for electron-
proton scattering, and physical values for electron-positron annihilation to give
proton anti-proton; (d) a diagram giving an anomalous threshold for the deuteron.

The physical interpretation of a vertex part becomes clearer from
considering its three-dimensional Fourier transform. Taking a negative
value for t we can write

$$t = -\mathbf{q}^2,$$ (4.3.6)

where $\mathbf{q}$ is a three-vector. The transform is called a form factor and is given by,

$$F(r) = \int d^3q \, F(-\mathbf{q}^2) \exp(i\mathbf{q}.\mathbf{r}). \qquad (4.3.7)$$

Substituting from (4.3.3) we obtain

$$F(r) = c \int_{4m^2}^{\infty} F_1(p^2) \frac{\exp(-pr)}{r} d(p^2). \qquad (4.3.8)$$

This is a superposition of Yukawa functions with weight function $F_1(p^2) = \operatorname{Im} F(p^2)$. The Yukawa function of longest range in (4.3.8) is

$$\frac{\exp(-2mr)}{r}. \qquad (4.3.9)$$

Thus the range is of order $(2m)^{-1}$ and is determined by the lowest possible physical intermediate state for the crossed reaction (4.3.4), namely the two-pion state.

The branch point at $4m^2$ lies at the threshold for creation of two physical pions. It is called a 'normal threshold'. More generally a normal threshold will occur in any scattering amplitude when the appropriate energy variable is just large enough for a new physical state to be allowed. Note that in this example, when t is sufficiently positive it becomes the energy of the system of electron and positron (or proton and antiproton).

The form factor (4.3.8) can be regarded as describing the charge distribution at a distance r from the centre of the proton. On this interpretation the charge distribution will have a radius of about 7.10^{-14} cm provided the function $F_1(p^2)$ becomes fairly large near the threshold of the two-pion cut. The size of $F_1(p^2)$ depends on the strength of the pion-pion interaction. To some extent the effect of this interaction can be approximated by resonances in the pion-pion system.

A more complicated situation occurs in electron-deuteron scattering, where it is found that the branch point at the smallest positive value of t comes from the diagram shown in Fig. 4.3.1 (d). This branch point occurs at

$$t = 16MB, \qquad (4.3.10)$$

where M is the mass and B is the deuteron binding energy. Its location is well below the value of t for which the two-proton intermediate state in Fig. 4.3.1 (d) could be physical. It is called an anomalous threshold and its location can be found from the perturbation diagram corresponding to Fig. 4.3.1 (d). An anomalous threshold will occur if

$$M_d^2 > M_n^2 + M_p^2, \tag{4.3.11}$$

which is easily satisfied in this example. For the deuteron, the position of the anomalous threshold largely determines the observed charge distribution. However, for a larger nucleus the size of the charge distribution is primarily due to the fact that the relevant weight function, $F_1(p^2)$ in (4.3.8) develops strong oscillations (Eden & Goldstone, 1963).

4.4 Dispersion relations and the Mandelstam representation

In this section we will state results on the analytic properties of the scattering amplitude $F(s,t)$ as a function of two complex variables. These results have been proved for potential scattering following the conjecture by Mandelstam for the relativistic case. For relativistic scattering the Mandelstam representation will be stated for equal mass particles and in § 4.6 we will indicate some of the modifications required for more general masses. The methods for deriving these results in perturbation theory will be outlined in § 4.5.

(a) Potential scattering

We will consider scattering by a Yukawa potential assuming that no subtractions are required for t real and negative. Then $F(s,t)$ satisfies the dispersion relation (4.2.29) with the addition of the pole (4.2.27), giving

$$F(s,t) = \frac{g^2}{t-\mu^2} + \frac{1}{\pi}\int_0^\infty \frac{dx\, F_1(x,t)}{x-s}, \tag{4.4.1}$$

where F_1 is the imaginary part of $F(x+i0,t)$, taken when x and t are both real,

$$2iF_1(x,t) = F(x+i0,t) - F(x-i0,t). \tag{4.4.2}$$

In this formula we can now allow t to be complex, although only for x and t real is F_1 equal to Im F. Neglecting possible subtractions we can write a dispersion relation for $F_1(x,t)$ in the variable t (Blankenbecler et al. 1960),

$$2iF_1(x,t) = \frac{1}{2\pi i}\int_{4\mu^2}^\infty dy\, \frac{F(x+i\epsilon, y+i\epsilon') - F(x+i\epsilon, y-i\epsilon')}{y-t}$$

$$-\frac{1}{2\pi i}\int_{4\mu^2}^\infty dy\, \frac{F(x-i\epsilon, y+i\epsilon') - F(x-i\epsilon, y-i\epsilon')}{y-t}, \tag{4.4.3}$$

in the limit ϵ, $\epsilon' \to 0$. $F_1(x,t)$ is real for t real $< 4\mu^2$, and satisfies the dispersion relation

$$F_1(x,t) = \frac{1}{\pi} \int_{4\mu^2}^{\infty} dy \frac{F_{12}(x,y)}{y-t},$$ (4.4.4)

where

$$F_{12}(x,y) = \lim_{\epsilon \to 0} \tfrac{1}{4}[F(x+i\epsilon, y+i\epsilon) - F(x+i\epsilon, y-i\epsilon)$$

$$- F(x-i\epsilon, y+i\epsilon) + F(x-i\epsilon, y-i\epsilon)].$$ (4.4.5)

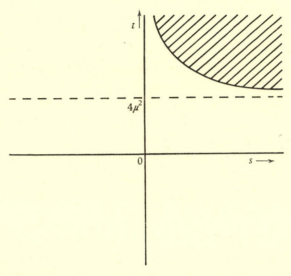

Fig. 4.4.1. The real s, t plane for non-relativistic scattering by a Yukawa potential. The shaded area shows where the Mandelstam spectral function is non-zero.

Substituting (4.4.4) into (4.4.1) we obtain a double dispersion relation for $F(s,t)$, which is the Mandelstam representation for potential scattering,

$$F(s,t) = \frac{g^2}{t-\mu^2} + \frac{1}{\pi^2} \int_0^{\infty} dx \int_{4\mu^2}^{\infty} dy \frac{F_{12}(x,y)}{(x-s)(y-t)}.$$ (4.4.6)

There may be subtractions required analogous to those in (4.2.32); these will be considered later in Chapter 5. The real function $F_{12}(x,y)$ is called the (Mandelstam) spectral function. It is in fact zero over part of the range of integration. This is due to the fact that $t = 4\mu^2$ is a singularity of the real part of F, and not of F_1, so there is cancellation in (4.4.5) near the boundaries of the region of integration. The region where F_{12} is non-zero can be obtained by a direct but difficult calculation from the second-order term in perturbation theory (see Mandelstam, 1958), or more simply by the methods described in § 4.5. It is the shaded area shown in Fig. 4.4.1 that lies above the curve

$$st = \mu^2(4s + \mu^2). \tag{4.4.7}$$

Potential scattering is often used as a model for discussing special aspects of the relativistic theory. Then one defines s as the relativistic energy invariant as in § 3.1. With particles of equal mass m, and with the mass in the Yukawa potential μ put equal to m, we have instead of (4.2.24) and (4.4.7),

$$s = 4(m^2 + k^2), \tag{4.4.8}$$

with a boundary curve for the double spectral function $F_{12}(s, t)$ given by

$$(s - 4m^2)(t - 4m^2) = 4m^4. \tag{4.4.9}$$

The above result does not include crossing symmetry.

(b) Relativistic theory and crossing symmetry

The scattering of identical bosons is represented by Fig. 4.4.2 (a). Suppose for simplicity that we can label each particle. Then three collision processes (and their inverse) are represented,

$$\left.\begin{array}{ll} \text{I:} & 1 + 2 \to 3 + 4, \\ \text{II:} & 1 + \bar{4} \to \bar{2} + 3, \\ \text{III:} & 1 + \bar{3} \to \bar{2} + 4. \end{array}\right\} \tag{4.4.10}$$

For simplicity we assume that particles and antiparticles are identical, $\bar{4} = 4$ etc., so we will not distinguish them. As in § 3.1 all the four-momenta are directed inwards so as to bring out the symmetry. Thus in process I, $-p_3$ and $-p_4$ denote the energy-momentum vectors for outgoing particles 3 and 4. The invariants defined in § 3.1 are

$$s = (p_1 + p_2)^2, \quad t = (p_1 + p_4)^2, \quad u = (p_1 + p_3)^2. \tag{4.4.11}$$

From total energy-momentum conservation, and the fact that initial and final particles are free,

$$\sum_1^4 p_i = 0; \quad p_i = m^2; \quad (i = 1, 2, 3, 4) \tag{4.4.12}$$

and

$$s + t + u = 4m^2. \tag{4.4.13}$$

The scattering amplitude for process I can be expressed as a function of s and t, $F(s, t)$. In general we will allow s and t to be complex. The function $F(s,t)$ corresponds to the physical scattering amplitude for process I when s and t are real, and

$$s > 4m^2, \quad t \leqslant 0, \quad u \leqslant 0. \tag{4.4.14}$$

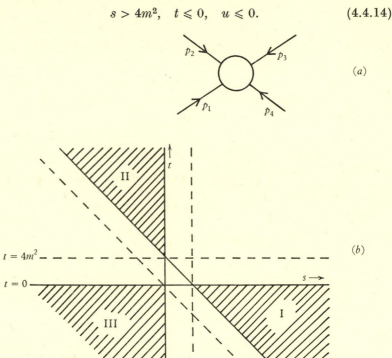

Fig. 4.4.2. (*a*) The scattering diagram for processes I, II and III of (4.4.10); (*b*) the physical scattering regions for processes I, II and III (shaded areas).

As we noted in § 4.2, the physical amplitude is obtained by taking the limit from the upper half s-plane on to the region (4.4.14)

$$F(\text{physical}) = \lim_{\epsilon \to 0} F(s+i\epsilon, t). \tag{4.4.15}$$

The physical region I, is shown in Fig. 4.4.2 (*b*), where we have also shown the regions where II and III can be physical processes. In region II the variable t becomes the centre of mass energy squared, and in III u is the centre of mass energy squared.

It has been proved from the axioms of quantum field theory (Bros, Epstein, Glaser, 1965) that the scattering amplitude $F(s,t)$ can be continued analytically from region I to regions II and III. Thus all three scattering processes are described by boundary values of one analytic function. When assuming the Mandelstam representation we will go beyond what has been proved from quantum field theory, although for equal mass scattering this representation is very plausible (Eden, 1961; Landshoff, Polkinghorne & Taylor, 1961; and Eden *et al.* 1961).

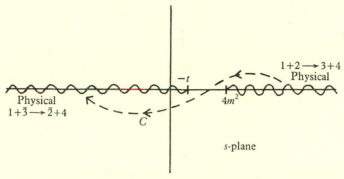

Fig. 4.4.3. The complex s-plane showing the path of analytic continuation that gives the crossing relation from the s channel (process I) to the u channel (process III).

The relation between the processes I, II and III obtained by analytic continuation is called crossing symmetry. It is especially simple for identical bosons. In Fig. 4.4.3 we show the complex s-plane for the function $F(s,t)$ with t real and negative, $0 > t > -4m^2$. We have, for process I,

$$F_{\rm I}(s,t) = \lim_{\epsilon\to 0} F(s+i\epsilon,t) \quad (s > 4m^2). \tag{4.4.16}$$

If the same function is continued analytically along the path C in Fig. 4.4.3, to a value of s such that u is greater than $4m^2$, with

$$u = 4m^2 - s - t > 4m^2, \tag{4.4.17}$$

we get

$$F_{\rm III}(u,t) = \lim_{\epsilon\to 0} F(4m^2 - u - t - i\epsilon, t), \tag{4.4.18}$$

where $F_{\rm III}$ denotes the physical scattering amplitude for process III. However with identical bosons, the process III is physically the same as I, so must be described by the same function. Putting $u = s$ in (4.4.18), we obtain

$$F_{\rm I}(s,t) \equiv F_{\rm III}(s,t) \equiv F(s,t). \tag{4.4.19}$$

(c) *The crossing relation*

Equation (4.4.18) can therefore be written

$$F(s+i\epsilon,t) = F(4m^2 - s - t - i\epsilon, t). \tag{4.4.20}$$

If the particles are not identical bosons, the process III may differ from process I, for example if we have

$$\text{I:} \quad \pi^- + p \to \pi^- + p, \tag{4.4.21}$$

the crossed process III becomes,

$$\text{III:} \quad \pi^+ + p \to \pi^+ + p. \tag{4.4.22}$$

In this example the crossing relation becomes

$$F_{(\pi^+, p)}(s + i\epsilon, t) = F_{(\pi^-, p)}(2m^2 + 2M^2 - s - t - i\epsilon, t). \tag{4.4.23}$$

We will consider the use of these crossing relations in more detail in Chapter 7. For the simple case (4.4.20) there is one more aspect to note here that follows from the reality of F in the gap between the branch cuts in Fig. 4.4.2. Using the Schwarz reflection principle, the reality of $F(s, t)$ in $-t < s < 4m^2$, gives for real t,

$$F(s - i\epsilon, t) = F^*(s + i\epsilon, t). \tag{4.4.24}$$

An amplitude satisfying this condition is said to have 'hermitean analyticity'. Combining (4.4.24) with (4.4.20) we obtain the crossing relation using hermitean analyticity,

$$F(s + i\epsilon, t) = F^*(4m^2 - s - t + i\epsilon, t). \tag{4.4.25}$$

Similar crossing relations can be established that involve process II for which t is the energy, for the case of identical bosons. More generally one obtains relations between different processes. Corresponding to $\pi^+ p \to \pi^+ p$ for process I, we get for process II,

$$\text{II:} \quad p + \bar{p} \to \pi^+ + \pi^-. \tag{4.4.26}$$

If we neglect the possibility of subtractions, the hermitean analyticity of $F(s, t)$ in the cut s-plane leads to the dispersion relation,

$$F(s, t) = \frac{1}{\pi} \int_{4m^2}^{\infty} ds' \frac{F_1(s', t)}{s' - s} + \frac{1}{\pi} \int_{-\infty}^{-t} ds' \frac{F_1(s', t)}{s' - s}. \tag{4.4.27}$$

This can be written more symmetrically in the form

$$F(s, t) = \frac{1}{\pi} \int_{4m^2}^{\infty} ds' \frac{F_1(s', t)}{s' - s} + \frac{1}{\pi} \int_{4m^2}^{\infty} du' \frac{F_3(u', t)}{u' - u}, \tag{4.4.28}$$

where we have made a change of variable,

$$s' = 4m^2 - t - u', \tag{4.4.29}$$

and where u is given by (4.4.17).

(d) The Mandelstam representation

The double dispersion relation proposed by Mandelstam (1958) was based originally on the evaluation of the fourth-order Feynman diagram. For this diagram, one can evaluate $F_1(s',t) = \operatorname{Im} F(s',t)$ explicitly, and show that it is analytic in the complex t-plane cut from $4m^2$ to ∞. This diagram (shown in Fig. 4.4.4 (a)) leads to single and double dispersion relations analogous to those for potential scattering,

$$F_1^A(s',t) = \frac{1}{\pi} \int_{4m^2}^{\infty} dt' \, \frac{F_{12}^A(s',t')}{t'-t}, \tag{4.4.30}$$

$$F^A(s,t) = \frac{1}{\pi^2} \int_{4m^2}^{\infty} ds' \int_{4m^2}^{\infty} dt' \, \frac{F_{12}^A(s',t')}{(s'-s)(t'-t)}, \tag{4.4.31}$$

where the superscript A denotes that only the single diagram in Fig. 4.4.4 (a) is included. The spectral function F_{12}^A is non-zero in the shaded region of Fig. 4.4.4 (b). The other two fourth-order diagrams shown in Fig. 4.4.4 (c) correspond to square diagrams for processes III and II. When they are included in $F(s,t)$, contributions like (4.4.31) come from each of the three shaded regions in Fig. 4.4.4 (d). The Mandelstam representation is based on the assumption that the resulting double dispersion relation can be generalised to the full amplitude giving,

$$\begin{aligned} F(s,t) = \frac{1}{\pi^2} \iint_{4m^2}^{\infty} ds' \, dt' \, \frac{F_{12}(s',t')}{(s'-s)(t'-t)} \\ + \frac{1}{\pi^2} \iint_{4m^2}^{\infty} dt' \, du' \, \frac{F_{23}(t',u')}{(t'-t)(u'-u)} + \frac{1}{\pi^2} \iint_{4m^2}^{\infty} du' \, ds' \, \frac{F_{31}(u',s')}{(u'-u)(s'-s)}. \end{aligned} \tag{4.4.32}$$

The regions of integration in the s, t plane (or s', t') are shown shaded in Fig. 4.4.4 (d). Note that

$$\left. \begin{aligned} s'+t'+u' &= 4m^2, \\ s+t+u &= 4m^2. \end{aligned} \right\} \tag{4.4.33}$$

Using these relations, we can rewrite the last term in (4.4.32) in the form

$$-\frac{1}{\pi^2} \iint_{4m^2}^{\infty} du' ds' \, \frac{F_{31}(u',s')}{(t'-t)(s'-s)} - \frac{1}{\pi^2} \iint_{4m^2}^{\infty} du' ds' \, \frac{F_{31}(u',s')}{(t'-t)(u'-u)}. \tag{4.4.34}$$

The first term of (4.4.34) combined with the first term on the right of (4.4.32) is equal to the first term on the right of the single variable dispersion relation (4.4.28). Similarly the last term of (4.4.34) combines with the (t', u') integral in (4.4.32) to give the u' integral of (4.4.28). This establishes consistency of (4.4.32) and the single variable

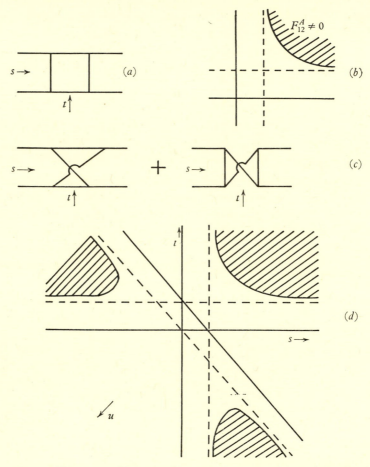

Fig. 4.4.4. (a) The square Feynman diagram whose singularities are considered; (b) the region in the real (s, t) plane for which the square diagram (a) gives a non-zero spectral function F_{12}^A; (c) diagrams that become square in terms of the crossed channel variables u, t and s, u, respectively; (d) the shaded areas show where the Mandelstam spectral function is non-zero, for the sum of the three diagrams in (a) and (c).

dispersion relation (4.4.28). For example the spectral function $F_1(s, t)$ in (4.4.28) can be written, with the aid of (4.4.32) and (4.4.34), in the form

$$F_1(s', t) = \frac{1}{\pi} \int_{4m^2}^{\infty} dt' \frac{F_{12}(s', t')}{t' - t} + \frac{1}{\pi} \int_{-\infty}^{-s'} dt' \frac{F_{31}(4m^2 - s' - t', s')}{t' - t}. \quad (4.4.35)$$

Note that $F_{12}(s', t')$ and $F_{31}(4m^2 - s' - t', s')$ are zero outside the shaded areas shown in Fig. 4.4.4 (d).

For the scattering of identical bosons, crossing symmetry shows that there must be full symmetry between the variables s, t and u. This means that F_{12}, F_{23} and F_{31} are the same function. We will denote this common spectral function by $\rho(s', t'), \rho(t', u'), \rho(u', s')$ in the integrands of the Mandelstam representation (4.4.32).

(e) Subtractions in the Mandelstam representation

We will see later, in Chapter 6, that for fixed real $t < 4m^2$, the scattering amplitude $F(s, t)$ satisfies

$$\left| \frac{F(s, t)}{s^2} \right| \to 0 \quad \text{as} \quad |s| \to \infty. \tag{4.4.36}$$

Hence the single variable dispersion relation in s requires no more than two subtractions with $t < 4m^2$. However there is no rigorous information about the number of subtractions required in the dispersion relation for $\text{Im} \, F(s, t)$, when $t > 4m^2$. There is even some indication (but no certainty) that as $t \to \infty$, the number of subtractions required may also tend to infinity (Mandelstam, 1963). The Mandelstam representation *assumes* that this inconvenient possibility does not occur. Thus it assumes that no more than N subtractions are required in the dispersion relation that generalises (4.4.35) to include necessary subtractions,

$$F_1(s', t) = \sum_{n=0}^{N-1} a_n(s') \, t^n + \frac{t^N}{\pi} \int_{4m^2}^{\infty} dt' \, \frac{\rho(s', t')}{t'^N(t' - t)} + \frac{t^N}{\pi} \int_{-\infty}^{-s} dt' \, \frac{\rho(u', s')}{t'^N(t' - t)}. \tag{4.4.37}$$

It is assumed that there is a finite value of N that suffices for all s'. In this equation the subtractions are introduced at $t = 0$; it may sometimes be more convenient to make them at some other fixed point. With the subtraction point s_0, t_0 and $u_0 = 4m^2 - s_0 - t_0$, the Mandelstam representation with $N = 1$ will have the form,

$$F(s, t) = F(s_0, t_0) + \frac{s - s_0}{\pi} \int_{4m^2}^{\infty} ds' \, \frac{\rho_1(s')}{(s' - s_0)(s' - s)}$$

$$+ \frac{t - t_0}{\pi} \int_{4m^2}^{\infty} dt' \, \frac{\rho_2(t')}{(t' - t_0)(t' - t)} + \frac{u - u_0}{\pi} \int_{4m^2}^{\infty} du' \, \frac{\rho_3(u')}{(u' - u_0)(u' - u)}$$

$$+ \frac{(s - s_0)(t - t_0)}{\pi^2} \iint_{4m^2}^{\infty} \frac{ds' \, dt' \, \rho(s', t')}{(s' - s_0)(s' - s)(t' - t_0)(t' - t)}$$

$$+ \text{cyclic permutation.} \tag{4.4.38}$$

In practice the Mandelstam representation is used mainly as a convenient method for analytic continuation of a single variable dispersion relation, in s for example, to finite values of $t > 4m^2$. In this connection one requires to know only that for bounded t, the single dispersion relation in s requires only a finite number of subtractions; it is very reasonable to assume that this is true.

4.5   Singularities from perturbation theory

In this section we give a brief account of the methods for studying the singularities of Feynman integrals. A more complete account of this subject is given by Drell & Bjorken (1965, Chapter 18) and by Eden, Landshoff, Olive & Polkinghorne (1966) in *The Analytic S-Matrix*. The latter book will be referred to by the abbreviation ELOP (1966) in this section.

(a) *Singularities of integral transforms*

In perturbation theory, scattering amplitudes are expressed as the sum of a series of integrals that correspond to Feynman diagrams. The analytic properties of these Feynman integrals are more easily obtained by studying the integrand and the relation of its singularities to the path of integration, rather than by attempting to carry out the full integration. To illustrate the method we begin by considering a simple integral transform.

Consider a function $f(z)$, which is defined as the integral of a given function $g(z, w)$ over a contour C in the complex w-plane between two end points A and B shown in Fig. 4.5.1 (*a*),

$$f(z) = \int_C g(z, w)\, dw. \qquad (4.5.1)$$

Let the singularities of the integrand $g(z, w)$ in the w-plane be located at the points

$$w = w_r(z) \quad (r = 1, 2, \ldots). \qquad (4.5.2)$$

In general w_r will vary as z is varied. If there are no singularities w_r between the contours C and C' in Fig. 4.5.1 (*a*), using Cauchy's theorem we can write

$$f(z) = \int_{C'} g(z, w)\, dw. \qquad (4.5.3)$$

In fact C can be varied to any contour C', provided it does not cross over any singularity (pole or branch point) of the integrand (this is called an allowed deformation of C). This provides a method of analytic continuation in z of the integral (4.5.1) in the following manner:

The integral (4.5.1) defines a function $f(z)$ that is regular except when a singularity $w_r(z)$ of the integrand $g(z, w)$ moves on to the integration contour C. However, if as z varies from z_1 to z_2 a singularity $w_1(z)$ of $g(z)$ moves across C as shown in Fig. 4.5.1 (b), we can usually deform C

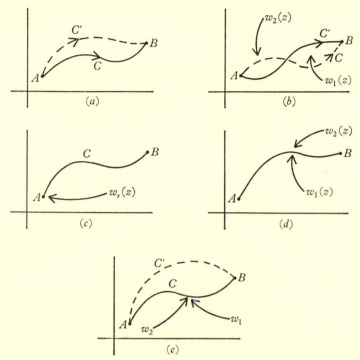

Fig. 4.5.1. The complex w-plane. (a) The contours C and C' in the complex w-plane that are used in the integral transforms (4.5.1) and (4.5.3); (b) deformation of C to avoid advancing singularities of the integrand as z varies; (c) variation of z giving an end-point singularity; (d) variation of z giving a pinch singularity; (e) *not* a pinch singularity, since w_1 and w_2 remain on the same side of C so it can undergo an allowed deformation to C' (in general).

to C', without crossing over any singularity w_r, before $w_1(z)$ reaches C, and we can work with the integral representation (4.5.3) instead of (4.5.1). This remains obviously regular when w_1 moves as shown in Fig. 4.5.1 (b). More generally as z varies we can analytically continue $f(z)$ by successively deforming C to C' to C'' etc. to avoid any approaching singularities w_r of the integrand $g(z, w)$.

This analytic continuation of $f(z)$ may be prevented for one of three reasons that arise from special positions of the singularities $w_r(z)$;

(i) *End-point singularities*. One of the singularities $w_r(z)$ may, for a particular value z_1 of z, meet one of the end points A or B of the integration contour C. Then no allowed variation of C can avoid the singularity w_r of the integrand $g(z, w)$, and the integral $f(z)$ may be singular at $z = z_1$. An end-point singularity is shown in Fig. 4.5.1 (c).

(ii) *Pinch singularities*. If two (or more) of the singularities $w_r(z)$ approach C from opposite sides and coincide when $z = z_2$, there is no allowed deformation that can avoid them. The point z_2 may be a singularity of $f(z)$. A pinch singularity is shown in Fig. 4.5.1 (d). Note however that the situation shown in Fig. 4.5.1 (e) is *not* a pinch of the contour C, and one can in general deform C to C' away from w_1 and w_2.

(iii) *Infinite deformations*. If a singularity $w_r(z)$ moves between the end points A and B, and goes to infinity as $z \to z_3$, an infinite deformation of the contour will result. This may cause a singularity, due to the possible divergence of the integral $f(z)$ along an infinite contour. By transforming with $\zeta = z^{-1}$, this can be studied in the same way as a pinch singularity.

The problem of finding singularities of $f(z)$ falls into two parts. One part is to locate possible singularities by finding values of z for which $w_r(z)$ meets an end point, or for which $w_1(z)$ and $w_2(z)$ coincide. The other part is to find whether a singularity actually occurs. For example, one must discover whether the coincidence of w_1 and w_2 corresponds to the pinch (singularity) situation of Fig. 4.5.1 (d), or to the non-pinch (non-singularity) situation shown in Fig. 4.5.1 (e). The first part of the problem can be solved in principle; the possible singularities correspond to solutions of a set of algebraic equations—the Landau equations, which will be formulated below.

The second part of the problem essentially concerns the sheet structure; on which Riemann sheet if any does a singularity lie? The standard technique here is to locate a pinch situation, when a contour is trapped as in Fig. 4.5.1 (d), and study the conditions under which the singularities $w_1(z)$ and $w_2(z)$, while remaining coincident, cease to pinch the contour. For this to occur, another parameter must be involved in addition to the variable z; let this parameter be t. Then

$$w_1 = w_1(z, t), \quad w_2 = w_2(z, t), \qquad (4.5.4)$$

are singularities of the integrand $g(z, t, w)$, in a generalisation of (4.5.1),

$$f(z, t) = \int_C g(z, t, w) \, dw. \qquad (4.5.5)$$

Now suppose, for a given value of z, we can choose t so that

$$w_1(z, t) = w_2(z, t). \tag{4.5.6}$$

Suppose also that (by methods to be discussed later) we find that these values of z and t correspond to the pinch situation of Fig. 4.5.1 (d). The solution of (4.5.6) can in principle be written

$$t = t(z). \tag{4.5.7}$$

Thus, as z varies we can choose new values of t so that (4.5.6) is retained; that is to say the two singularities w_1 and w_2 of $g(z, t, w)$ remain coincident. If it happens that, when we vary z through real values, the function $t(z)$ is also real (given by (4.5.7)), then our variation of z and t corresponds to moving along a real curve Γ in the (real z, real t) plane as shown in Fig. 4.5.2 (a). At the same time the points w_1 and w_2, while remaining coincident (4.5.6), will themselves move in the complex w-plane (in general). They cannot cease to pinch the contour unless the pinch slides off the end of C as shown in Fig. 4.5.2 (b).

Let (z_0, t_0) or P be the point on Γ that corresponds to w_1 and w_2 sliding off the contour at A. Then at P we will have

$$w_1(z_0, t_0) = w_2(z_0, t_0) = A. \tag{4.5.8}$$

Hence, if we can ensure the coincidence of two particular singularities w_1, w_2, then the corresponding points $z, t(z)$, will be singularities of $f(z, t)$ defined by (4.5.5), until we reach an end point while on the curve Γ in Fig. 4.5.2 (a). But this means that P *also* corresponds to an end-point singularity (see condition (i) above). To maintain the end-point singularity we need only require that either w_1 or w_2 remains at A, say

$$w_1(z, t) = A. \tag{4.5.9}$$

This can be solved to give a curve γ_1, whose equation is

$$t = t_1(z). \tag{4.5.10}$$

Thus P also lies on the curve γ_1 in the variables z and t, as indicated in Fig. 4.5.2 (a). In general, γ_1 may not be real when z is real but for simplicity we assume that this is so. The general solution of (4.5.10) will be a two-dimensional surface in the four-dimensional space of the two complex variables z and t. In general this will meet the, real z real t, plane in isolated points. However, in practice, many of the curves relating to singularities in perturbation theory do have real sections like those shown in Fig. 4.5.2 (a).

We cannot always identify which two singularities of $g(z, t, w)$ are coincident, but can only be sure that we have

$$w_r(z, t) = w_s(z, t), \qquad (4.5.11)$$

for some pair w_r and w_s out of $w_1, w_2, \ldots, w_n$. This means that there is another way in which a pair of singularities can cease to pinch the

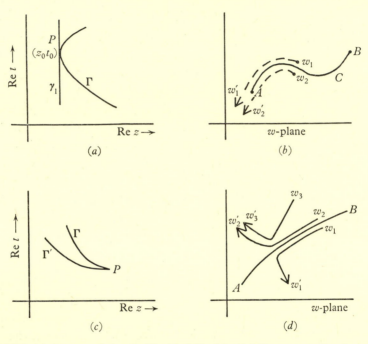

Fig. 4.5.2. (a) A curve of singularities Γ of $f(z, t)$ in (4.5.5) corresponding to a pinch singularity, and a curve of singularities γ corresponding to an end-point singularity. At P, the pinch reaches an end point; (b) the complex w-plane showing how a pinch can slide off a contour at an end point; (c) a curve of singularities Γ and Γ', that itself has a cusp singularity; (d) a multiple pinch that has been found sometimes to give rise to a cusp as in (c). The diagram shows how the pinch may cease to trap the contour C as z, t (and w_r) vary.

contour C in (4.5.5). This is illustrated by the point P on Γ in Fig. 4.5.2 (c), which corresponds to the coincidence of three singularities w_1, w_2 and w_3 on the integration contour C. Such a situation gives a multiple pinch as shown in Fig. 4.5.2 (d). It appears likely in practice that when a multiple pinch occurs the corresponding point P on Γ is a singular point of Γ, like a cusp as shown in Fig. 4.5.2 (c). On the curve Γ there is a pinch as with w_1, w_2 in Fig. 4.5.2 (d), but on Γ' the pinch is lost since now w_1 and w_3 are coincident but do not pinch the contour. The transition occurs at a multiple pinch. An example of this situation is discussed by Eden, Landshoff, Polkinghorne & Taylor (1961), (see also ELOP (1966)).

The above methods can be generalised to multiple integrals. The generalisation is quite difficult to make rigorous; it was at first assumed to be obvious (Eden, 1952), it was discussed more generally by Polkinghorne & Screaton (1960) and by Stapp (1962a), but its rigorous development requires the use of homology theory (Fotiadi, Froissart, Lascoux & Pham, 1964; Hwa & Teplitz, 1966).

(b) The Landau equations

These equations determine the conditions for a Feynman integral to have end-point or coincident singularities of its integrand. The Feynman integral is a term in the perturbation series for a scattering amplitude and corresponds to a Feynman diagram (see Bjorken & Drell (1965), Schweber (1961), Feynman (1949) Dyson (1949, 1951)).

A general Feynman integral has the form

$$F_\epsilon(s,t) = \int d^4k_1 \dots d^4k_l \frac{N}{\prod\limits_1^n (q_j^2 - m_j^2 + i\epsilon)}, \qquad (4.5.12)$$

with $\epsilon > 0$. The integration is over l independent loop four-momenta, and q_j $(j = 1, \dots, n)$ denotes the four-momentum in an internal line of the diagram. The numerator N does not lead to singularities except those avoided by renormalisation. It may sometimes cause cancellations of singularities, and in calculations from perturbation theory, the spin matrices that it contains may cause considerable difficulty. However, N does not affect the location of possible singularities in most situations that can be solved, so we will make the simplifying assumption that $N = 1$. Introducing Feynman parameters, $\alpha_1, \dots, \alpha_n$, and using the Feynman relation,

$$\frac{1}{u_1 u_2 \dots u_n} = (n-1)! \int_0^1 d\alpha_1 \dots d\alpha_n \frac{\delta(1 - \Sigma\alpha_j)}{\left[\sum\limits_1^n \alpha_j u_j\right]^n}, \qquad (4.5.13)$$

we can rewrite (4.5.12) as

$$F_\epsilon(s,t) = c_1 \int_0^1 d\alpha_1 \dots d\alpha_n \int d^4k_1 \dots d^4k_l \frac{\delta(1 - \Sigma\alpha)}{[\psi(k_1 \dots k_l) + i\epsilon]^n}, \qquad (4.5.14)$$

where
$$\psi(k_1 \dots k_l) = \sum\limits_1^n \alpha_j(q_j^2 - m_j^2). \qquad (4.5.15)$$

Since each q_j is linear in the variables k_i, the function ψ is quadratic in $k_1, k_2, \ldots, k_l$. The integral over the four-momenta $k_1 \ldots k_l$ in (4.5.14) can therefore be carried out. The general method for doing this is due to Chisholm (1952). The procedure is to change variables $k \to k'$, so that ψ becomes $\Sigma k'^2 + C$. Then one can use the relation

$$\int d^4k \frac{1}{(k^2+L)^3} = \frac{i\pi^2}{2L}, \qquad (4.5.16)$$

and its derivatives with respect to L. The result is

$$F_\epsilon(s,t) = c_2 \int_0^1 d\alpha_1 \ldots d\alpha_n \frac{\delta(1-\Sigma\alpha)[C(\alpha)]^{n-2l-2}}{[D(\alpha,s,t)+i\epsilon C(\alpha)]^{n-2l}}. \qquad (4.5.17)$$

$D(\alpha,s,t)$ is the discriminant of the full quadratic form ψ, regarded as a function of k_i^2, k_i, and terms not involving k. The function $C(\alpha)$ is the discriminant of only the second-order part of ψ, namely of the terms involving k_i^2, $k_i k_j$.

The physical Feynman amplitude consists of terms like (4.5.17), where s and t have real physical values, with $\epsilon > 0$. The limit of this amplitude as $\epsilon \to 0$ will lead to a distortion of the contour of integration from the real axis for at least one α variable. Thus

$$F(\text{physical}) = \lim_{\epsilon \to 0+} [F_\epsilon(s,t)]. \qquad (4.5.18)$$

The integral (4.5.17) also defines a function $F(s,t)$ when $\epsilon = 0$ and s (and/or t) is complex. In a simple situation with real s replaced by $s+i\epsilon$, the α contour need not be distorted from the real axis. So we have

$$F(s+i\epsilon,t) = c_2 \int_0^1 d\alpha_1 \ldots d\alpha_n \frac{\delta(1-\Sigma\alpha)[C(\alpha)]^{n-2l-2}}{[D(\alpha,s+i\epsilon,t)]^{n-2l}}. \qquad (4.5.19)$$

In such a situation it can be shown that as $\epsilon \to 0$ one obtains the same contour distortion as in the limit (4.5.18). Thus

$$F(\text{physical}) = \lim_{\epsilon \to 0+} [F(s+i\epsilon,t)]. \qquad (4.5.20)$$

This relation expresses the statement made earlier, that the physical amplitude is the boundary value of a function $F(s,t)$ of the complex variable s, on the upper side of the cut along the real s axis.

The conditions for $F(s,t)$ to be singular can be obtained either from the form (4.5.14) or from (4.5.19), or (less conveniently) from (4.5.12). These conditions are usually called the Landau equations. They were derived independently by Landau (1959), Bjorken (1959), Mathews

(1959) and Nakanishi (1959). Details of their derivation can be found in ELOP 1966 and only the results will be stated here. These give the conditions for zeros of the denominator in the integrals (4.5.14) and (4.5.19) to be coincident or end-point. There are two forms of these conditions that are particularly useful:

First form: Landau equations based on (4.5.14). The amplitude $F(s, t)$ may be singular when,

$$\text{either} \quad q_i^2 = m_i^2, \tag{4.5.21}$$

$$\text{or} \quad \alpha_i = 0, \tag{4.5.22}$$

for each value of $i = 1, 2, ..., n$. *In addition* we must have

$$\frac{\partial}{\partial k_j} \sum_1^n \alpha_i (q_i^2 - m_i^2) = 0 \quad (j = 1, 2, ..., l). \tag{4.5.23}$$

Since each q_i is linear in k, (4.5.23) is equivalent to

$$\sum_{(j)} \pm \alpha_i q_i = 0, \quad j = 1, ..., l. \tag{4.5.24}$$

where (j) denotes the sum round the loop of the Feynman diagram round which k_j runs, and the signs depend on the relative directions of k_j and q_i.

Second form: Landau equations based on (4.5.19). The amplitude $F(s, t)$ may be singular when,

$$D(\alpha, s, t) = 0, \tag{4.5.25}$$

and $$\text{either} \quad \alpha_i = 0, \tag{4.5.26}$$

$$\text{or} \quad \frac{\partial D}{\partial \alpha_i} = 0, \tag{4.5.27}$$

for each $i = 1, 2, ..., n$. In practice D is homogeneous in the α variables so (4.5.25) follows from the other equations.

(c) *The fourth-order diagram*

As an example we consider the square diagram with equal masses shown in Fig. 4.5.3 (a). The first form of the Landau equations gives the result that F may be singular when

$$\text{either} \quad q_i^2 = m^2 \quad \text{or} \quad \alpha_i = 0 \quad (i = 1, 2, 3, 4); \tag{4.5.28}$$

and $$\alpha_1 q_1 + \alpha_2 q_2 + \alpha_3 q_3 + \alpha_4 q_4 = 0. \tag{4.5.29}$$

We begin by considering $\alpha_1 = \alpha_3 = 0$. Then we must have

$$q_2^2 = m^2, \quad q_4^2 = m^2, \tag{4.5.30}$$

and

$$\alpha_2 q_2 + \alpha_4 q_4 = 0. \tag{4.5.31}$$

These equations correspond to the 'reduced' diagram shown in Fig. 4.5.3(b). There is an obvious electric circuit analogy in which α_i

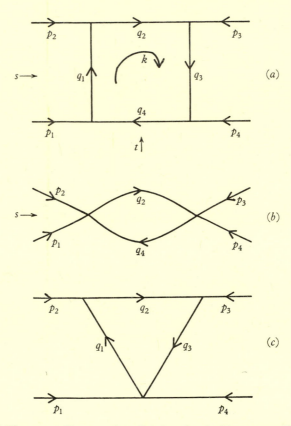

Fig. 4.5.3. (a) The square diagram whose Landau singularities are considered; (b) the reduced diagram giving the normal threshold $s = 4m^2$; (c) a reduced diagram giving no physical sheet singularities for equal masses in all lines.

denotes a resistance and q_i a current (of magnitude m). Then (4.5.29) and (4.5.31) correspond to Kirchhoff's second law. Conservation of the total four-momentum at each vertex corresponds to the first law. In this analogy one changes, in Fig. 4.5.3, from diagram (a) to diagram (b) by 'short-circuiting' lines 1 and 3. This electric circuit analogy has been extensively used by Bjorken (1959) and Bjorken & Drell (1965).

From (4.5.31) it is evident that the four vectors q_2 and q_4 must be parallel or antiparallel. The physical singularity has $\alpha_i > 0$ $(0 \leqslant \alpha_i \leqslant 1)$. Hence $q_2 = -q_4$, and at the singularity

$$s = (q_2 - q_4)^2 = 4q_2^2 = 4m^2. \tag{4.5.32}$$

This is the lowest normal threshold singularity of the scattering amplitude $F(s, t)$ with energy s. For this diagram there is also a normal threshold at $t = 4m^2$, which is the physical threshold for processes in which t is the energy.

Attached to the normal threshold in s is a branch cut from $s = 4m^2$ to $s = \infty$. For real t $(t < 4m^2)$ and complex s, there are no singularities of the Feynman integral for the square diagram. This can be seen from the form (4.5.19) in this case, since

$$D(\alpha, s_1 + is_2, t) = is_2 \alpha_2 \alpha_4 + D(\alpha, s_1, t). \tag{4.5.33}$$

The second term is real and the first term is non-zero if $\alpha_2 \neq 0$, $\alpha_4 \neq 0$, so that D is non-zero and F is regular. If $\alpha_2 = \alpha_4 = 0$, we obtain the fixed singularity at $t = 4m^2$, (with its branch cut $4m^2$ to ∞), but this is excluded if we require $t < 4m^2$.

More detailed consideration shows that for the square diagram, $F(s, t)$ is regular for complex s and t in the product:

{complex s-plane cut $(4m^2, \infty)$}

$$\times \text{\{complex } t\text{-plane cut } (4m^2, \infty)\}. \tag{4.5.34}$$

This domain is described as the 'physical Riemann sheet', or the physical sheet, of the scattering amplitude $F(s, t)$. We may also refer to the physical sheet in s (or in t), which means the cut s-plane in (4.5.34), (or the cut t-plane).

On the (real) boundary of the physical sheet (i.e. on the real s, real t, axes), there are no singularities of $F(s, t)$ from the conditions (4.5.28), (4.5.29) with only one α_i zero (to see this verify that $D \neq 0$ with α_j positive $(j \neq i)$, when the conditions are satisfied). However, if no α_i is zero, there are solutions with $0 < \alpha_i < 1$, and

$$q_i^2 = m^2 \quad (i = 1, 2, 3, 4), \tag{4.5.35}$$

$$\alpha_1 q_1 + \alpha_2 q_2 + \alpha_3 q_3 + \alpha_4 q_4 = 0. \tag{4.5.36}$$

Multiply (4.5.36) in turn by q_1, q_2, q_3, q_4. This gives four linear homogeneous equations in α_1, α_2, α_3, α_4. If there is to be a solution with $\alpha_i \neq 0$, the determinant associated with these four equations must vanish,

$$\det (q_i . q_j) = 0. \tag{4.5.37}$$

Now use the relations

$$s = (q_2 - q_4)^2, \quad t = (q_1 - q_3)^2, \quad q_i^2 = m^2. \tag{4.5.38}$$

Eliminating $(q_i \cdot q_j)$ from (4.5.37) by (4.5.38), we obtain

$$(s - 4m^2)(t - 4m^2) = 4m^4, \tag{4.5.39}$$

or $st = 0$. The latter is not a physical sheet singularity, but is reached only by going through a branch cut on to an unphysical sheet.

The branch of (4.5.39) lying in $s > 4m^2, t > 4m^2$ gives a singularity of $F(s,t)$ on the boundary of the physical sheet, where s and t are both real.

If we take $\operatorname{Im} F(s,t)$ for $t < 4m^2$ and $s > 4m^2$, to give a function $F_1(s,t)$, then on continuation in t we find that this function is singular for values of $t = t(s)$ on the curve (4.5.39) in $t > 4m^2$. This curve gives the boundary of the Mandelstam spectral function for equal masses, shown in Fig. 4.4.3(c). Further details are given by ELOP (1966).

After considering the sum of the square diagram shown, and the other fourth-order diagrams (with crossed rungs), Mandelstam (1958) proposed the symmetric double dispersion relation described in §4.4. The Mandelstam representation implies that there are no singularities of the scattering amplitude $F(s,t)$ on the physical sheet, but only on its boundary. The physical sheet is the product of the three complex planes of s, t and u, each cut along the real axis from $4m^2$ to ∞ (for equal masses), with the restriction that $s + t + u = 4m^2$. We consider next the form of the singularities of the scattering amplitude when the masses associated with the external and internal lines of a Feynman diagram are not equal.

4.6   General masses and anomalous thresholds

When the masses of the particles involved in a two-body reaction are unequal, two new features arise. One is the change in kinematics that causes the physical boundaries to be less simple and changes the mapping $(s,t) \leftrightarrow (q^2, \cos\theta)$. The second involves changes in the type of singularity that may occur on the boundary of the physical sheet or even in the complex part of the physical sheet. We will state the results in these two situations, illustrating the first by pion-nucleon scattering and the second by using general masses.

(a) *Pion-nucleon scattering*

Denote the pion and nucleon four-momenta by q_1, q_2, p_1, and p_2. We consider three processes,

$$\text{I}\,(s): \quad \pi_1 + N_1 \to \pi_2 + N_2,$$
$$\text{III}\,(u): \quad \bar{\pi}_2 + N_1 \to \bar{\pi}_1 + N_2,$$
$$\text{II}\,(t): \quad \pi_1 + \bar{\pi}_2 \to \bar{N}_1 + N_2, \quad\quad (4.6.1)$$

where a bar denotes the antiparticle. There are of course other processes having the same physical regions as I, III and II. For example the following three processes have similar kinematics,

$$\text{I}': \quad \pi_2 + N_2 \to \pi_1 + N_1,$$
$$\text{III}': \quad \bar{\pi}_1 + N_2 \to \bar{\pi}_2 + N_1,$$
$$\text{II}': \quad \bar{N}_1 + N_2 \to \pi_1 + \bar{\pi}_2. \quad\quad (4.6.2)$$

Let s, u, and t denote the invariant energies (squared) for processes I, III and II, respectively. These variables can all be expressed in terms of the three-momenta of π and N relative to their centre of mass in their initial state ($\mathbf{q}$ and $-\mathbf{q}$) and their final state ($\mathbf{q}'$, $-\mathbf{q}'$) for process I. From the definitions in § 3.1, we obtain,

$$s = (p_1 + q_1)^2 = [(M^2 + q^2)^{\frac{1}{2}} + (m^2 + q^2)^{\frac{1}{2}}]^2, \quad\quad (4.6.3)$$

$$u = (p_1 + q_2)^2 = [(M^2 + q^2)^{\frac{1}{2}} - (m^2 + q^2)^{\frac{1}{2}}]^2 - 2q^2(1 + \cos\theta), \quad\quad (4.6.4)$$

$$t = (p_1 + p_2)^2 = -2q^2(1 - \cos\theta), \quad\quad (4.6.5)$$

where $\quad q^2 = \mathbf{q}'^2 = \mathbf{q}^2 \quad$ and $\quad \mathbf{q}.\mathbf{q}' = q^2 \cos\theta.$ $\quad\quad (4.6.6)$

The boundaries of the physical regions I and III are given by $\cos\theta = \pm 1$. These give the line and curve shown in Fig. 4.6.1, in the $[s + u > 0]$ region of the (s, u) plane. Their equations are $t = 0$, which gives,

$$s + u = 2M^2 + 2m^2 \quad (\cos\theta = +1), \quad\quad (4.6.7)$$

$$su = (M^2 - m^2)^2 \quad (\cos\theta = -1). \quad\quad (4.6.8)$$

For I to be physical we must have $s > (M+m)^2$, and for III to be physical we must have $u > (M+m)^2$. Similarly by choosing a centre of mass frame for process II, the boundary of this physical region can be shown to be given by the other branch of the curve (4.6.8) (in $[s + u < 0]$). For II to be physical, we require $t > 4M^2$. The physical regions are shaded in Fig. 4.6.1 (recall that $s + t + u = 2m^2 + 2M^2$).

The lowest normal threshold in the s channel (I) is at the physical threshold. Thus $F(s,t)$ has a fixed branch point at

$$s = (M+m)^2. \qquad (4.6.9)$$

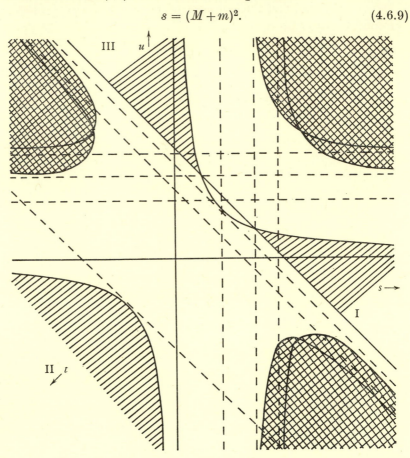

Fig. 4.6.1. Kinematics for pion-nucleon processes (not drawn to scale). Physical scattering regions are shown shaded with diagonal lines, I for $\pi N \to \pi N$, II for $\pi N \to \pi N$, and III for $\pi\pi \to N\overline{N}$. The cross-hatched areas are regions where the Mandelstam spectral functions are non-zero.

There is a similar normal threshold in the u channel (III). In the t channel (noting conservation rules) there will be normal thresholds at

$$t = (nm)^2 \quad (n = 2, 4, 6, ...), \qquad (4.6.10)$$

in addition to that at $t = 4M^2$. There are further normal thresholds in each channel whenever a new physical particle is allowed, for example at

$$s = (M+nm)^2 \quad (n = 2, 3, ...). \qquad (4.6.11)$$

There will also be a pole at $s = M^2$ and one at $u = M^2$.

Some of the normal thresholds are shown as broken lines in Fig. 4.6.1. The Mandelstam representation is believed to hold for the πN scattering amplitudes. It takes a form similar to (4.4.38) for each invariant amplitude (see §4.9), but the regions where the spectral functions are non-zero are more complicated. These regions can be obtained by considering the fourth-order diagrams having the lightest internal masses allowed by conservation laws. One can then apply the formulae that are given below for the general mass case. The resulting spectral functions $\rho_{12}(s,t)$ etc. are non-zero in the regions shown double shaded in Fig. 4.6.1.

For pion-nucleon scattering there are six possible charge states. Allowing for isospin invariance there are two isospin amplitudes, having $T = \frac{3}{2}$ and $T = \frac{1}{2}$. In addition there are two independent invariant amplitudes A and B for each charge state. These spin and charge considerations complicate the crossing conditions. For example if in (4.6.1) process I is

$$\pi^- + p \rightarrow \pi^- + p, \tag{4.6.12}$$

then the crossed process III is

$$\pi^+ + p \rightarrow \pi^+ + p. \tag{4.6.13}$$

Invariant amplitudes and crossing relations for the pion-nucleon system will be considered in §4.9. We proceed now to the evaluation of the singularities of a scalar amplitude with general masses.

(b) Mandelstam representation with general masses

Our discussion will be confined to the fourth-order square diagram shown in Fig. 4.5.3 (a) except that now we allow general internal masses $m_1, \ldots, m_4$ and general external masses $M_1, \ldots, M_4$. As before there are normal thresholds when the energy reaches a value at which the intermediate states become physical. This gives branch points of the amplitude $F(s,t)$ at

$$s = (q_2 - q_4)^2 = (m_2 + m_4)^2, \tag{4.6.14}$$

$$t = (q_1 - q_3)^2 = (m_1 + m_3)^2. \tag{4.6.15}$$

These singularities come from reduced diagrams analogous to that shown in Fig. 4.5.3 (b), with $q_2^2 = m_2^2$ and $q_4^2 = m_4^2$. It is useful to introduce new variables y_{ij} defined by

$$y_{ij} = \frac{(q_i - q_j)^2 - m_i^2 - m_j^2}{2 m_i m_j}. \tag{4.6.16}$$

Then

$$y_{24} = \frac{s - m_2^2 - m_4^2}{2 m_2 m_4}, \tag{4.6.17}$$

$$y_{13} = \frac{t - m_1^2 - m_3^2}{2 m_1 m_3}, \tag{4.6.18}$$

$$y_{41} = \frac{M_1^2 - m_4^2 - m_1^2}{2 m_4 m_1} \quad \text{etc.} \tag{4.6.19}$$

The reduced diagram shown in Fig. 4.5.3 (c) now may have singularities on the boundary of the physical sheet. This is ascertained by finding whether there are solutions, with $0 < \alpha_i < 1$, of the corresponding Landau equations. From (4.5.21, 24) these equations are

$$q_i^2 = m_i^2 \quad (i = 1, 2, 3), \tag{4.6.20}$$

$$\alpha_1 q_1 + \alpha_2 q_2 + \alpha_3 q_3 = 0. \tag{4.6.21}$$

Multiply (4.6.21) successively by q_1, q_2 and q_3 to give three linear equations in $\alpha_1, \alpha_2, \alpha_3$. There can be non-zero solutions only if

$$\det (q_i \cdot q_j) = 0. \tag{4.6.22}$$

Using (4.6.17, 18, 19) this equation can be solved for t or y_{13} (it does not depend on s). There are two solutions:

$$y_{13} = a_{13} \quad \text{and} \quad y_{13} = a_{13}'. \tag{4.6.23}$$

Both lie in $(-1 < y_{13} < 1)$, so that t is below the normal threshold (4.6.15) for both values. One can now ascertain whether these solutions give $\alpha_i > 0$ $(i = 1, 2, 3)$. It is found (Karplus, Sommerfeld & Wichmann (1958)) that the solution nearest to the normal threshold has $\alpha_i > 0$ $(i = 1, 2, 3)$ if

$$(\arccos y_{12} + \arccos y_{23}) > \pi. \tag{4.6.24}$$

Then the corresponding point $t = t_a$ is a singularity of the scattering amplitude $F(s, t)$ on the boundary of the physical sheet.

When (4.6.24) holds, the point $t = t_a$ is called an *anomalous threshold* of the amplitude $F(s, t)$. 'Anomalous' because it lies below the first allowed physical (normal) threshold.

A similar analysis can be done of the singularities of the fourth-order diagram 4.5.3 (a) itself, (see, for example, ELOP (1966)). Full information can be evaluated only by studying the connection between singularities on different Riemann sheets, reached by going through the (real) branch cuts. For practical use the most important results concern the physical sheet. The possible situations on the physical sheet and its boundary are summarised in Fig. 4.6.2. The successive diagrams (a) to (e) correspond to the situation of singularities on the physical sheet as the external masses M_i are successively increased relative to the internal masses m_i.

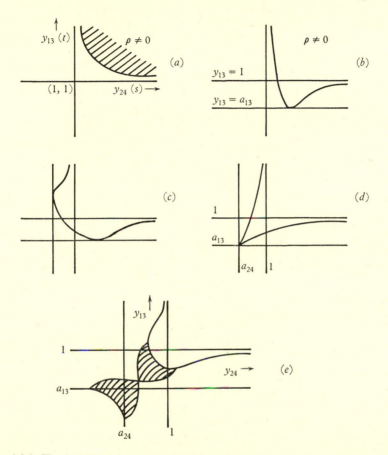

Fig. 4.6.2. The boundary of the Mandelstam spectral function for equal or nearly equal masses is shown in diagram (a). The other diagrams show how the boundary curve changes as the external masses are increased relative to the internal masses. In diagram (b) there is one anomalous threshold, in (c) and (d) there are two. In diagram (e) there is a complex surface coming on to the physical sheet.

In Fig. 4.6.2(a) the scattering amplitude for the single square diagram satisfies the simplest form of Mandelstam representation,

$$F(s,t) = \frac{1}{\pi^2} \int_a^\infty ds' \int_b^\infty dt' \frac{\rho(s',t')}{(s'-s)(t'-t)}, \qquad (4.6.25)$$

where a and b denote the normal thresholds, corresponding to $y_{24} = 1$ and $y_{13} = 1$ respectively.

In diagram (b) of Fig. 4.6.2, one anomalous threshold has developed, as the condition (4.6.24) is satisfied. The region where ρ is non-zero is bounded by the curve shown in diagram (b) (in y_{13}, y_{24} variables). In diagram (c), a further increase in the external masses has produced an anomalous threshold in s as well as in t. In diagram (d) the boundary curve of the region $\rho \neq 0$ has developed a singular point—a 'crunode'. Finally in (e) the real section of the curve has split into two parts, and they are connected by a complex surface that enters the physical sheet of the variables s and t (or y_{24}, y_{13}). Then there are complex singularities of $F(s, t)$ on the physical sheet and the Mandelstam representation (4.6.25) no longer holds. One can instead, write down an integral transform with a complex contour but it is much less convenient to use. Further details of these and other complex singularities of scattering amplitudes are given in *The Analytic S-Matrix* (ELOP, 1966).

There is one aspect of the kinematics of unequal mass collisions that causes a particular difficulty in Regge theory which we discuss later (see §5.6). This arises from a singularity in the mapping between different variables.

(c) Mapping singularities

As an example consider pion-nucleon scattering. From (4.6.3, 4, 5)

$$\cos\theta = 1 + \frac{t}{2q^2}, \tag{4.6.26}$$

$$4q^2 = s - 2(M^2 + m^2) + \frac{(M^2 - m^2)^2}{s}. \tag{4.6.27}$$

If s is held fixed and $t \to \infty$ then $\cos\theta \to \infty$. However by taking s of order m^4/t, we find that $\cos\theta$ is no longer large when t is large. In terms of u we have

$$\cos\theta = 1 + \frac{2M^2 + 2m^2 - s - u}{2q^2}. \tag{4.6.28}$$

The same remarks apply now as $u \to \infty$. We can keep $\cos\theta$ finite by choosing s of order $m^4/(2M^2 + 2m^2 - s - u)$, i.e. of order m^4/u, for large u.

If we consider channel III for which u is the energy variable (see Fig. 4.6.1), small values of s and large values of u correspond to backward pion-nucleon scattering at high energy. Thus the singular point $s = 0$ in the change of variable s to q^2 lies in the physical region III.

For equal masses, instead of (4.6.27) we have

$$4q^2 = s - 4m^2. \tag{4.6.29}$$

In this case there is no singularity in the transformation s to q^2, at $s = 0$, and the mapping difficulty in Regge theory mentioned above does not arise.

4.7 Partial wave dispersion relations

Partial wave dispersion relations can be derived most conveniently from the Mandelstam representation. As we will see below, for unequal masses the branch cuts are not in general real in the energy plane, so the corresponding dispersion relation is a generalisation from the Hilbert transforms that we have so far considered. For equal masses the partial wave dispersion relations can be proved for every term in the perturbation series for a scattering amplitude (Eden, 1960). Although they have not yet been proved from axiomatic field theory, a large but finite domain of analyticity has been established (Martin, 1966 a and §4.8). The problem of subtractions in partial wave dispersion relations has not been solved although interesting progress towards this end has been made by Kinoshita (1966).

(a) Equal masses

The kinematics for equal mass scattering with real invariants s and t is shown in Fig. 4.4.2 *(b)*. We define the partial wave amplitude $g_l(s)$ by

$$g_l(s) = \frac{W}{2k} f_l(s) = \frac{1}{4} \int_{-1}^{1} d\,(\cos\theta) \frac{F(s,t)}{8\pi} P_l(\cos\theta), \tag{4.7.1}$$

where
$$s = 4(m^2 + k^2) = W^2, \quad \cos\theta = 1 + \frac{2t}{s - 4m^2}. \tag{4.7.2}$$

Using t as the variable of integration,

$$g_l(s) = \frac{1}{64\pi k^2} \int_{4m^2 - s}^{0} dt\, F(s,t)\, P_l\!\left(1 + \frac{2t}{s - 4m^2}\right). \tag{4.7.3}$$

Assuming the Mandelstam representation, $F(s,t)$ will be analytic in the product of the cut s-plane and the cut t-plane. The branch cut from $s = 4m^2$ to ∞, leads to a corresponding branch cut for $g_l(s)$. Similarly the branch cut for $F(s,t)$, for $-\infty < s \leqslant 0$, will also be a branch cut for $g_l(s)$.

The other possible branch cuts that we must consider, arise from the fact that the path of integration (and one end point), in the variable t, is a function of s. There will be a singularity of $g_l(s)$ if the path of integration is pinched by singularities of $F(s,t)$; for equal masses this can be shown not to occur in perturbation theory, and it does not occur if $F(s,t)$ satisfies the Mandelstam representation. However, $g_l(s)$ will have a singularity (and an attached branch cut) whenever the end-point $t = 4m^2 - s$ meets a singularity of $F(s,t)$. We must therefore consider singularities of the function,

$$F(s,t(s)) \equiv F(s, 4m^2 - s). \tag{4.7.4}$$

This has the normal threshold singularities for $s \geqslant 4m^2$, and $s \leqslant 0$, that we have already considered. From Fig. 4.4.2 (b) it is evident also that the normal threshold singularities for $t \geqslant 4m^2$ will also give branch points (and cuts) along $s \leqslant 0$. The coincidence of singularities from $t = 4m^2$ and $s = 0$ is special to the equal mass case as we noted in §4.6. If the scattering amplitude F has a pole in s, say at $s = m^2$, this will appear as a pole in $g_l(s)$. However a pole in t,

$$F(s,t) = \frac{C}{t - m^2} + \dots, \tag{4.7.5}$$

leads to a distortion of the integration path $(0, 4m^2 - s)$ in (4.7.3) whenever $s < 3m^2$, so that the end-point singularity, $4m^2 - s = m^2$, gives a branch point. We conclude that the left-hand branch cut is along $(-\infty, 3m^2)$ when there is a pole in t, and $(-\infty, 0)$ when there is no pole. If we assume that there are no poles and that

$$g_l(s) \to 0 \quad \text{as} \quad |s| \to \infty, \tag{4.7.6}$$

the dispersion relation for $g_l(s)$ will have the form

$$g_l(s) = \frac{1}{\pi} \int_{4m^2}^{\infty} \frac{ds' \, g_l^{(1)}(s')}{s' - s} + \frac{1}{\pi} \int_{-\infty}^{0} \frac{ds' \, g_l^{(1)}(s')}{s' - s}. \tag{4.7.7}$$

(b) *Unitarity and N over D solutions*

We consider a definite value of l, and write $g_l(s) = g(s)$ for the partial wave amplitude in this angular momentum state. The discontinuity $g^{(1)}(s)$ of g on its left-hand cut can be interpreted as a generalisation of the potential. To see this we will follow the discussion by Chew (1961). Assume that the left-hand discontinuity is a given function $v(s)$,

$$\operatorname{Im} g(s) = g^{(1)}(s) = v(s) \quad \text{for} \quad s < 0. \tag{4.7.8}$$

It can be shown for potential scattering (including the possibility of coupled channels, Omnes 1958, 1961; Frye & Warnock, 1962), that the partial wave amplitude $g(s)$ can be expressed in the form,

$$g(s) = \frac{N(s)}{D(s)}, \tag{4.7.9}$$

where $N(s)$ has no singularities in the complex s-plane except along the left-hand cut, and $D(s)$ is singular only on the right-hand cut (see (3.5.35)). It is often assumed that a similar separation can be made in relativistic theory.

On the left-hand cut, using (4.7.8, 9)

$$\operatorname{Im} N(s) = v(s)\, D(s) \quad \text{for} \quad s < 0. \tag{4.7.10}$$

On the right-hand cut,

$$\operatorname{Im} N(s) = 0 \quad \text{for} \quad s > 4m^2. \tag{4.7.11}$$

In the elastic region of the right-hand cut ($4m^2 < s < 9m^2$), unitarity gives a simple relation between $\operatorname{Im} D$ and N. From (4.7.1),

$$g(s) = \frac{W \exp i\delta \sin \delta}{2k}, \tag{4.7.12}$$

so, when δ is real (i.e. in the elastic region),

$$\operatorname{Im} [g(s)]^{-1} = -2k/W. \tag{4.7.13}$$

Then

$$\operatorname{Im} D(s) = -2kN(s)/W, \quad 4m^2 < s < 9m^2, \tag{4.7.14}$$

$$= -2kR(s)\, N(s)/W, \quad 9m^2 < s, \tag{4.7.15}$$

$$= 0, \quad s < 4m^2, \tag{4.7.16}$$

where $R(s)$ denotes the effect of inelasticity. From §3.5,

$$g_l(s) = \frac{(\eta_l \exp 2i\delta_l - 1)\, W}{4ik}, \tag{4.7.17}$$

$$R_l(s) = \frac{\sigma_l(\text{total})}{\sigma_l(\text{elastic})} = -\frac{W}{2k} \operatorname{Im} [g_l(s)]^{-1}. \tag{4.7.18}$$

The functions N and D are both analytic except on part of the real axis, and each is real on part of the real axis, so they satisfy dispersion relations. The dispersion relation for N is (using 4.7.10, 11),

$$N(s) = \frac{1}{\pi} \int_{-\infty}^{0} dx\, \frac{v(x)\, D(x)}{x - s}. \tag{4.7.19}$$

It is convenient to introduce a subtraction in the dispersion relation for D, and to assume that $D(s) \to 1$ as $s \to \infty$. This gives

$$D(s) = 1 - \frac{(s-s_0)}{\pi} \int_{4m^2}^{\infty} dx \left(\frac{x-4m^2}{x}\right)^{\frac{1}{2}} \frac{R(x)\,N(x)}{(x-s)\,(x-s_0)}. \qquad (4.7.20)$$

Now we make two simplifying assumptions: (i) take $v(x)$ to have the form,

$$v(x) = -\pi G\,\delta(x-s_1) \quad \text{where} \quad s_1 < 4m^2, \qquad (4.7.21)$$

and (ii) approximate the resulting integral for D by assuming dominance by the integrand near $4m^2$. Then, $R \approx 1$, and, choosing $s_0 = s_1$,

$$N(s) = \frac{G}{s-s_1}. \qquad (4.7.22)$$

Write $s_1 = 4(m^2 - y_1^2)$, then

$$D(s) = 1 - \frac{G(y_1 - ik)}{32my_1(y_1 + ik)}. \qquad (4.7.23)$$

For small $k^2 > 0$, from (4.7.12),

$$\mathrm{Re}\left(\frac{D}{N}\right) = \frac{2k\cot\delta}{W} \approx \frac{k\cot\delta}{m}. \qquad (4.7.24)$$

Using the effective range expansion for $\cot\delta$ we obtain,

$$\mathrm{Re}\left(\frac{D}{N}\right) = \frac{1}{am} + \frac{rk^2}{2m}. \qquad (4.7.25)$$

We can make also an expansion of D/N from (4.7.22, 23) in powers of k. Comparing this expansion with (4.7.25) we obtain the effective range and scattering length in terms of the parameters G and s_1. For example the scattering length is given by,

$$\frac{1}{ma} = \frac{4y_1^2}{G}\left[1 - \frac{G}{32my_1}\right]. \qquad (4.7.26)$$

If the strength G of the discontinuity on the left-hand cut is small, then a is small and has the same sign as G. Thus G represents a weak attraction if positive, and a weak repulsion if negative. For large negative G, the scattering length a never exceeds $8/y_1$, as one would expect for a strongly repulsive potential.

If G is positive, an increase will cause the scattering length a to become infinite when

$$G = 32my_1. \qquad (4.7.27)$$

When G exceeds this value, the scattering amplitude $g = g_l(s)$ becomes singular at a point given by

$$s = 4(m^2 + k^2) = 4m^2 - \frac{2}{ar}, \qquad (4.7.28)$$

where a is given by (4.7.26) and r is the effective range obtained as described above. This value of s corresponds to a bound state and is a pole of $g(s)$ on the physical sheet. Thus zeros of D correspond to bound states.

In some circumstances $D(s)$ may have poles for real or complex values of s. In the single-channel problem these may arise as an inherent ambiguity in the solution of the scattering equation. They do not correspond to poles of the amplitude. They are called CDD poles (Castillejo, Dalitz & Dyson, 1956). In a many-channel problem, as the coupling is varied (or by variation of the angular momentum), it appears that the right-hand branch cut (4.7.20) may become distorted by singularities of D coming on to the physical sheet. This occurrence can, by a change of both N and D, be replaced by a CDD situation. It is discussed in detail by Hartle & Jones (1965).

Bootstraps

If crossing symmetry is used then, in principle, a knowledge of partial waves in the s channel permits one to evaluate the full scattering amplitude in the t channel and the u channel. From this, one can in principle evaluate the discontinuity across the left-hand cut of any partial wave in the s channel. Thus, knowing all $g_l(s)$, $l = 0, 1, \ldots$ for $s > 4m^2$, we can determine F and hence the discontinuity $v_l(s)$ across the left-hand cut. Then we can in principle solve the dispersion relation for $g_l(s)$ with $v_l(s)$ as the driving term (or 'potential') to regain $g_l(s)$ on the right-hand cut.

In practice the following approximation procedure can be considered. For pion-pion scattering, assume a ρ resonance for $l = 1$ giving a definite form for $g_1(t)$. Take this to determine the full amplitude $F(s,t)$ in the t channel, neglecting all other partial waves. This will produce a branch cut in the partial wave $g_1(s)$ for the s channel for $s < s_0$, where s_0 depends on the assumed mass of the ρ. Approximate this branch cut by a pole near s_0 and a residue at the pole related to the width of the ρ resonance. This determines a potential $v_1(s)$ like that introduced in (4.7.8). It can be used to calculate the form of $g_1(s)$ for $s > 4m^2$. With a suitable choice of parameters $g_1(s)$ is found to have a resonance. If one requires this resonance to be the same as that assumed for $g_1(t)$ in position and in width, one obtains a self-consistency condition. This is the bootstrap condition—that the ρ meson produces an attractive force when exchanged by two pions, and that this exchange force is strong enough to produce a nearly bound state of the pions—namely the ρ meson.

Unfortunately the bootstrap procedure seems to be unstable after the first iteration. Only with some form of cut-off on the left-hand cut does it give reasonable answers. This cut-off may come from terms that are neglected in the very drastic approximations involved. The idea of strongly interacting particles generating themselves is an attractive one in spite of the considerable difficulties that have prevented any self-consistent bootstrap up to the present.

The bootstrap method is described by Zachariasen (1964) and Balazs (1963). A discussion of its possible place in a future theory of strong interactions is given by Chew 1966).

(c) Pion-nucleon partial waves

We conclude this section by outlining for πN scattering, the form of the singularities of the partial wave amplitude $g_l(s)$ in the complex s-plane. They are obtained from the singularities of the amplitude discussed earlier in § 4.6, using the kinematics for πN scattering and the methods discussed in part (a) of this section. Essentially one assumes that no pinching singularities arise as a result of variations of the partial wave range of integration, but one includes all singularities encountered by the end points. This assumption is correct if the Mandelstam representation is valid.

The partial wave amplitude that we are using is given by,

$$g_l(s) = \frac{1}{64\pi q^2} \int_{t_2}^{t_1} dt\, F(s,t)\, P_l(\cos\theta), \qquad (4.7.29)$$

where t_1 and t_2 are given by $\cos\theta = +1$ and -1 in equation (4.6.5), namely
$$t_1 = 0, \qquad (4.7.30)$$

$$t_2 = -4q^2 = -s + 2(m^2 + M^2) - \frac{(m^2 - M^2)^2}{s}. \qquad (4.7.31)$$

Assuming that $g_l(s)$ is singular only when $F(s,t)$ is singular at the endpoints, we consider the singularities of $F(s,t_1)$ and $F(s,t_2)$ for complex s when $t_1(s)$ and $t_2(s)$ are given by (4.7.30, 31). The singularities of $F(s,t)$ are shown in Fig. (4.6.1), using $s + t + u = 2m^2 + 2M^2$.

The resulting singularities of $g_l(s)$ are shown in Fig. 4.7.1. They arise from singularities of $F(s,t)$ in the following manner:

(i) The Born term in s gives a pole (the nucleon pole) at
$$s = M^2. \qquad (4.7.32)$$

(ii) The Born term in u (a nucleon pole in $F(s,t)$) gives a branch cut, from
$$s = M^2 - 2m^2 + \frac{m^4}{M^2} \quad \text{to} \quad s = M^2 + 2m^2. \qquad (4.7.33)$$

(iii) The normal threshold in s gives a branch cut from,

$$s = (M+m)^2 \quad \text{to} \quad +\infty. \tag{4.7.34}$$

(iv) The normal threshold in u gives a branch cut from,

$$s = (M-m)^2 \quad \text{to} \quad -\infty. \tag{4.7.35}$$

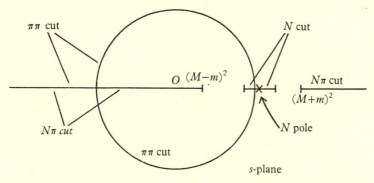

Fig. 4.7.1. Singularities of the partial wave amplitude for pion-nucleon scattering (not drawn to scale), shown in the complex s-plane.

(v) The normal threshold in t at $t = 4m^2$ gives a branch point at

$$s = M^2 - m^2, \tag{4.7.36}$$

with which we associate a branch cut so that it goes through the multipion thresholds $(2n\pi)^2$; this is a branch cut along the intersection of (4.7.31) with the line

$$t_2 > 4m^2 \quad \text{with } t_2 \text{ real}. \tag{4.7.37}$$

We must consider both $t_2 + i0$ and $t_2 - i0$, and we find that the branch cut satisfies the equation

$$\frac{s}{(M^2-m^2)} + \frac{(M^2-m^2)}{s} = 2b, \tag{4.7.38}$$

where b is real and less than one. For $-1 < b < 1$ we obtain,

$$s = (M^2 - m^2)\exp i\theta, \tag{4.7.39}$$

giving the circular branch cut in Fig. 4.7.1, when $-\pi < \theta < \pi$. For $b < -1$, s is real and the branch cut is from

$$s = -(M^2 - m^2) \quad \text{to} \quad -\infty. \tag{4.7.40}$$

The discontinuity across these branch cuts has to be obtained by considering their origin via the Mandelstam representation from physical discontinuities. There is a very full account of these questions in papers by Hamilton and collaborators (Hamilton, 1964, 1966).

4.8   Rigorous results on analyticity

In this section we will describe the regions of (complex) s and t, for which the scattering amplitude $F(s,t)$ has been shown rigorously to be analytic. The assumptions that lead to these results are the axioms of quantum field theory, which were given qualitatively in §4.1 with references for further details. The impetus for studying dispersion relations came from Goldberger's work (1955), and the simpler results on analytic properties in one variable were proved soon afterwards by Symanzik (1957), Bogoliubov (1958), Bremermann, Oehme & Taylor (1958) and Lehmann (1958). Analyticity in two complex variables was established in successively larger regions by Mandelstam (1960), Lehmann (1964), Hepp (1964), Bros, Epstein & Glaser (1964, 1965) and Martin (1966a). The techniques in the (1966a) papers by Martin are important as an illustration of the practical application of analyticity to high energy scattering and they will be described in some detail in Chapter 7. There is also a review of results on analyticity by Martin (1966b), which gives more complete details than are given here.

(a)   The Lehmann ellipse

It has been proved by Lehmann (1958), with certain restrictions on the masses of participating particles, that the scattering amplitude $F(s, t(\cos\theta))$ is regular for $\cos\theta$ inside an ellipse in the complex $(\cos\theta)$-plane. The imaginary part of F is analytic inside a slightly larger ellipse. The energy squared, s, is physical and fixed, and θ is the scattering angle in the centre of mass system.

The Lehmann ellipse in the $(\cos\theta)$-plane has foci at $\cos\theta = \pm 1$, and a semi-major axis of length

$$b(s) = \left[1 + \frac{(m_1^2 - M_1^2)\,(m_2^2 - M_2^2)}{k^2\{s - (M_1 - M_2)^2\}}\right]^{\frac{1}{2}}, \qquad (4.8.1)$$

where m_1 and m_2 are the masses of the incident particles and where k is the momentum in the centre of mass system. M_1, M_2 are the lowest mass states such that

$$\langle 0|j_1(x)|M_1\rangle \neq 0, \quad \langle 0|j_2(x)|M_2\rangle \neq 0, \qquad (4.8.2)$$

where j_1 and j_2 are the current operators associated with particles 1 and 2. Thus for equal mass scalar bosons $M_1 = M_2 = 2m$, whilst for pion-pion scattering $M_1 = M_2 = 3\mu$, where μ is the pion mass.

It will be recalled that,

$$s = [(m_1^2 + k^2)^{\frac{1}{2}} + (m_2^2 + k^2)^{\frac{1}{2}}]^2 \tag{4.8.3}$$

and from (3.1.19),

$$t = \sum_1^4 m_r^2 - s + \frac{\lambda(s, m_1^2, m_2^2)\,\lambda(s, m_3^2, m_4^2)\cos\theta - (m_1^2 - m_2^2)(m_3^2 - m_4^2)}{2s}. \tag{4.8.4}$$

The imaginary part, $\operatorname{Im} F(s, t)$ is regular inside the 'large' Lehmann ellipse having foci $(+1, -1)$ and semi-major axis $(2b^2 - 1)$.

The Lehmann ellipse 'shrinks' as $s \to \infty$, so that asymptotically it approaches the real line

$$-1 < \cos\theta < 1. \tag{4.8.5}$$

Of more direct significance, the end of the major axis corresponds to a point $t = t_0(s)$ such that

$$t_0(s) \sim \frac{C}{s} \to 0 \quad \text{as} \quad s \to \infty. \tag{4.8.6}$$

This result is much weaker than the result $t_0 = m^2$ (or $4m^2$) that one expects intuitively from crossing symmetry to the t channel. The important advance by Martin (1966 a) that we discuss in detail in § 7.3, shows that the ellipse in which $F(s, t)$ is regular can indeed be enlarged so that $t_0(s)$ is replaced by m^2 (or $4m^2$ if the pole is removed).

(b) The Martin domain of analyticity

The amplitude $F(s, t)$, for s below threshold (between the branch cuts), has been shown in the papers quoted above, to be an analytic function of t regular in a neighbourhood of $t = 0$, $|t| < R_0$, say. For s above threshold, but not at any higher threshold singularity, $F(s, t)$ has been shown by Bros, Epstein & Glaser (1964, 1965) to be regular in both s and t, for t near to zero and s near to physical values. For small values of t, they show also that $F(s, t)$ is regular in the cut s-plane, and when t is small and real F satisfies a dispersion relation in s.

The work of Martin, which we discuss in § 7.3, shows that for s near to its physical values (no matter how large), $F(s, t)$ is regular inside the region

$$|t| < R, \tag{4.8.7}$$

where R is a positive constant. The important point of this result, in relation to the behaviour of amplitudes and cross-sections at high energy, is that R is independent of the variable s. There are a number of consequent enlargements of the analyticity domain of $F(s, t)$ as a function of the two complex variables, but we will indicate these only briefly and refer the reader to the papers by Martin (1966a, b) for further details.

The size of the domain (4.8.7) has been established by Sommer (1966). It satisfies,

$$R \geqslant \max. R(s) \quad (\text{threshold} \leqslant s < \infty), \qquad (4.8.8)$$

$$R(s) = 2k^2[b(s) - 1], \qquad (4.8.9)$$

where $b(s)$ is the semi-major axis of the Lehmann ellipse (4.8.1). In terms of the pion mass μ, the value of R for $\pi\pi$ scattering is $4\mu^2$, which was expected from perturbation theory. It seems likely that, in practice, $R(s)$ will be the perturbation value, although there are some difficulties in proving this when intermediate single particle states can occur, as with πN scattering.

From the above results it is known that $F(s, t)$ is regular in an ellipse (the Lehmann ellipse), and in a region (4.8.7), which extends outside and ellipse and includes points where $\cos\theta$ is real and greater than one. The theory of expansion in terms of Legendre polynomials shows that the ellipse can be enlarged so as to include the largest real (positive) value of $\cos\theta$, which corresponds to $t = 4\mu^2$. This enlarged ellipse is of great importance in establishing high energy bounds on the amplitude and on cross-sections.

The range of t, for which $F(s, t)$ satisfies a dispersion relation in s, can be extended using Martin's enlargement of the Lehmann ellipse. It seems probable that the only limitations on extending this range arise from singularities of $\operatorname{Im} F(s, t)$ for s above the first inelastic threshold. In the $\pi\pi$ case it has been proved by Martin (1966a, b) that $F(s, t)$ satisfies a dispersion relation in s when t is in the range

$$-28\mu^2 < t < 4\mu^2. \qquad (4.8.10)$$

The lower limit arises from the inelastic thresholds at $16\mu^2$ in s and in u. With $u = s = 16\mu^2$, the relation $s + t + u = 4\mu^2$ gives $t = -28\mu^2$. Similar results have been proved for πK scattering, and slightly less satisfactory results have been proved for πN scattering (due to difficulties from the nucleon pole).

The work of Martin (1966a, b) also extends the domain of analyticity of $F(s, t)$ as a function of two variables. The real section (real s, real t) of this domain for $\pi\pi$ scattering goes up to the boundary of the Mandelstam spectral functions $\rho(s, t)$, $\rho(t, u)$, $\rho(u, s)$, except for the small indentations (see Fig. 4.6.1). The complex section of the domain of analyticity is somewhat larger than the product of the Martin–Lehmann ellipse in t, times the cut s-plane, because, (1) crossing symmetry ($st \to tu \to us$) extends the domain, and (2) analytic completion can be used to cut off any projecting corners. However, the domain of analyticity is still much less than the full cut-plane analyticity in both s and t that is assumed by the Mandelstam representation.

(c) *Polynomial boundedness*

It has been shown by Hepp (1964) that dispersion relations for a scattering amplitude $F(s,t)$ exist and require only a finite number of subtractions when t is real and negative, ($-t_1 < t \leqslant 0$), thus

$$|F(s,t)| < s^N \quad \text{for} \quad -t_1 < t \leqslant 0, \tag{4.8.11}$$

where N is finite. This is often called a temperedness condition, in analogy with the temperedness assumption of quantum field theory. However it appears (Jaffe, 1966) that temperedness in quantum field theory is not always necessary for the derivation of polynomial boundedness of a scattering amplitude on the physical sheet.

Using the results of Bros, Epstein & Glaser (1964, 1965), it has been shown by Martin (1966 a, see also § 7.3) that in (4.8.11), $N \leqslant 2$, and the domain of the variable t for which it holds can be extended to $|t| < R$, where R is the constant that occurs in (4.8.7).

4.9 Crossing conditions with charge and spin

In the earlier sections of this chapter we have considered scattering amplitudes for scalar bosons and have normally ignored questions of charge. Then a single amplitude $F(s,t)$ was sufficient to describe the three (formally different) scattering processes (4.4.10). By analytic continuation (see Fig. 4.4.3) from one physical region to another the same function F gives all three physical scattering amplitudes. In this section we consider the additional complications that occur when the particles have charge and when they have spin.

(a) *Crossing matrices for charged particles*

A general discussion of crossing matrices for particles of arbitrary isospin has been given by Carruthers & Krisch (1965) and Taylor (1966), and for SU 3 by de Swart (1964) and Carruthers (1966). Consider the associated reactions:

$$s \text{ channel:} \quad a+b \to c+d, \tag{4.9.1}$$

$$t \text{ channel:} \quad a+\bar{c} \to \bar{b}+d, \tag{4.9.2}$$

$$u \text{ channel:} \quad a+\bar{d} \to c+\bar{b}. \tag{4.9.3}$$

Denote the corresponding scattering amplitudes, defined initially as the physical amplitudes in the s, t and u channels, by

$$(cd|\, M_s\, |ab), \quad (\bar{b}d|\, M_t\, |a\bar{c}), \quad (c\bar{b}|\, M_u\, |a\bar{d}). \tag{4.9.4}$$

Crossing symmetry asserts that when these amplitudes are analytically continued in the manner described in § 4.4, there are relations between them that depend on the isospin quantum numbers. For example, continuing M_s at fixed t from $(s+i\epsilon)$ to $(-s-i\epsilon)$ giving

$$u = \sum_1^4 m_i^2 - t + s + i\epsilon, \tag{4.9.5}$$

we have a relation between M_s and M_u evaluated at the same (s, t, u) value,

$$(cd|\, M_s\, |ab)_c = \xi_{su}(c\bar{b}|\, M_u\, |a\bar{d}), \tag{4.9.6}$$

where the suffix c denotes that the amplitude has been analytically continued. Similarly,

$$(cd|\, M_s\, |ab)_c = \xi_{st}(\bar{b}d|\, M_t\, |a\bar{c}), \tag{4.9.7}$$

where the coefficients ξ_{su} and ξ_{st} are phase factors that depend on the isospin quantum numbers and on whether the particles concerned are self-conjugate (e.g. pions) or pair-conjugate (e.g. nucleon and anti-nucleon). The rules for evaluating ξ_{su} and ξ_{st} have been derived by Carruthers & Krisch (1965 denoted CK in the following, see also Carruthers, 1966 & Taylor, 1966).

CK adopt the Condon & Shortley (1957) phase convention within an isomultiplet, which gives

$$(cd\, |M\, |ab) = (-1)^{a+b-c-d}(ab|\, M\, |cd), \tag{4.9.8}$$

where a, b, c, d denote the isospin of particles a, b, c, d. The third components of isospin will be denoted by greek letters α, β, γ, δ.

The phase factors ξ_{su}, ξ_{st}, depend on the types of particle or field. Following Carruthers (1966) we denote a self-conjugate particle by ϕ, and a pair-conjugate one by ψ. Then there are three, apparently different, types of crossing,

    I: both particles involved in the crossing of type ϕ,

    II: both particles of type ψ,

    III: one of type ϕ and one of type ψ.

Corresponding to the above possibilities, Carruthers (1966) gives the following phases.

$$I_u: \qquad (c\phi_d|\, M_s\, |a\phi_b) = \xi_{su}^{I}(c\overline{\phi}_b|\, M_u\, |a\overline{\phi}_d), \qquad (4.9.9a)$$

$$\xi_{su}^{I} = (-1)^{\gamma-\alpha-b-d}, \qquad (4.9.9b)$$

$$I_t: \qquad (\phi_c d|\, M_s\, |a\phi_b) = \xi_{st}^{I}(\overline{\phi}_b d|\, M_t\, |a\overline{\phi}_c), \qquad (4.9.10a)$$

$$\xi_{st}^{I} = (-1)^{\delta-\alpha-b-c}, \qquad (4.9.10b)$$

$$II_u: \qquad (c\psi_d|\, M_s\, |a\psi_b) = \xi_{su}^{II}(c\overline{\psi}_b|\, M_u\, |a\overline{\psi}_d), \qquad (4.9.11a)$$

$$\xi_{su}^{II} = (-1)^{\gamma-\alpha+b-d}, \qquad (4.9.11b)$$

$$II_t: \qquad (\psi_c d|\, M_s\, |a\psi_b) = \xi_{st}^{II}(\overline{\psi}_b d|\, M_t\, |a\overline{\psi}_c), \qquad (4.9.12a)$$

$$\xi_{st}^{II} = (-1)^{\delta-\alpha+b-c}, \qquad (4.9.12b)$$

$$III_u: \qquad (c\psi_d|\, M_s\, |a\phi_b) = \xi_{su}^{III}(c\overline{\phi}_b|\, M_u\, |a\overline{\psi}_d), \qquad (4.9.13a)$$

$$\xi_{su}^{III} = (-1)^{\gamma-\alpha-b-d}, \qquad (4.9.13b)$$

$$III_t: \qquad (\psi_c d|\, M_s\, |a\phi_b) = \xi_{st}^{III}(\overline{\phi}_b d|\, M_t\, |a\overline{\psi}_d), \qquad (4.9.14a)$$

$$\xi_{st}^{III} = (-1)^{\delta-\alpha-b-c}. \qquad (4.9.14b)$$

In each channel the scattering amplitudes can be expressed in terms of isospin amplitudes in the usual way. The isospin amplitudes in the s channel will be written

$$M_s(s'), \qquad (4.9.15)$$

where s' denotes the isospin. Then,

$$(cd|\, M_s\, |ab) = \sum_{s'} M_s(s')\, C(abs';\alpha,\beta)\, C(cds';\gamma,\delta). \qquad (4.9.16)$$

Similar expansions hold relating

$$(c\overline{b}|\, M_u\, |a\overline{d}) \quad \text{to} \quad M_u(u'), \qquad (4.9.17)$$

$$(\overline{b}d|\, M_t\, |a\overline{c}) \quad \text{to} \quad M_t(t'), \qquad (4.9.18)$$

with Clebsch–Gordan coefficients given by Rose (1957). These equations can be solved to give relations between the isospin amplitudes which CK write using the Racah recoupling coefficients.

These relations between the isospin amplitudes take the form,

$$M_u(u') = \sum_{s'} X(u',s')\, M_s(s'), \qquad (4.9.19)$$

$$M_t(t') = \sum_{s'} X(t',s')\, M_s(s'). \qquad (4.9.20)$$

The matrices $X(u', s')$ and $X(t', s')$ are called the u channel and the t channel crossing matrices respectively. They are given by

$$X(u', s') = \xi_{su}^{-1}(-1)^{\gamma-\alpha+a-c} (2s'+1) W(abdc; s'u'), \quad (4.9.21a)$$

$$= (-1)^{a-b-c+d} (2s'+1) W(abdc; s'u'), \quad (4.9.21b)$$

$$X(t', s') = \xi_{st}^{-1}(-1)^{\delta-\alpha+a+d} (2s'+1) W(as't'd; bc), \quad (4.9.22a)$$

$$= (-1)^{a-b+c+d} (2s'+1) W(as't'd; bc), \quad (4.9.22b)$$

where the Racah coefficients W are defined by Rose (1957). The formulae (4.9.21b) and (4.9.22b) are derived using the values of ξ_{su}^{II} and ξ_{st}^{II} given in (4.9.11b) and (4.9.12b) respectively. These formulae for type II also give the correct answers for the crossing types I and III because self-conjugate particles have integer isospin so $(-1)^{2b} = 1$. Thus the apparent difference in types is not a real difference. For an alternative derivation of this result see Taylor (1966).

For pion-nucleon scattering there are just two isospin amplitudes in each channel. We take the channels to be:

s channel: $N\pi \to N\pi$, isospin $\frac{1}{2}$ and $\frac{3}{2}$.

t channel: $N\overline{N} \to \overline{\pi}\pi$, isospin 0 and 1.

u channel: $N\overline{\pi} \to N\overline{\pi}$, isospin $\frac{1}{2}$ and $\frac{3}{2}$.

Then writing $X(\frac{1}{2}, \frac{1}{2})$ as the top left matrix element,

$$X(u', s') = \begin{pmatrix} -\frac{1}{3}, & \frac{4}{3} \\ \frac{2}{3}, & \frac{1}{3} \end{pmatrix}, \quad (4.9.23)$$

Writing $X(0, \frac{1}{2})$ in the top left of the matrix,

$$X(t', s') = \begin{pmatrix} (\frac{2}{3})^{\frac{1}{2}}, & (\frac{8}{3})^{\frac{1}{2}} \\ \frac{2}{3}, & -\frac{2}{3} \end{pmatrix}. \quad (4.9.24)$$

These crossing conditions apply to each of the invariant amplitudes A and B for pion-nucleon scattering, which have the analytic properties required for the continuation that has been used. One can obtain analogous crossing relations for the isospin states using helicity amplitudes.

(b) Crossing relations for helicity amplitudes

General crossing relations for helicity amplitudes have been derived by Trueman & Wick (1964). Their general relations will not be given here but we will follow their illustrative example for pion-nucleon scattering. The isospin indices for the invariant amplitudes A and B will be omitted (see § 3.6 and earlier in this section).

The pion-nucleon scattering amplitude is the matrix element between Dirac spinors (for the final and initial nucleon states) of the amplitude (3.6.3)

$$T = A(st) - \tfrac{1}{2}iB(s,t)\,\gamma\,.(q_1 + q_2). \tag{4.9.25}$$

The helicity amplitudes (§ 3.6) F_{++} and F_{+-} are obtained from T by a suitable choice of the Dirac spinors, which makes the spin of a nucleon parallel to its direction of motion in the centre of mass system. From (3.6.8, 9) and (3.6.13, 14), the helicity amplitudes in the s channel (θ_s = scattering angle in this channel which describes $\pi N \rightarrow \pi N$), satisfy

$$F'_{++} = \cos\left(\tfrac{1}{2}\theta_s\right)\left[A + \left(\frac{WE - m^2}{m}\right)B\right], \tag{4.9.26}$$

$$F'_{+-} = \sin\left(\tfrac{1}{2}\theta_s\right)\left[\frac{E}{m}A + (W - E)B\right], \tag{4.9.27}$$

where $F' = (4\pi W/m)\,F$, using the F defined in § 3.6.; and E denotes the centre of mass energy of the nucleon.

Define the variable x by

$$x^2 = [s - (m + \mu)^2]\,[s - (m - \mu)^2]. \tag{4.9.28}$$

Then if k denotes the centre of mass momentum, we have the following relations,

$$k^2 = x^2/4s; \quad \sin\left(\tfrac{1}{2}\theta_s\right) = (-st)^{\frac{1}{2}}/x; \tag{4.9.29}$$

$$\cos\left(\tfrac{1}{2}\theta_s\right) = (x^2 + st)^{\frac{1}{2}}/x = [(m^2 - \mu^2)^2 - su]^{\frac{1}{2}}/x; \tag{4.9.30}$$

$$E = (s + m^2 - \mu^2)/(4s)^{\frac{1}{2}}; \tag{4.9.31}$$

$$W - E = (s - m^2 + \mu^2)/(4s)^{\frac{1}{2}}. \tag{4.9.32}$$

In the physical region of the s channel all the square roots are taken to be positive (πN scattering). In the t channel ($\pi\pi \rightarrow N\bar{N}$) the helicity amplitudes G, are given by (Frazer & Fulco, 1960),

$$G_{++} = -(p/m)\,A(s,t) + (q\cos\theta)\,B(s,t), \tag{4.9.33}$$

$$G_{+-} = (Eq/m)\cos\theta\,.B(s,t), \tag{4.9.34}$$

where

$$p = (\tfrac{1}{4}t - m^2)^{\frac{1}{2}}, \quad q = (\tfrac{1}{4}t - \mu^2)^{\frac{1}{2}}, \tag{4.9.35}$$

$$\cos\theta = (s - u)/4pq. \tag{4.9.36}$$

The functions A and B in (4.9.33, 34) have been analytically continued, as described in § 4.4, from the s channel to the t channel.

The coefficients of A and B in (4.9.26, 27) have branch points at

$$s = 0; \quad s = (m \pm \mu)^2, \tag{4.9.37}$$

$$t = 0 \quad \text{and} \quad su = (m^2 - \mu^2)^2. \tag{4.9.38}$$

The analytic continuation from the s channel to the t channel must avoid these singularities. There is a choice of path. The convention adopted by Trueman & Wick (1964) is that the path should cross the real (s, t) plane within the hyperbolic segment bounded by $t = 0$ and $su = (m^2 - \mu^2)^2$, (see Fig. 4.6.1 and (4.6.7, 8)). Then

$$\cos \left(\tfrac{1}{2}\theta_s \right) = -\frac{2ipq \sin \theta}{x}, \tag{4.9.39}$$

$$s^{-\frac{1}{2}} \sin \left(\tfrac{1}{2}\theta_s \right) = \frac{it^{\frac{1}{2}}}{x}. \tag{4.9.40}$$

This gives the following relation between the helicity amplitudes in the two channels (note that our F and G are interchanged when compared with those of Trueman & Wick),

$$F_{++} = \frac{2i}{x} [(mq \sin \theta) G_{++} + \tfrac{1}{2}t^{\frac{1}{2}}(p - q \cos \theta) G_{+-}], \tag{4.9.41}$$

$$F_{+-} = \frac{2i}{x} [\tfrac{1}{2}t^{\frac{1}{2}}(p - q \cos \theta) G_{++} - (mq \sin \theta) G_{+-}]. \tag{4.9.42}$$

Trueman & Wick observe that this is essentially an orthogonal transformation. Generalising from this observation they obtain crossing relations with arbitrary helicity.

The singularities in the helicity amplitudes at (4.9.37, 38) are called kinematic singularities. They are inconvenient for the partial wave analysis used in Regge theory and a number of authors have devised methods for constructing singularity-free helicity amplitudes (Wang, 1966; Hara, 1964; Williams, 1963; Fox, 1966). It has been proved by Hepp (1963) that the full scattering amplitude can be expressed in terms of invariant amplitudes that (like A and B for pion-nucleon scattering) are free from kinematic singularities. These papers are concerned primarily with the problem of general spin. For pion-nucleon scattering the problem was first solved by Chew, Goldberger, Low & Nambu (1957), and for nucleon-nucleon scattering by Goldberger, Grisaru, MacDowell & Wong (1960).

4.10 Experimental comparison of forward dispersion relations

In this section we will outline the comparison of forward dispersion relations with experiment which has been carried out by a number of authors (Hamilton & Woolcock, 1963; Lindenbaum, 1965 and 1967; Lautrup & Olesen, 1965; Barashenkov, 1966; Hohler *et al.* 1966; Carter 1966). We will use the example of pion-nucleon scattering, and the notation of Lautrup & Olesen.

Forward dispersion relations are used for a symmetric amplitude $F_S(E)$ and an antisymmetric amplitude $F_A(E)$, where E is the laboratory energy of the pion,

$$F_S = \tfrac{1}{2}[F(\pi^-p) + F(\pi^+p)], \quad F_A = \tfrac{1}{2}[F(\pi^-p) - F(\pi^+p)]. \quad (4.10.1)$$

The dispersion relations, using crossing symmetry ((4.4.20) see also Chapter 8) can be written

$$F_S(E) = F_S(m) + \frac{8\pi f^2(E^2 - m^2)}{E^2 - (m^2/2M)^2}$$

$$+ \frac{4M(E^2 - m^2)}{\pi} P \int_m^\infty \frac{dx\, x\sigma_S(x)}{(x^2 - m^2)^{\frac{1}{2}}(x^2 - E^2)}, \quad (4.10.2)$$

$$F_A(E) = \frac{E}{m}\left[F_A(m) - \frac{16\pi^2 f^2 M(E^2 - m^2)}{m\{E^2 - (m^2/2M)^2\}}\right]$$

$$+ \frac{4ME(E^2 - m^2)}{\pi} P \int_m^\infty \frac{dx\sigma_A(x)}{(x^2 - m^2)^{\frac{1}{2}}(x^2 - E^2)}, \quad (4.10.3)$$

where m, M are the pion and nucleon masses; P denotes a principal value integral and

$$\sigma_S = \tfrac{1}{2}\sigma_t(\pi^-p) + \tfrac{1}{2}\sigma_t(\pi^+p); \quad \sigma_A = \tfrac{1}{2}\sigma_t(\pi^-p) - \tfrac{1}{2}\sigma_t(\pi^+p). \quad (4.10.4)$$

In the above dispersion relations, the amplitude has been normalised so that it is dimensionless, and the optical theorem has the form

$$\operatorname{Im} F(E) = 2M(E^2 - m^2)^{\frac{1}{2}}\, \sigma(E). \quad (4.10.5)$$

The subtraction has been made at the threshold $E = m$. It can be argued that it is more convenient to subtract at a high energy (Lindenbaum, 1965).

Experimental data on $\sigma(\text{total})$ for pion-nucleon scattering are available only up to 20 Gev (Galbraith *et al.* 1965; Foley *et al.* 1967).

There are several ways to consider σ above 20 Gev. One is to extrapolate from existing data assuming that $\sigma(\pi^+p)$ and $\sigma(\pi^-p)$ tend to constant values in some smooth manner. The other is to separate off from the dispersion integrals (4.10.2, 3) the parts above 20 Gev and treat them as unknown functions $X_S(E)$ and $X_A(E)$.

For the first method a suitable assumption could be

$$\sigma_S(E) = a + b(E/E_0)^{-\frac{1}{2}}, \tag{4.10.6}$$

$$\sigma_A(E) = c(E/E_0)^{-\frac{1}{2}}, \tag{4.10.7}$$

with the constants a, b, c, E_0 chosen to fit the data on total cross-sections in 10 to 20 Gev. Alternatively, (and probably better), one could take the best fit to σ_S and σ_A given by Regge theory (see § 9.3), though even then the scale parameter E_0 is usually fixed somewhat arbitrarily around 1 Gev. With chosen formulae for σ_S and σ_A above 20 Gev, and experimental data below 20 Gev, the real parts of F_S and F_A can be calculated from the dispersion relations, giving

$$\alpha_+(E) = \frac{\operatorname{Re} F(\pi^+p)}{\operatorname{Im} F(\pi^+p)}, \quad \alpha_-(E) = \frac{\operatorname{Re} F(\pi^-p)}{\operatorname{Im} F(\pi^-p)}. \tag{4.10.8}$$

The values obtained for $\alpha_+(E)$ and $\alpha_-(E)$ at different energies can then be compared with the experimental values obtained by Coulomb interference (see Chapters 2 and 3 and Lindenbaum, 1967; Foley *et al.* 1967). By varying the parameters in (4.10.6, 7) one can estimate the uncertainty introduced by the high energy region that has not been studied experimentally. The data available in 1965 give agreement with experimental values of α_+ and α_-, using data at 8, 10, 12 Gev, but do not give very precise information about what values of the parameters are compatible with this data. More satisfactory information is given by the new experiments of Lindenbaum *et al.* (1967) (see § 2.1).

The second method introduces the unknown functions X_S and X_A, for example (Lautrup & Olesen, 1965),

$$X_S(E) = \frac{4M(E^2 - m^2)}{\pi} P \int_{20\,\mathrm{Gev}}^{\infty} \frac{x\,dx\,\sigma_S(x)}{(x^2 - m^2)^{\frac{1}{2}}(x^2 - E^2)}. \tag{4.10.9}$$

The dispersion relations, together with the direct measurements of $\alpha_+(E)$ and $\alpha_-(E)$, can be used to obtain values for $X_S(E)$ and $X_A(E)$ at different energies. It has been suggested that these could then be used to test the validity of the dispersion relations since the formula (4.10.9) implies certain inequalities on $X_S(E)$ at different values. However, it is probably more profitable to regard the dispersion relations as sacrosanct. Then the resulting $X_S(E)$ and $X_A(E)$ give information about the high energy behaviour of the total cross-sections. This can be used to test theories like Regge theory that have a less secure foundation than forward dispersion relations.

CHAPTER 5

COMPLEX ANGULAR MOMENTUM AND REGGE THEORY

The use of complex angular momentum to study high energy collisions has led to many theoretical and experimental developments. The analytic continuation of partial wave amplitudes into the complex angular momentum plane was applied to potential scattering by Regge (1959, 1960). In particular, he noted and used the relation between poles of the partial wave amplitude in the l-plane and the asymptotic behaviour of the full amplitude for large momentum transfer. The subsequent application of the method of complex angular momentum to high energy scattering was proposed independently by Blankenbecler & Goldberger (1961, 1962), and by Chew & Frautschi (1961, 1962) (see also Gribov, 1962a). The general theory has been reviewed by Chew (1962, 1966), Gell-Mann (1962b), Squires (1963), Frautschi (1963), Omnes & Froissart (1963), Oehme (1964), and potential theory has been reviewed by Bottino, Longoni & Regge (1962), Alfaro & Regge (1963) and Newton (1964).

In this chapter we will consider the formulation of Regge theory, beginning in § 5.1 with a general description and summary of the ideas involved. Next, in § 5.2, potential scattering theory will be described using complex angular momentum; this is a useful model both because the theoretical results can be rigorously proved and because explicit numerical solutions are possible. The main part of the chapter, in §§ 5.3–5.6, will be concerned with the formulation of the theory for relativistic scattering. Applications of Regge theory and its comparison with experiment will be described in Chapter 9, and will also be used in parts of Chapters 6, 7 and 8, where general results on high energy behaviour will be described.

5.1 Outline of the method

(a) Analytic continuation of partial waves

We will begin by considering the scattering of equal mass spinless bosons. Then

$$s = 4(k^2 + m^2), \tag{5.1.1}$$

$$t = -2k^2(1 - \cos\theta), \tag{5.1.2}$$

where k is the magnitude of the momentum and θ is the scattering angle in the centre of mass system. The partial wave series for the scattering amplitude is

$$F(s,t) = \frac{8\pi s^{\frac{1}{2}}}{k} \sum_{0}^{\infty} (2l+1) f_l(s) P_l(\cos\theta). \tag{5.1.3}$$

This series converges within the Martin–Lehmann ellipse (see §4.8), and $F(s,t)$ is regular in this region. If information is required about analytic properties of $F(s,t)$ for t outside the ellipse, one method is to replace the sum in (5.1.3) by an integral over a contour in the complex l-plane. This method requires an extension of $f_l(s)$, which is needed in (5.1.3) only for integer values of l, to a function $f(l,s)$ where l may be complex, such that

$$f(l,s) = f_l(s) \quad \text{for} \quad l = 0, 1, 2, \dots . \tag{5.1.4}$$

The partial wave amplitude $f_l(s)$ in potential theory satisfies an explicit differential equation involving l as a parameter, so its properties can be studied directly when l is complex; this method was used by Regge (1959) (see §5.2). Alternatively, if the Mandelstam representation is assumed, one can define $f(l,s)$ by an integral over the double spectral function $\rho(s,t)$; this method was used by Froissart (1961a) and Gribov (1962b). Either way it is necessary to establish uniqueness of $f(l,s)$.

The relation (5.1.4) by itself does not establish uniqueness since, for example, to any solution we could add $g(s)\sin l\pi$, which is zero at all integer values of l. However, the relation (5.1.4) does define $f(l,s)$ uniquely if we impose the additional conditions that,

$$f(l,s) \quad \text{is regular for} \quad \text{Re}\,(l) > l_0, \tag{5.1.5}$$

and in $\text{Re}\,(l) > l_0$, as $l \to \infty$,

$$|f(l,s)| < \exp(a|l|) \quad \text{with} \quad a < \pi. \tag{5.1.6}$$

If $f(l,s)$ satisfies these conditions, a theorem due to Carlson (see Titchmarsh, 1939) establishes that it is uniquely defined by (5.1.4).

Carlson's theorem

Given a function $f(z)$ that (i) is regular in $\text{Re}\,(z) > A$, and (ii) $f(z) < \exp(a|z|)$, $a < \pi$, in $\text{Re}\,(z) > A$, and (iii) $f(z) = 0$, for an infinite sequence $z = N+1, N+2, \dots$, then $f(z)$ is identically zero.

The sequence need not be equally spaced but the separation of discrete points must not increase too fast. For a proof of Carlson's theorem see Titchmarsh (1939) or Boas (1954). The conditions for Carlson's theorem can be proved to hold for potential scattering by a Yukawa potential, and for more general potentials (Challifour & Eden 1963*b*); but in relativistic theory they require assumptions about convergence. The latter assumptions follow from the assumption that the Mandelstam representation is valid with only a finite number of subtractions.

(b) The Sommerfeld–Watson transform

We will discuss methods for defining the unique extension of $f_l(s)$ to complex l in §5.2 and §5.3. In order to indicate the motivation, we will continue now by assuming that we have a suitable definition of $f(l, s)$. We assume also that, for some physical values of s, $f(l, s)$ is regular on and near the real l axis from $l = 0$ to ∞. Then we can rewrite the partial wave series (5.1.3) as an integral representation used by Watson (1918) and Sommerfeld (1949),

$$F(s, t) = -\frac{a}{2i} \int_C \frac{dl(2l+1)f(l, s)\,P_l(-z)}{\sin \pi l}, \qquad (5.1.7)$$

where

$$a = \frac{8\pi s^{\frac{1}{2}}}{k}, \quad z = \cos\theta = 1 + \frac{t}{2k^2}. \qquad (5.1.8)$$

The contour C is shown in Fig. 5.1.1. It surrounds the real l axis from 0 to ∞. The integral therefore picks out the residues of the integrand at the zeros of $\sin \pi l$, for $l = 0, 1, 2, \ldots$. This reproduces the partial wave series since (by assumption) $f(l, s)$ has no singularities inside C for our chosen value of s. The integral can be shown to converge for physical values of s and t under reasonable assumptions about $f(l, s)$ as $l \to \infty$ along the real axis.

The basic step in the Regge method is to deform the contour C into the contour C' shown in Fig. 5.1.1, along $\mathrm{Re}\,(l) = -\frac{1}{2}$, together with several contours C'' around singularities of $f(l, s)$ in $\mathrm{Re}\,(l) > -\frac{1}{2}$. In Fig. 5.1.1 a single such contour C'' is shown surrounding a pole of $f(l, s)$ at $l = \alpha_n(s)$. In order to achieve this step it is necessary to assume, or to prove, (i) that the analytic partial wave amplitude $f(l, s)$ is meromorphic (i.e. it has no singularities except poles) in $\mathrm{Re}\,(l) > -\frac{1}{2}$, and (ii) that its behaviour, as $|l| \to \infty$, in $\mathrm{Re}\,(l) > -\frac{1}{2}$, is such that there is no contribution from the curved contour at ∞ that joins C and C'.

Evaluating the integrals round C'', we obtain from (5.1.7), with C replaced by $C' + C''$,

$$F(s,t) = -\frac{a}{2i} \int_{-\frac{1}{2}-i\infty}^{\frac{1}{2}+i\infty} \frac{dl(2l+1)f(l,s)\,P_l(-z)}{\sin l\pi}$$

$$-\sum_n \frac{a\pi(2\alpha_n+1)\,r(\alpha_n,s)\,P_{\alpha_n}(-z)}{\sin \pi\alpha_n} \quad (5.1.9)$$

where $\alpha_n(s) = \alpha_n$ denotes a Regge pole (a pole of $f(l,s)$) in $\mathrm{Re}\,(l) > -\frac{1}{2}$, and the sum is over all such poles. The function $r(\alpha_n, s)$ denotes the residue of $f(l,s)$ at the pole.

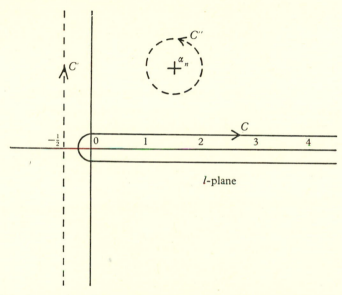

Fig. 5.1.1. The complex l-plane showing the contour C used in the Sommerfeld–Watson transform, and the equivalent contours C' and C'' of the Regge representation.

If, in addition to poles, $f(l,s)$ has branch points in $\mathrm{Re}\,(l) > -\frac{1}{2}$, there will be terms additional to those in (5.1.9) coming from contours C''' that surround the branch cuts attached to these branch points. These may occur in the relativistic theory of collisions (Mandelstam, 1963), but for the present we will assume that if such branch points exist they can be neglected, (but see § 5.6).

The expression (5.1.9) for the scattering amplitude is often described as the Regge representation. As noted above, it implies assumptions about the asymptotic behaviour of $f(l,s)$ as well as its meromorphy. The usefulness of the Regge representation comes from the information it gives about $F(s,t)$ as $t \to \infty$. Since in general this implies $z \to \infty$, we note that for $\mathrm{Re}\,(l) > -\tfrac{1}{2}$,

$$|P_l(-z)| \sim |z|^{\mathrm{Re}(l)} \quad \text{as} \quad z \to \infty. \tag{5.1.10}$$

It will be assumed that the background integral along C' can be neglected as $z \to \infty$, and that it behaves like $|z|^{-\frac{1}{2}}$. Then the terms in the sum in (5.1.9) will dominate over the integral for large z.

(c) Physical interpretation of Regge poles

Consider the contribution to $F(s,t)$ from a particular Regge pole $\alpha_n = \alpha(s)$ in (5.1.9). This is

$$-\frac{a\pi[2\alpha(s)+1]\,r(\alpha,s)\,P_\alpha(-z)}{\sin \pi \alpha}. \tag{5.1.11}$$

In deriving the formula (5.1.9) we assumed that $\alpha(s)$ was not an integer or zero. If in fact $f(l,s)$ does have a pole (for s real) at an integer value l_0 of l, then we shall see in §5.2 for potential scattering, that the pole corresponds to a bound state and can only occur for $s = s_0 \leqslant 4m^2$. Then for $s \neq s_0$ but with $|s - s_0|$ small, and $l_0 = \alpha(s_0)$, we can write

$$f(l,s) \approx \frac{r(l_0,s)}{l-\alpha(s)}. \tag{5.1.12}$$

Assuming that $\alpha(s)$ is regular at s_0, and writing l_0' for the derivative of $\alpha(s)$ at l_0, we have

$$\alpha(s) = l_0 + l_0'(s - s_0), \tag{5.1.13}$$

$$f(l_0,s) \approx \frac{r(l_0,s)}{l_0'(s_0 - s)} \approx f_{l_0}(s). \tag{5.1.14}$$

Thus the partial wave amplitude $f_{l_0}(s)$ will have a real pole at $s = s_0$, as we would expect for a bound state.

Resonance poles can be studied from (5.1.12), but (see §5.2) $\alpha(s)$ must be complex with $\mathrm{Im}\,\alpha > 0$ when $s > 4m^2$. Suppose the pole $\alpha(s_0)$ is near to an integer l_0, so that for small $|s - s_0|$, we can write

$$\alpha(s_0) = l_0 + i\gamma_0, \tag{5.1.15}$$

$$\alpha(s) = l_0 + i\gamma_0 + l_0'(s - s_0), \tag{5.1.16}$$

where s_0 is real. Then

$$f_{l_0}(s) = f(l_0,s) \approx \frac{r(l_0,s)}{-i\gamma_0 - l_0'(s-s_0)}. \tag{5.1.17}$$

This has a pole at $s = s_0 - i\gamma_0/l_0',$ (5.1.18) which is a resonance pole and must lie in the unphysical sheet of the s-plane, below the real cut (see § 5.2). This implies that l'_0 is approximately real and has a positive real part. This is a general property of the function,

$$l = \alpha(s), \tag{5.1.19}$$

that can be proved in potential scattering. The path of $l = \alpha(s)$, as s varies along the real axis (above the real branch cut), is called a Regge trajectory. We will consider Regge trajectories further in § 5.2 and § 5.3, and again when we consider applications of Regge theory in Chapter 9.

(d) Regge poles and large values of t

From (5.1.1 and 2), with $z = \cos\theta$,

$$z = 1 + \frac{2t}{s - 4m^2}. \tag{5.1.20}$$

For fixed s, as $|t| \to \infty$, we have $|z| \to \infty$. From the assumptions stated after (5.1.10), we see that the scattering amplitude $F(s, t)$, when expressed by the Regge representation (5.1.9), will be dominated for large $|t|$ or $|z|$ by the $P_{\alpha_n}(-z)$ term for which Re (α_n) is largest. We can label the poles $\alpha_n(s)$ for given s, so that

$$\operatorname{Re}\alpha_1 > \operatorname{Re}\alpha_2 > \ldots \tag{5.1.21}$$

and we will call α_1 the right-most pole. Then for $|z| \to \infty$,

$$F(s, t) \sim -\frac{a\pi(2\alpha_1 + 1)\, r(\alpha_1, s)}{\sin \pi\alpha_1}(-z)^{\alpha_1}, \tag{5.1.22}$$

$$\sim C_1(s)\, t^{\alpha_1(s)} \quad \text{as} \quad t \to \infty, \tag{5.1.23}$$

where the symbol $\sim$ denotes asymptotic equality, and where

$$C_1 = -\frac{a\pi(2\alpha_1 + 1)\, r(\alpha_1, s)}{\sin \pi\alpha_1}\left(\frac{-2}{s - 4m^2}\right)^{\alpha_1}. \tag{5.1.24}$$

(e) Regge trajectories

Since each $\alpha_n(s)$ is a function of s, the ordering (5.1.21) of the poles will change as s changes. It is therefore of considerable importance to ascertain how each Regge pole, $l = \alpha_n(s)$, moves along its corresponding trajectory as s varies. This problem has been solved numerically for a Yukawa potential; its general solution is not known in relativistic theory, although some information about trajectories can be obtained from experimental evidence on reasonable assumptions (Chapter 9). A typical Regge trajectory for a single Yukawa potential is shown in Fig. 5.1.2 (a). This behaviour is typical for the right-most trajectory with an attractive potential. The behaviour of other trajectories in the left half l-plane is much more complicated (Ahmadzadeh, Burke & Tait, 1963 and Lovelace & Masson, 1963).

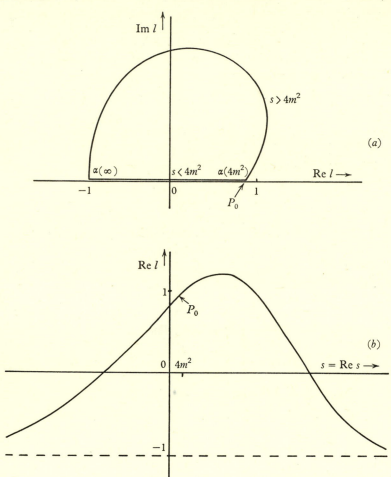

Fig. 5.1.2. Regge trajectories for a Yukawa potential, (a) the path of a Regge pole $l = \alpha(s)$ as s varies through real values from $-\infty$ to $+\infty$; (b) the projection on to the (real l, real s) plane of the Regge trajectory $l = \alpha(s)$; (to the right of P_0, Im $\alpha(s)$ will be non-zero).

In Fig. 5.1.2(b) another useful representation is shown for a Regge trajectory. In this, the real part of $l = \alpha(s)$ is plotted against the value of s (s real). Note that this trajectory has a maximum real value of l, and in general one would expect there to be a similar maximum for any trajectory if the amplitude $F(s,t)$ is bounded by a polynomial.

(f) Signature

When there are exchange forces in potential scattering, and always in relativistic scattering, it is necessary to separate odd and even l values of the partial wave amplitude $f_l(s)$ before making an analytic extension in l. This is because exchange forces introduce a factor, $(-1)^l = \exp il\pi$, into part of $f_l(s)$. Since $|(-1)^l| \sim \exp(\operatorname{Im} l\pi)$, one finds that this part of f_l does not satisfy the uniqueness conditions required for Carlson's theorem (5.1.6), when $l \to \infty$ parallel to the imaginary axis.

It is therefore necessary to consider two partial wave amplitudes f^+ and f^-, defined so that

$$f^+(l,s) = f_l(s) \quad \text{for} \quad l = 0, 2, 4, \ldots, \tag{5.1.25}$$

$$f^-(l,s) = f_l(s) \quad \text{for} \quad l = 1, 3, 5, \ldots. \tag{5.1.26}$$

These are called the partial wave amplitudes of even and odd signature respectively. They will have different sets of Regge poles, which must be distinguished, $\alpha_n^+(s)$ and $\alpha_n^-(s)$.

The physical interpretation of Regge poles discussed above applies to α_n^+ only near even l values, and to α_n^- only near odd l values. Thus a pole of $f^+(l,s)$ at $l =$ even integer, for real $s < 4m^2$, will correspond to a physical bound state, but a pole of f^+ at $l =$ odd integer does not correspond to a physically observed state. The symbol τ is often used to denote the signature ($\tau = +$ for even and $\tau = -$ for odd signature).

(g) Relativistic theory, crossing and high energy behaviour

In §4.4 we discussed the relation between scattering amplitudes for different processes through crossing symmetry. We considered the following channels:

$$\left. \begin{aligned} s \text{ channel:} &\quad 1 + 2 \to 3 + 4, \\ t \text{ channel:} &\quad 1 + \bar{4} \to \bar{2} + 3, \\ u \text{ channel:} &\quad 1 + \bar{3} \to \bar{2} + 4, \end{aligned} \right\} \tag{5.1.27}$$

in which s, t and u respectively denote the squared centre of mass energies for each process.

In the s channel, the variable t is the square of the momentum transfer. If we assume that the methods, proved by Regge for potential scattering, apply also to relativistic scattering, the behaviour for the s-channel amplitude as $t \to \infty$ will be given by (5.1.23), that is by,

$$F(s,t) \sim C_1(s)\, t^{\alpha_1(s)}. \tag{5.1.28}$$

It will be recalled that α_1 denotes the right-most pole in the complex l-plane of the partial wave amplitude $f(l, s)$ in the s channel (omitting for the present the question of signature). Although this s channel partial wave amplitude was originally defined for s above threshold, it can be continued to $s \leqslant 0$. Let $\alpha_1(s)$ be the pole with the largest real part when s is fixed at some negative value (in practice we will apply the theory when $0 > s > -1 \,(\mathrm{Gev/c})^2$).

From Fig. 4.4.2(b) it will be seen that when we take

$$s \leqslant 0 \quad \text{and} \quad t \to \infty, \tag{5.1.29}$$

we obtain $F(s, t)$ in the physical scattering region for the t channel. In the simplest case of scalar bosons there is just one invariant amplitude, so the asymptotic form (5.1.28) determines the high energy behaviour of the physical scattering amplitude in the t channel. We have seen, in subsection (c) above, that the Regge pole $\alpha_1(s)$ may correspond to a resonance for scattering in the s channel, or to a bound state in this channel. Assuming that this correspondence is always valid it appears that, by observing resonances in the s channel, we can predict high energy behaviour in the t channel. This requires also that we can determine experimentally or theoretically the s dependence of the resonance poles $\alpha_n(s)$ of $f(l, s)$ in the complex l-plane.

In the relativistic case, with simplifying theoretical assumptions, one can deduce this s dependence of $\alpha_n(s)$ by observing the high (t) energy behaviour of $F(s, t)$ in (5.1.28). Then one can reverse the above argument and predict the Regge trajectories on which resonances, or stable states, in the s channel should be observed. Including the requirement of positive and negative signatures, one does not necessarily have a physically observable resonance each time $\alpha_n^+(s)$ or $\alpha_n^-(s)$ goes near an integer; this happens only when the integer is even or odd respectively (and the derivative of $\alpha^\pm(s)$ has a positive real part).

(h) *Factorisation*

The relativistic form of Regge theory will be described in more detail in §5.3 and in Chapter 9, but there is one more aspect that should be mentioned in this outline; this is the factorisation hypothesis, which is obtained by analogy with nuclear reaction theory. When a resonance can decay into several different channels, the cross-section for its formation and decay depends on the product of the partial widths for its formation from the incident channel and its decay into another channel (here the word 'channel' is used in its conventional sense as meaning an allowed decay product, and not in its extended sense using analytic continuation as in (5.1.27)). The close relation between Regge poles and resonances suggests that near a Regge pole the partial wave amplitudes will have a similar property. For multi-channel scattering the partial wave amplitudes form a matrix with elements $f_{mn}(l,s)$ corresponding to each reaction $m \to n$. The factorisation hypothesis states that near a Regge pole,

$$f_{mn}(l,s) = \frac{\gamma_m \gamma_n}{l - \alpha(s)} + R_{mn}(l,s),\qquad(5.1.30)$$

where $R_{mn}(l,s)$ is regular at the pole. This simple situation may become more complicated if there are neighbouring or coincident Regge poles, but it does form a valuable initial hypothesis.

Summary

The relativistic Regge theory contains the following main features:

(i) High energy behaviour in the t channel for fixed s, is given by resonances or bound states in the s channel, and conversely.

(ii) Trajectories $l = \alpha(s)$ can sometimes be determined for $s < 0$ from observed high energy behaviour in the t channel.

(iii) Trajectories $l = \alpha(s)$ can be determined for $s > 0$ from observed resonances in the s channel having different masses and angular momenta; resonances on the same trajectory are called Regge recurrences.

(iv) States of odd and even signature lie on different Regge trajectories in general.

(v) The factorisation hypothesis can be used to give relations between the effects of Regge poles on different reactions.

(vi) With spin, one should use helicity amplitudes, or the conventional amplitudes f_1 and f_2 that were defined in §3.6.

There are a number of hazards and complications in putting the theory to practical use. Some of these involve theoretical difficulties such as possible branch cuts in the complex l-plane; others are more practical like the experimental difficulty of reaching the asymptotic region where the leading (right-most in l) Regge pole can dominate in $F(s,t)$, as was implied in (5.1.28). These difficulties will be discussed later in this chapter and also in Chapter 9. Before developing the relativistic Regge theory in detail, we will outline the non-relativistic potential theory.

5.2   Potential scattering and complex angular momentum

(a) Coulomb scattering

The partial wave scattering amplitudes can be obtained explicitly for Coulomb scattering, and give a simple illustration of the properties of Regge poles for complex angular momentum. However, there are certain special characteristics that arise from the long range r^{-1} potential, which are not present with a Yukawa potential, and probably not in relativistic scattering with strong interactions. In this subsection will follow the treatment of Coulomb scattering given by Singh (1962).

The Schrodinger equation for scattering by a Coulomb potential (e^2/r) is,

$$\nabla^2 \Psi + \left(k^2 + \frac{e^2}{r}\right) \Psi = 0, \tag{5.2.1}$$

where $k^2 = E$ is the energy. The corresponding partial wave $\psi(l,r)$ satisfies

$$\frac{d^2}{dr^2}\psi(l,r) + \left[k^2 + \frac{e^2}{r} - \frac{l(l+1)}{r^2}\right]\psi(l,r) = 0. \tag{5.2.2}$$

Although physical solutions must have l = integer, this equation can be used to define solutions for other values of l. The asymptotic solutions define the phase $\delta_l(k)$, after removal of the infinite part

$$\exp\left[(ie^2/2k)\log r\right],$$

which is special to Coulomb scattering and has no observable effect. From the asymptotic phase $\delta_l(k)$ one obtains the S-matrix element, for $\mathrm{Re}\,(l) > -\tfrac{1}{2}$,

$$S(l,k) = \exp\left[2i\delta_l(k)\right] = 1 + 2if(l,k). \tag{5.2.3}$$

This is given explicitly by

$$S(l,k) = \frac{\Gamma(l+1 - ie^2/2k)}{\Gamma(l+1 + ie^2/2k)}. \tag{5.2.4}$$

For real k, real l, unitarity is obviously satisfied,

$$S^*(l,k)\,S(l,k) = 1. \tag{5.2.5}$$

Although the solution (5.2.4) was obtained for $\mathrm{Re}\,(l) > -\tfrac{1}{2}$, (where the wave function is normalisable), it can be continued to any value of l. The gamma function $\Gamma(z)$ is meromorphic in the entire z-plane, with poles at

$$z = 0, \quad -1, \quad -2, \quad \ldots \tag{5.2.6}$$

and it has no zeros for finite z. Hence these poles determine the Regge poles, which are singularities of $S(l, k)$. The $(n+1)$th pole is given by

$$l + 1 - \frac{ie^2}{2k} = -n, \tag{5.2.7}$$

which gives

$$l = \alpha_n(k) = -n - 1 + \frac{ie^2}{2k}. \tag{5.2.8}$$

The Regge trajectory is the path in the l-plane of α_n, moving as a function of $E = k^2$. For an attractive Coulomb potential, it varies through real values from $l = -n - 1$ to $+\infty$, as E varies from $-\infty$ to 0. Each positive integer (or zero) value of l corresponds to a physical bound state of energy,

$$E_{l_0} = k^2 = -\frac{e^4}{4(l_0 + n + 1)^2}. \tag{5.2.9}$$

These are the well known Coulomb bound state energies. The degeneracy of these levels is, of course, a special feature of the Coulomb potential.

In order to obtain the trajectory for $E > 0$, we must avoid the origin by a small detour in the upper half E-plane so as to arrive above the branch cut $(0 < E < \infty)$. This small detour causes the trajectory to move to the positive imaginary part of the l-plane as shown in Fig. 5.2.1 (a). As $E \to \infty$, the trajectory completes the cycle and returns to $l = -n - 1$. For a Yukawa potential the trajectory does not go to infinity at $E = 0$ (see Fig. 5.1.2). A repulsive Coulomb potential has its trajectory in the lower left-hand quadrant of the l-plane as shown also in Fig. 5.2.1 (a).

The other type of curve, that is also called a Regge trajectory, is the plot of Re (l) against Re (E). The three leading Coulomb trajectories are shown in Fig. 5.2.1 (b). For Re $(E) < 0$, they are curves; but for Re $(E) > 0$, they correspond to the straight line Re $(l) = -n - 1$, $(n = 0, 1, 2)$. The connection between these two parts of the trajectory is shown for the leading trajectory as a broken line in Fig. 5.2.1 (b); it is obtained by giving E a small imaginary part. It is evident on physical grounds that a short range (Yukawa) potential cannot bind states of arbitrarily large l. This means that the Yukawa trajectory cannot go to infinity as does the trajectory for the Coulomb potential. In the l-plane the Yukawa–Regge trajectory will become complex just above the highest bound state, that is to say $\alpha(E)$ becomes complex at $E = 0$, (see Fig. 5.1.2).

In the Coulomb solution (5.2.8), the position $\alpha_n(E)$ of a Regge pole satisfies the dispersion relation

$$\alpha_n(E) = \alpha_n(\infty) + \frac{1}{\pi}\int_0^\infty \frac{dE'\,\mathrm{Im}\,\alpha_n(E')}{E'-E}, \qquad (5.2.10)$$

where $\mathrm{Im}\,\alpha_n(E) = e^2/(2E^{\frac{1}{2}})$ and $\alpha_n(\infty) = -n-1$.

The asymptotic behaviour for large momentum transfer is unexpectedly simple for Coulomb scattering, as the Legendre polynomials

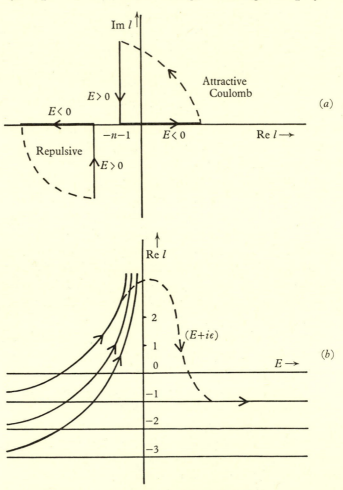

Fig. 5.2.1. Regge trajectories for the Coulomb potential, (a) the path of the leading Coulomb Regge pole $l(E)$ as E varies from $-\infty$ to $+\infty$, for both attractive and repulsive potentials; (b) the projection on to the (real l, real E) plane of Coulomb Regge poles; the broken line shows how the projection of a pole moves as E avoids the singular point $E = 0$ in going from negative to positive values.

in (5.1.9) coming from the sum over Regge poles combine with the background integral to give a simple sum. This sum is more easily obtained from the exact asymptotic solution of (5.2.1) which gives,

$$F(E,t) = \frac{\Gamma(1 - ie^2/2k)}{\Gamma(ie^2/2k)} \left(\frac{i}{2k}\right) \left(-\frac{t}{4k^2}\right)^{\alpha_0(E)}, \qquad (5.2.11)$$

where
$$\alpha_0(E) = -1 + ie^2/2k.$$

(b) Yukawa potential

A detailed account of the Regge theory of potential scattering using complex angular momentum is given by Bottino, Longoni & Regge (1962) and by Regge (1963). I will summarise here parts of these papers, since they not only provide the historical basis for Regge theory, but they are also valuable in providing a model for investigating surmised properties of relativistic scattering.

Our initial objective is to make a transformation from the partial wave series

$$F(E, \theta) = \frac{1}{k} \sum_0^\infty (2l + 1) g_l(k) P_l(\cos\theta), \qquad (5.2.12)$$

to the Sommerfeld–Watson formula,

$$F(E, \theta) = -\frac{1}{2ik} \int_C \frac{dl(2l+1)}{\sin\pi l} g(l, k) P_l(-\cos\theta), \qquad (5.2.13)$$

where $S(l, k) = 1 + 2ig(l, k)$ is the non-relativistic partial wave S-matrix element when l is an integer. The contour C surrounds the positive integers as in Fig. 5.1.1. We use g to denote the partial wave amplitude in this section so as to distinguish it from the Jost function f which we introduce below.

In order to achieve this transformation, we must first define $g(l, k)$ when l and k have general (complex) values, and secondly investigate the convergence of (5.2.13). The latter requires information about the behaviour of $g(l, k)$ when l is large. These objectives can be achieved for the Yukawa potential,

$$V(z) = \frac{\exp(-mz)}{z}. \qquad (5.2.14)$$

When $z = r$ is real it denotes distance from the scattering centre. The partial wave Schrodinger equation for $\psi = \psi_l$, is

$$\psi''(z) + k^2\psi(z) - \frac{(\lambda^2 - \frac{1}{4})}{z^2}\psi(z) - V(z)\psi(z) = 0, \qquad (5.2.15)$$

where $\lambda = l + \frac{1}{2}$.

We require that $V(z)$ converges as $|z| \to \infty$ in Re $(z) > 0$, and that on any ray with $|\arg z| = |\theta| < \frac{1}{2}\pi$,

$$\int^{\infty} d|z| \cdot |z| \cdot |V(z)| < M < \infty. \tag{5.2.16}$$

These conditions are satisfied by (5.2.14).

We need to study the equation (5.2.15), and associated functions like $g(l, k)$, when l and k are both complex. We will give the relevant formulae without detailed derivations, and refer the reader to the original papers by Regge *et al.* (loc. cit.) for details. There are two types of solution of the equation (5.2.15). One type is defined by its behaviour near $z = 0$,

$$\phi(\lambda, k, z) \sim z^{\lambda + \frac{1}{2}} \quad \text{as} \quad z \to 0. \tag{5.2.17}$$

The second type is defined by its behaviour as $z \to \infty$,

$$f(\lambda, k, z) \sim \exp(-ikz) \quad \text{as} \quad z \to \infty. \tag{5.2.18}$$

More precisely ϕ and f are defined by means of an integral equation, using the following lemma:

Lemma (to derive an integral equation from a differential equation). Let ψ satisfy the differential equation,

$$D(\lambda, k, z)\,\psi(\lambda, k, z) \equiv \left[\frac{d^2}{dz^2} + g(\lambda, k, z)\right] \psi(\lambda, k, z)$$

$$= h(\lambda, k, z)\,\psi(\lambda, k, z). \tag{5.2.19}$$

Let ψ_1 be a solution when h is replaced by zero. Then, if we require $\psi \to \psi_1$ as $z \to z_1$, the equivalent integral equation is,

$$\psi(\lambda, k, z) = \psi_1(\lambda, k, z) + \frac{1}{W[\phi_1, \phi_2]} \int_{z_1}^{z} dx [\phi_1(x)\,\phi_2(z) - \phi_2(x)\,\phi_1(z)]$$

$$\times h(\lambda, k, x)\,\psi(\lambda, k, x), \tag{5.2.20}$$

where ϕ_1 and ϕ_2 are independent solutions of the 'free' equation,

$$D(\lambda, k, z)\,\phi_i(\lambda, k, z) = 0, \tag{5.2.21}$$

and the Wronskian is defined by

$$W[\phi_1, \phi_2] \equiv \phi_1 \frac{\partial \phi_2}{\partial z} - \phi_2 \frac{\partial \phi_1}{\partial z}. \tag{5.2.22}$$

From the above lemma, we obtain ϕ satisfying (5.2.15) and having the form (5.2.17) when it is defined by the integral equation,

$$\phi(\lambda, k, z) = \phi_0(\lambda, k, z)$$
$$-\frac{\pi z^{\frac{1}{2}}}{2\sin\pi\lambda}\int_0^z dx\, x^{\frac{1}{2}}[J_\lambda(kx)\, J_{-\lambda}(kz) - J_{-\lambda}(kx)\, J_\lambda(kz)]\, V(x)\, \phi(\lambda, k, x),$$

$$(5.2.23)$$

where ϕ_0 denotes the free solution for partial waves with $V = 0$,

$$\phi_0 = \Gamma(\lambda+1)\left(\frac{2}{k}\right)^\lambda z^{\frac{1}{2}}J_\lambda(kz) \tag{5.2.24}$$

$$\approx z^{\lambda+\frac{1}{2}} \quad \text{as} \quad z \to 0. \tag{5.2.24a}$$

From the form as $z \to 0$, it can be seen that

$$W[\phi(\lambda, k, z), \phi(-\lambda, k, z)] = -2\lambda, \tag{5.2.25}$$

which shows that we have two independent solutions of (5.2.15), namely $\phi(\lambda, k, z)$ and $\phi(-\lambda, k, z)$.

The second type of solution, the scattering solution f, is defined by an analogous integral equation, using the free solution f_0 for $V = 0$,

$$f_0(\lambda, k, z) = \exp\left[-i\pi(\lambda+\tfrac{1}{2})/2\right](\pi kz/2)^{\frac{1}{2}}H_\lambda^{(2)}(kz), \tag{5.2.26}$$

$$\sim \exp(-ikz). \tag{5.2.26a}$$

The scattering solution of (5.2.15) is,

$$f(\lambda, k, z) = f_0(\lambda, k, z)$$
$$-\frac{i\pi z^{\frac{1}{2}}}{4}\int_z^\infty dx\, x^{\frac{1}{2}}[H_\lambda^{(1)}(kx)\, H_\lambda^{(2)}(kz) - H_\lambda^{(2)}(kx)\, H_\lambda^{(1)}(kz)]\, V(x)f(\lambda, k, x).$$

$$(5.2.27)$$

Since there is a branch cut at $k = 0$, we must define $f(\lambda, -k, z)$ by specifying a rotation (in the lower half plane),

$$f(\lambda, -k, z) \equiv f(\lambda, k\exp(-i\pi), z). \tag{5.2.28}$$

Then, from the form as $z \to \infty$,

$$W[f(\lambda, k, z), \quad f(\lambda, -k, z)] = 2ik, \tag{5.2.29}$$

showing that we have two more independent solutions, namely $f(\lambda, k, z)$ and $f(\lambda, -k, z)$.

Analyticity of the Jost functions

Define the Jost function (Jost, 1947; Bargman, 1949), by

$$f(\pm\lambda, \pm k) \equiv W[f(\lambda, \pm k, z), \quad \phi(\pm\lambda, k, z)]. \tag{5.2.30}$$

Note that in (5.2.13) we used $g(l, k)$ for the partial amplitude to distinguish it from $f(\lambda, k)$, which in this section denotes the Jost function. We have four solutions, but only two can be independent. Hence each ϕ can be expressed as a linear combination of the two f's,

$$\phi(\lambda, k, z) = Af(\lambda, k, z) + Bf(\lambda, -k, z), \qquad (5.2.31)$$

$$\phi(-\lambda, k, z) = Cf(\lambda, k, z) + Df(\lambda, -k, z), \qquad (5.2.32)$$

where A, B, C and D depend on λ and k, but not on z.

Substitute from (5.2.31, 32) for ϕ in (5.2.30), and use (5.2.25, 29), giving,

$$\left.\begin{aligned}
f(\lambda, -k) &= -2ikA, \\
f(\lambda, k) &= 2ikB, \\
f(-\lambda, -k) &= -2ikC, \\
f(-\lambda, k) &= 2ikD.
\end{aligned}\right\} \qquad (5.2.33)$$

Now substitute for both $\phi(\lambda, k, z)$ and $\phi(-\lambda, k, z)$ in (5.2.25). Using (5.2.33) we obtain the identity

$$f(\lambda, -k)f(-\lambda, k) - f(\lambda, k)f(-\lambda, -k) \equiv 4i\lambda k. \qquad (5.2.34)$$

The free Jost function ($V = 0$) can be calculated explicitly, and is

$$f_0(\lambda, k) = \left(\frac{2}{\pi}\right)^{\frac{1}{2}} 2^\lambda \Gamma(\lambda + 1) k^{-\lambda + \frac{1}{2}} \exp\left[-i\pi(\lambda - \tfrac{1}{2})/2\right]. \qquad (5.2.35)$$

From the asymptotic form of the Bessel functions, we obtain the asymptotic form of the free solution ϕ_0, as $z \to \infty$, which is

$$\phi_0(\lambda, k, z) \sim \exp\left[i\pi(\lambda - \tfrac{1}{2})/2\right] \frac{f_0(\lambda, k)}{k} \sin\left[kz - \pi(\lambda - \tfrac{1}{2})/2\right]. \qquad (5.2.36)$$

The phase shift $\delta(\lambda, k)$ is obtained by comparing the perturbed solution ($V \neq 0$) with its asymptotic form in terms of δ, namely

$$\phi(\lambda, k, z) \sim \exp\left[\tfrac{1}{2}i\pi(\lambda - \tfrac{1}{2}) - i\delta\right] \frac{f_0(\lambda, k)}{k} \sin\left[kz - \frac{\pi(\lambda - \tfrac{1}{2})}{2} + \delta\right]. \qquad (5.2.37)$$

But for large z we can also use (5.2.31, 33) to obtain an expression for the perturbed regular solution ϕ in terms of the Jost functions,

$$\phi(\lambda, k, z) \sim \frac{1}{2ik}[f(\lambda, k)\exp(ikz) - f(\lambda, -k)\exp(-ikz)]. \qquad (5.2.38)$$

Hence, comparing the two forms (5.2.37, 38) for $\phi(\lambda, k, z)$, as $z \to \infty$, the S-matrix element is,

$$S(\lambda, k) = \exp(2i\delta) = \frac{f(\lambda, k)}{f(\lambda, -k)} \exp\left[i\pi(\lambda - \tfrac{1}{2})\right]. \qquad (5.2.39)$$

We now quote the analytic properties that follow from the general theory of differential equations applied to (5.2.15), or alternatively that can be obtained by iterating the integral equations and studying convergence properties, (for details see Bottino, Longoni & Regge, 1962, and Challifour & Eden, 1963b).

(i) All four solutions $\phi(\pm\lambda, k, z), f(\lambda, \pm k, z)$ have the same analyticity domain as $V(z)$. If V is real for z real we have Hermiticity,

$$\phi(\lambda, k, z) = \phi^*(\lambda^*, k^*, z), \tag{5.2.40}$$

$$f(\lambda, k, z) = f^*(\lambda^*, -k^*, z). \tag{5.2.41}$$

(ii) $\phi(\lambda, k, z)$ and $\phi'(\lambda, k, z) = \partial\phi/\partial z$, are integral or entire functions of k, (i.e. regular for all k except for an essential singularity at ∞) and are analytic in λ for $\mathrm{Re}\,(\lambda) > 0$. Thus $\phi(\lambda, k, z)$ is analytic in the region,

$$\{k | k \neq \infty\} \times \{\lambda | \mathrm{Re}\,(\lambda) > 0\}, \tag{5.2.42}$$

and it is continuous on $\mathrm{Re}\,(\lambda) = 0$.

(iii) $f(\lambda, k, z)$ is analytic in the region

$$\{k | \mathrm{Im}\,(k) < 0\} \times \{\lambda | \lambda \neq \infty\}, \tag{5.2.43}$$

and it is continuous on $\mathrm{Im}\,(k) = 0$. From these results and the definition (5.2.30) of the Jost function $f(\lambda, k)$, namely

$$f(\lambda, k) \equiv W[f(\lambda, k, z), \phi(\lambda, k, z)],$$

we see that $f(\lambda, k)$ is analytic in

$$\{k | \mathrm{Im}\,(k) < 0\} \times \{\lambda | \mathrm{Re}\,(\lambda) > 0\}, \tag{5.2.44}$$

and it is continuous on $\mathrm{Im}\,k = 0$ and $\mathrm{Re}\,\lambda = 0$. There is a branch point at $k = 0$. Results for other domains of λ and k follow similarly from (5.2.30).

Enlargement of the domain of analyticity of $f(\lambda, k)$ can be obtained by writing

$$z = \rho\exp i\sigma \quad \text{with} \quad |\sigma| < \tfrac{1}{2}\pi. \tag{5.2.45}$$

Then, $f_1(\lambda, k_1, \rho)$ with $k_1 = k\exp i\sigma$ satisfies the equation,

$$\frac{d^2\psi}{d\rho^2} - \frac{(\lambda^2 - \tfrac{1}{4})}{\rho^2}\,\psi + k_1^2\,\psi - \exp\,(2i\sigma)\,V(\rho\exp i\sigma)\,\psi = 0, \tag{5.2.46}$$

with $\psi \equiv f_1(\lambda, k_1, \rho) \sim \exp\,(-ik_1\rho)$ as $\rho \to \infty$. Since V still converges if $|\sigma| < \tfrac{1}{2}\pi$, it follows as before that $f_1(\lambda, k_1, \rho)$ is analytic in

$$\{k_1 | \mathrm{Im}\,k_1 < 0\} \times \{\lambda | \lambda \neq \infty\}. \tag{5.2.47}$$

Since $f_1(\lambda, k_1, \rho)$ and the analytic continuation of $f(\lambda, k, z)$ have the same asymptotic form and satisfy the same equation, they are equal. Similarly, the solution of (5.2.46) satisfies

$$\phi_1(\lambda, k_1, \rho) = \exp\left[-i\sigma(\lambda + \tfrac{1}{2})\right]\phi(\lambda, k, z), \qquad (5.2.48)$$

giving
$$f_1(\lambda, k_1) = \exp\left[-i\sigma(\lambda + \tfrac{1}{2})\right]f(\lambda, k). \qquad (5.2.49)$$

Thus $f(\lambda, k)$ is analytic in the sum of all the domains of regularity (5.2.47) obtained with $|\sigma| < \tfrac{1}{2}\pi$, which is the region,

$$\{k \,|\, k \text{ not positive imaginary}\} \times \{\lambda \,|\, \mathrm{Re}\,\lambda > 0\}. \qquad (5.2.50)$$

Similarly, $f(\lambda, k \exp(-i\pi))$ is analytic in

$$\{k \,|\, k \text{ not negative imaginary}\} \times \{\lambda \,|\, \mathrm{Re}\,\lambda > 0\}. \qquad (5.2.51)$$

It follows that the S-matrix element,

$$S(\lambda, k) = \frac{f(\lambda, k)}{f(\lambda, -k)} \exp\left[i\pi(\lambda - \tfrac{1}{2})\right], \qquad (5.2.52)$$

is analytic in the region

$$\{k \,|\, k \text{ not pure imaginary}\} \times \{\lambda \,|\, \mathrm{Re}\,\lambda > 0\}, \qquad (5.2.53)$$

except at the zeros of $f(\lambda, -k)$ which are poles of S.

The branch cut of $f(\lambda, k)$ from 0 to im can be shown to be purely kinematic (when the potential is $V = [\exp(-mz)]/z$). This is done by showing that $f_0(\lambda, k)$ has the same branch cut. In fact

$$f_0(\lambda, ke^{-2i\pi}) = f_0(\lambda, k) - 2i \cos \lambda\pi f_0(\lambda, ke^{-i\pi}). \qquad (5.2.54)$$

This gives the corresponding change in S and shows that

$$S(\lambda, ke^{-i\pi}) = \frac{\exp\left[2i\pi(\lambda - \tfrac{1}{2})\right]}{S(\lambda, k) - 2i \cos \lambda\pi \exp\left[i\pi(\lambda - \tfrac{1}{2})\right]}. \qquad (5.2.55)$$

From this, it follows that the function $Z(\lambda, k)$ is analytic at $k = 0$ and is an even function of k, where

$$Z(\lambda, k) = \frac{ik^{2\lambda}[S(\lambda, k) - \exp(2i\pi\lambda)]}{S(\lambda, k) - 1}, \qquad (5.2.56)$$

$$= z_0 + z_1 k^2 + z_2 k^4 + \dots. \qquad (5.2.57)$$

This is the generalisation of the effective range expansion of

$$k^{2l+1} \cotan \delta(l, k)$$

for small k.

Uniqueness

In order to prove uniqueness of the extension from $S_l(k)$ to $S(l, k)$, it must be shown that the conditions of Carlson's theorem § 5.1 are satisfied. This has been done by Bottino *et al.* (1962).

Location of Regge poles

From the above results we obtain information on the allowed location of Regge poles. These are poles of $S(\lambda, k)$ arising from zeros of the Jost function $f(\lambda, -k)$, (see (5.2.52)).

(i) For $\lambda = l + \frac{1}{2}$, $l =$ integer or zero (i.e. physical values), and k complex; there may be poles at:

$k = ib$,   bound states for $b > 0$,

$k = -ib$,   anti-bound states or virtual states for $b > 0$,

$k = a - ib$,   resonance if $a > 0$, $b > 0$.

There are no poles in $\{\text{Im } k > 0, \text{Re } k \neq 0\}$.

(ii) For k real (physical) and λ complex. Poles of S occur at the zeros of $f(\lambda, -k)$. Strictly we can only discuss these in $\text{Re } \lambda > 0$ since only there has $f(\lambda, -k)$ been shown to be analytic, but in fact the domain of analyticity can be extended also to

$$-M < \text{Re } \lambda < 0 \quad \text{for} \quad M < \infty,$$

so one can also consider poles of S in $\text{Re } \lambda < 0$.

(iii) There are no poles in

$$[\text{Re } k > 0, \text{Im } \lambda < 0],$$

or in $$[\text{Re } k < 0, \text{Im } \lambda > 0].$$

Further details are given by Bottino & Longoni (1962).

This completes our description of the methods used by Regge and collaborators for establishing, in potential theory, that partial wave scattering amplitudes have the properties quoted in § 5.1. There is a further question that is of considerable importance in justifying the dominance of the right-most Regge pole when the momentum transfer becomes large. This concerns the convergence of the background integral along $\text{Re } \lambda = 0$, or $\text{Re } l = -\frac{1}{2}$, and the proof that it behaves like $P_{(-\frac{1}{2})}(z)$, for large values of $z = \cos \theta$. This result can be proved but it requires a complicated use of the WKB method and will not be considered here (but see Bottino, Longoni & Regge, 1962). It should also be noted that the Regge theory provides an elegant proof of the Mandelstam representation for scattering by a Yukawa potential (Regge, 1959).

(c) Asymptotic behaviour in perturbation theory

The general behaviour of Feynman integrals as the momentum transfer variable t tends to infinity has been considered by Polkinghorne (1963b) and by Federbush & Grisaru (1963) with particular reference to their relation to Regge theory. We follow their methods for a simple example. These methods are introduced here to illustrate an analogue of the potential scattering example used earlier; they apply also in relativistic theory.

Diagram (a) in Fig. (5.2.2) gives a Feynman integral (see Chapter 4, particularly §4.5),

$$f_4 = g^2\left(-\frac{g^2}{16\pi^2}\right)\int_0^1 \frac{d\alpha_1\,d\alpha_2\,d\beta_1\,d\beta_2\,\delta(\alpha_1+\alpha_2+\beta_1+\beta_2-1)}{[\alpha_1\alpha_2 t + d(\alpha,\beta,s)]^2}. \quad (5.2.58)$$

As $t \to \infty$, the integrand will become small except near $\alpha_1 = \alpha_2 = 0$. To obtain the leading behaviour for large t, only the integral near $\alpha_i = 0$ need be considered. Writing $d' = d(0,\beta,s)$, this gives an integral involving

$$\int_0^\epsilon d\alpha_1\,d\alpha_2\,\frac{1}{[\alpha_1\alpha_2 t + d']^2} \sim \frac{1}{d'}t^{-1}\log t \quad \text{as} \quad t \to \infty. \quad (5.2.59)$$

Hence (5.2.58) gives for large t,

$$f_4 \sim g^2 K(s)\,t^{-1}\log t, \quad (5.2.60)$$

where
$$K(s) = -\frac{g^2}{16\pi^2}\int_0^1 \frac{d\beta_1\,d\beta_2\,\delta(\beta_1+\beta_2-1)}{d'(\beta,s)}. \quad (5.2.61)$$

This expression for $K(s)$ is just the integral corresponding to the Feynman diagram in Fig. (5.2.2 (b))), but using a two-dimensional loop momentum k instead of the usual four-vector.

This procedure can be generalised to give contributions from each of the diagrams (c) in Fig. 5.2.2. For large t the sum of these terms has the form

$$\sum_{n=1}^\infty f_{2n} \sim \frac{g^2}{t}\left[1 + K(s)\log t + \dots + \frac{(K\log t)^n}{n!} + \dots\right], \quad (5.2.62)$$

$$\sim \frac{g^2}{t}\exp\left[K(s)\log t\right], \quad (5.2.63)$$

$$\sim g^2 t^{\alpha(s)}, \quad (5.2.64)$$

where
$$\alpha(s) = -1 + K(s), \quad (5.2.65)$$

For large s, the function $K(s)$ is small. This illustrates the feature of Yukawa potential scattering discussed earlier. The leading Regge trajectory tends to $l = -1$ as $s \to \infty$ (see Fig. 5.1.2). It can also be shown to have the expected singularities at normal thresholds (Polkinghorne, 1963c).

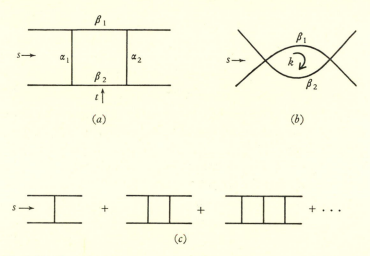

Fig. 5.2.2. Feynman diagrams used to illustrate how asymptotic behaviour can be obtained in perturbation theory.

Details of the above methods and their use in relativistic theory are described in *The Analytic S-Matrix* (Eden, Landshoff, Olive & Polkinghorne, 1966) where further references are also given.

5.3   Relativistic formulation of Regge theory

In this section we begin by defining analytic extensions of the partial wave amplitude to complex l. Two such extensions are required, one from even integer l and one from odd integer l. We then show how the use of crossing symmetry relates high energy behaviour in the t channel to Regge poles in the s channel.

(a) *The continued partial wave amplitudes for even and odd signature*

The Mandelstam representation given in §4.4 defines a dispersion relation in t for the scattering amplitude $F(s,t)$, when s is fixed above the elastic threshold. In addition to the analyticity required by this representation, we must make an assumption about the asymptotic behaviour of $F(s,t)$ for large t, so as to define the number of subtractions that are required to make the dispersion integral converge. We assume

$$|F(s,t)| < |t|^{N-\epsilon} \quad \text{as} \quad |t| \to \infty, \tag{5.3.1}$$

so that the dispersion relation requires only N subtractions in t. Assuming equal masses, we have

$$z = \cos\theta = 1 + \frac{t}{2k^2}. \tag{5.3.2}$$

We can write the dispersion relation in terms of z,

$$F(s,z) = \sum_0^{N-1} a_n z^n + \frac{z^N}{\pi}\int_{z_0}^{\infty}\frac{dx\,F_1(s,x)}{x^N(x-z)} + \frac{z^N}{\pi}\int_{-z_0}^{-\infty}\frac{dx\,F_2(s,x)}{x^N(x-z)}, \tag{5.3.3}$$

where $\pm z_0$ denotes the position of the branch point nearest to $z = \pm 1$. The minimum required number of subtractions N will, in general, change as s varies. If it did not change, then a paradox could be established by comparing the requirements of unitarity and a (nearly) constant cross-section for large t, in the t channel.

The partial wave amplitude is,

$$f_l(s) = \frac{1}{2}\int_{-1}^{1} dz\, P_l(z)\, F(s,z). \tag{5.3.4}$$

Substituting (5.3.3) into (5.3.4) and using the relation

$$\frac{1}{2}\int_{-1}^{1} dz\, P_l(z)\frac{z^N}{x^N(x-z)} = Q_l(x) \quad (l \geqslant N), \tag{5.3.5}$$

we obtain,

$$f_l(s) = \frac{1}{\pi}\int_{z_0}^{\infty} dx\, Q_l(x)\, F_1(s,x) + \frac{1}{\pi}\int_{-z_0}^{-\infty} dx\, Q_l(x)\, F_2(s,x). \tag{5.3.6}$$

The polynomial in (5.3.3) gives no contribution when $l \geqslant N$. These integrals converge since, by assumption, $l \geqslant N$ and

$$|F_1(s,x)| < x^N, \tag{5.3.7}$$

while from Bateman (1953) for large z,

$$Q_l(z) = \frac{\pi^{\frac{1}{2}}\,\Gamma(l+1)}{(2z)^{l+1}\,\Gamma(l+\frac{3}{2})}\,F\left(\frac{l}{2}+1,\frac{l}{2}+\frac{1}{2},l+\frac{3}{2},\frac{1}{z^2}\right)$$

$$\sim \frac{\pi^{\frac{1}{2}}}{(2z)^{l+1}}\frac{\Gamma(l+1)}{\Gamma(l+\frac{3}{2})} \quad \text{as} \quad z \to \infty. \tag{5.3.8}$$

However (5.3.6) is not yet in the right form for unique analytic continuation in l, since

$$Q_l(z) \sim \frac{C}{l^{\frac{1}{2}}}\exp\left[(l+\tfrac{1}{2})\log\{z-(z^2-1)^{\frac{1}{2}}\}\right] \quad \text{as} \quad |l| \to \infty. \tag{5.3.9}$$

Thus the second integral in (5.3.6) involving negative values of x, introduces terms that behave like $\exp |l| \, \pi$, for large l on the imaginary axis. These do not satisfy the conditions of Carlson's theorem (§ 5.1). The trick to avoid this difficulty is due to Froissart (1961 a) and Gribov (1962 b), (see also Challifour & Eden 1963 a, Squires, 1962 and Frautschi $et\ al.$ 1962). Change the variable $x \to -x$ in the second integral of (5.3.6). Then for even integer values of l, since $Q_l(-x) = (-1)^{l+1} Q_l(x)$,

$$f_l(s) = f^+(l, s) = \frac{1}{\pi} \int_{z_0}^{\infty} dx \, Q_l(x) \, [F_1(s, x) + F_2(s, -x)], \quad (5.3.10)$$

and for odd integer values of l,

$$f_l(s) = f^-(l, s) = \frac{1}{\pi} \int_{z_0}^{\infty} dx \, Q_l(x) \, [F_1(s, x) - F_2(s, -x)]. \quad (5.3.11)$$

The amplitudes, $f^+(l, s)$ and $f^-(l, s)$, each have the behaviour for large l that is required for unique analytic continuation by Carlson's theorem. They are called the partial wave amplitudes for even and odd signature ($\tau = +$ or $-$) respectively.

The continued partial wave amplitudes also satisfy extended unitarity (Froissart, 1963). For s on the real axis consider the function

$$H(l) \equiv f^+(l, s) - [f^+(l^*, s)]^* - \frac{4ik}{s^{\frac{1}{2}}} f^+(l, s) \, [f^+(l^*, s)]^*. \quad (5.3.12)$$

The function $H(l)$ vanishes when l is any even integer larger than N, and it satisfies the conditions of Carlson's theorem given in § 5.1. Hence $H(l)$ is identically zero, and (5.3.12) gives an extended form of unitarity. A similar equation holds for $f^-(l, s)$.

(b) Analytic properties of partial waves in the s-plane

In order to obtain the asymptotic behaviour for a physical process in the t channel we require $s \leqslant 0$. The continuation from physical values in the s channel above threshold $4m^2$, to below threshold involves

$$k^2 \to k^2 \exp i\pi; \quad s = 4(m^2 + k^2).$$

We will see that $f^{\pm}(l, s)$ is not real below threshold when l is not integer, and it is necessary to relate it to a real analytic function g in order to obtain its phase below threshold.

In the definitions (5.3.10) and (5.3.11) of f^+ and f^-, for complex l, the branch point z_0 depends on s. It is therefore more convenient to work with the variable t, which has a fixed position t_0 for its leading singularity, which is independent of s. Then

$$f^{\pm}(l,s) = \frac{1}{\pi} \int_{t_0}^{\infty} \frac{dt}{2k^2} Q_l\left(1+\frac{t}{2k^2}\right) [F_1(s,t) \pm F_2(s,t)], \quad (5.3.13)$$

where t_0 is the smaller of t_0 and u_0, the leading singularities in the t and u channels respectively.

The Legendre function $Q_l(z)$ has branch cuts along the intervals,

$$(-\infty < z < -1) \quad \text{and} \quad (-1 < z < +1). \quad (5.3.14)$$

For $z < -1$, the discontinuity is given by (Bateman, 1953),

$$Q_l(x \exp \pm i\pi) = -\exp(\pm i\pi l) Q_l(x) \quad (x > 1). \quad (5.3.15)$$

This shows that the function,

$$(k^2 \pm i\epsilon)^{-l} Q_l\left\{1+\frac{t}{2(k^2 \pm i\epsilon)}\right\} = -(-k^2)^{-l} Q_l\left(-1-\frac{t}{2k^2}\right), \quad (5.3.16)$$

has no discontinuity in,

$$-\tfrac{1}{4}t < k^2 < 0.$$

Using this result, we can construct partial amplitudes $g^{\pm}(l,s)$ having simpler properties than $f^{\pm}(l,s)$. Define $g^{\pm}$ by,

$$g^{\pm}(l,s) \equiv (k^2)^{-l} f^{\pm}(l,s). \quad (5.3.17)$$

Then $g^{\pm}(l,s)$ is real in the interval,

$$-\tfrac{1}{4}t_0 < k^2 < 0.$$

The discontinuity of $Q_l(z)$ in $(-1 < z < 1)$ can be evaluated in terms of $P_l(z)$ which has no branch cut in this interval,

$$\sin \pi l \cdot Q_l(-z \mp i\epsilon) = \tfrac{1}{2}\pi \exp(\pm i\pi l) P_l(-z) - \tfrac{1}{2}\pi P_l(z). \quad (5.3.18)$$

Hence for $-\infty < k^2 < -\tfrac{1}{4}t$,

$$\frac{Q_l\left(1+\dfrac{t}{2k^2+i\epsilon}\right)}{(k^2+i\epsilon)^l} - \frac{Q_l\left(1+\dfrac{t}{2k^2-i\epsilon}\right)}{(k^2-i\epsilon)^l} = \frac{-i\pi P_l\left(-1-\dfrac{t}{2k^2}\right)}{(-k^2)^l}. \quad (5.3.19)$$

This gives the following analytic properties of g defined by (5.3.17) and (5.3.13). The functions $g^{\pm}(l,s)$ are real analytic, regular in the s-plane except for branch cuts on the real axis, along $(-\infty < s < 4m^2-t_0)$ and

$(4m^2 < s < \infty)$. The discontinuities are obtained from (5.3.13) as noted below:

(i) For the right-hand cut $s > 4m^2$, the discontinuities of $g^{\pm}(l, s)$ come from the branch cuts in $F_1(s, t)$ and $F_2(s, t)$. Their discontinuities can be obtained from the Mandelstam representation, giving for $s > 4m^2$,

$$\text{disc.}\,[g^{\pm}(l, s)] = \frac{2i}{\pi(k^2)^l} \int_{t_0}^{\infty} \frac{dt}{2k^2} Q_l\left(1 + \frac{t}{2k^2}\right) [\rho_{12}(s, t) \pm \rho_{13}(s, t)]. \quad (5.3.20)$$

(ii) On the left-hand cut, $k^2 < -\frac{1}{4}t_0$, there is a discontinuity following from (5.3.19), and

(iii) There is a discontinuity due to disc. F_1 and disc. F_2 given by the Mandelstam representation.

Combining (ii) and (iii), the discontinuity on the left-hand cut $(-\infty < k^2 < -\frac{1}{4}t_0)$ is given by

$$\text{disc.}\,[g^{\pm}(l, s)] = \frac{-i}{(-k^2)^l} \int_{t_0}^{-4k^2} \frac{dt}{2k^2} P_l\left(-1 - \frac{t}{2k^2}\right)[F_1(s - i0, t) \pm F_2(s - i0, t)]$$

$$+ \frac{2i}{\pi(k^2 + i\epsilon)^l} \int_{t_0}^{\infty} \frac{dt}{2k^2} Q_l\left(1 + \frac{t}{2k^2 + i\epsilon}\right) \rho_{23}(s, t)\,[1 \mp e^{-i\pi l}]. \quad (5.3.21)$$

In the last term $\rho_{23}(s, u)$ has been eliminated by using the relation (5.3.15), which gives the factor $\exp(-i\pi l)$, (Squires (1963)).

Assuming convergence of the integrals, we can continue $g^{\pm}(l, s)$, and hence continue $f^{\pm}(l, s)$, below threshold in s. If we *assume* that $g^{\pm}(l, s)$ is meromorphic below $\text{Re}\,l = N$ down to $\text{Re}\,l = -\frac{1}{2}$, then we can separate out the poles in the Sommerfeld–Watson transform, and explicitly show their contribution to the scattering amplitude $F(s, t)$ for large t. We can also make the same assumption about $f^{\pm}(l, s)$ provided we take care in considering $k^2 < 0$.

(c) *Regge poles, crossing and high energy behaviour*

We assume now that the partial wave amplitudes for even and odd signature are meromorphic (no singularities except poles) in $\text{Re}\,l > -\frac{1}{2}$. The Sommerfeld–Watson transform can be applied to even and odd partial waves separately. We have

$$F(s, t) = \frac{8\pi s^{\frac{1}{2}}}{k} \sum_{0}^{\infty} (4l + 1) f_{2l}(s)\, P_{2l}(\cos\theta)$$

$$+ \frac{8\pi s^{\frac{1}{2}}}{k} \sum_{0}^{\infty} (4l + 3) f_{2l+1}(s)\, P_{2l+1}(\cos\theta). \quad (5.3.22)$$

Taking a contour C round the positive real l axis as shown in Fig. 5.1.1, we obtain

$$F(s,t) = -\frac{2\pi s^{\frac{1}{2}}}{ik}\int_C \frac{dl(2l+1)f^+(l,s)\,P_l(z)\exp\left(-i\pi l/2\right)}{\sin\left(\pi l/2\right)}$$

$$-\frac{2\pi s^{\frac{1}{2}}}{ik}\int_C \frac{dl(2l+1)f^-(l,s)\,P_l(z)\exp\left[-i\pi(l-1)/2\right]}{\sin\left[\pi(l-1)/2\right]}. \quad (5.3.23)$$

The exponentials are required so that the residues have the correct signs. The product fP_l in the integrands can be rewritten,

$$f^{\pm}(l,s)\,P_l\left(1+\frac{t}{2k^2}\right) = k^{2l}g^{\pm}(l,s)\,P_l\left(1+\frac{t}{2k^2}\right),$$

$$= (-k^2)^l g^{\pm}(l,s)\,P_l\left(-1-\frac{t}{2k^2}\right).$$

When $k^2 < 0$, and t is sufficiently large, the argument of the last Legendre polynomial is positive.

We now assume that $g^{\pm}(l,s)$ tends to zero, as $|l| \to \infty$, in Re $(l) > -\frac{1}{2}$, so that C can be replaced by C' and a sum over C'' round the poles of $g^{\pm}(l,s)$, as in Fig. 5.1.1. Evaluating the integrals over C'', which pick out the residues at the Regge poles in Re $(l) > -\frac{1}{2}$, we obtain

$$F(s,t) =$$
$$-\frac{2\pi s^{\frac{1}{2}}}{ik}\int_{-\frac{1}{2}-i\infty}^{-\frac{1}{2}+i\infty} dl \left[\frac{(2l+1)\,g^+(l,s)\,(-k^2)^l\,P_l\left(-1-\frac{t}{2k^2}\right)\exp\left(-i\pi l/2\right)}{\sin\left(\pi l/2\right)}\right.$$

$$\left.+\frac{(2l+1)\,g^-(l,s)\,(-k^2)^l\,P_l\left(-1-\frac{t}{2k^2}\right)\exp\left[-i\pi(l-1)/2\right]}{\sin\left[\pi(l-1)/2\right]}\right]$$

$$+\sum_n C_n^+(s)\,P_{\alpha_n^+}\left(-1-\frac{t}{2k^2}\right) + \sum_m C_m^-(s)\,P_{\alpha_m^-}\left(-1-\frac{t}{2k^2}\right). \quad (5.3.24)$$

The sum over n and m includes all the poles in Re $(l) > -\frac{1}{2}$, of $g^+(l,s)$ and $g^-(l,s)$ respectively. Denote the residues at these poles by $b_n^+(s)$ and $b_m^-(s)$. In the interval between the branch cuts these residues will be real, since g^+ and g^- are real analytic functions. It is often assumed also that they are real when $s < 4m^2 - t_0$.

The coefficients in (5.3.24) are given in terms of the real residues b_n^+ by

$$C_n^+(s) = -\frac{4\pi s^{\frac{1}{2}}}{k}(2\alpha_n^+ + 1)\,(-k^2)^{\alpha_n^+}\,b_n^+(s)\,\frac{\exp\left[-\frac{1}{2}i\pi\alpha_n^+\right]}{\sin\left(\frac{1}{2}\pi\alpha_n^+\right)},$$

$$= \frac{4\pi s^{\frac{1}{2}}}{k}(2\alpha_n^+ + 1)\,(-k^2)^{\alpha_n^+}\,b_n^+(s)\,[i - \cot\left(\frac{1}{2}\pi\alpha_n^+\right)]. \quad (5.3.25)$$

Similarly,

$$C_m^-(s) = \frac{4\pi s^{\frac{1}{2}}}{k}(2\alpha_m^- + 1)(-k^2)^{\alpha_m^-} b_m^-(s)\left[i + \tan\left(\tfrac{1}{2}\pi\alpha_m^-\right)\right]. \quad (5.3.26)$$

When s is below threshold (between the branch points), the phases of the Regge pole terms are given by the above equations. Note that all factors in C_n^+, C_m^- are then real except those in the square brackets. We will see in Chapter 7 that their phases follow from crossing symmetry. Thus once an expansion in terms of Regge pole contributions is assumed, the phases of each term in the even and odd series are determined by crossing symmetry. This means that the phase conditions cannot easily be relaxed without modifying the basic expansion. Note, however, that the situation is more complicated when the colliding particles have non-zero spin (see § 5.5).

We now assume that the representation (5.3.24) can be continued to negative values of s, and that for large values of t the background integral in (5.3.24) behaves like $t^{-\frac{1}{2}}$. Then the contribution from the sums over residues in (5.3.24) will dominate $F(s,t)$ for large t, in channel II, for which t is the energy squared and s is the momentum transfer variable. Since for large z,

$$|P_l(-z)| \sim |z|^{\mathrm{Re}(l)} \sim \left|1 + \frac{2t}{s - 4m^2}\right|^{\mathrm{Re}(l)}, \quad (5.3.27)$$

we can arrange the Regge pole contributions to the scattering amplitude $F(s,t)$ as $t \to \infty$ in order of decreasing importance. The dominant term will come from the pole in $f^+(l,s)$ or $f^-(l,s)$, that has the largest real value of l, for given $s \leqslant 0$.

(d) The Pomeranchuk pole

Let $\alpha(s)$ denote the position of the pole of f^+ or f^- that has the largest real part at $s = 0$. Then at $s = 0$,

$$F(s,t) = F(0,t) \sim C_0\left(\frac{t}{t_0}\right)^{\alpha(0)} \quad \text{as} \quad t \to \infty, \quad (5.3.28)$$

where t_0 gives an (arbitrary) normalisation, usually taken to be of the order of 1 (Gev)^2. The phase factor in C_0 will be determined by (5.3.25) or (5.3.26), depending on whether the pole is in the partial wave amplitude of even or odd signature respectively.

From the optical theorem, apart from a constant factor,

$$\mathrm{Im}\,F(0,t) \sim t\,\sigma(\text{total}) \quad \text{as} \quad t \to \infty. \quad (5.3.29)$$

But from (5.3.28),

$$\operatorname{Im} F(0,t) \sim \operatorname{Im} C_0 \left(\frac{t}{t_0}\right)^{\alpha(0)} \quad \text{as} \quad t \to \infty.$$

Hence, if we assume from experimental evidence that the total cross-section is asymptotically constant as the energy variable t tends to infinity, we must have

$$\alpha(0) = 1. \tag{5.3.30}$$

For σ(total) to be asymptotic to a non-zero constant, we must also require that the imaginary part of C_0 shall be finite. From the phase conditions, (5.3.25, 26) and (5.3.30), we see that the pole must therefore occur in the amplitude of even signature. Hence,

$$F(0,t) \sim i\,|C_0| \left(\frac{t}{t_0}\right) \quad \text{as} \quad t \to \infty. \tag{5.3.31}$$

Therefore σ(total) $\sim$ constant, requires that the scattering amplitude is dominated, at $s = 0$ and $t \to \infty$, by a Regge pole in the even signature amplitude $f^+(l,s)$, at

$$l = \alpha_P(s) \quad \text{with} \quad \alpha_P(0) = 1. \tag{5.3.32}$$

The suggestion that asymptotic cross-sections should be non-zero constants was made by Pomeranchuk in 1956. The Regge pole having $\alpha(0) = 1$ is called the Pomeranchuk pole and the corresponding trajectory is called the Pomeranchuk trajectory. The abbreviation Pomeron is also used. At $l = 1$ in the even signature amplitude the Pomeron does not correspond to a physical particle. However, if the trajectory (5.3.32) goes nearly through $l = 2$ when $s > 4m^2$, it will then correspond to a physical particle of spin 2, and it should be observable (subject to $(d/ds)\,\alpha(s)$ having a positive real part at $\alpha = 2$, see (5.1.18)).

(e) *Properties of Regge trajectories*

In our discussion of potential scattering in §5.2, it was noted that there are no Regge poles in the region

$$\operatorname{Re} k > 0, \quad \operatorname{Im} l < 0. \tag{5.3.33}$$

For k real, this region corresponds to the upper (physical) side $(s+i\epsilon)$ of the s branch cut. For $s < 4m^2$, k is pure imaginary, and for potential theory the Regge poles must be at real values of l. These results have not been proved for a relativistic theory but it is assumed that they hold. Hence for s real,

$$\operatorname{Im}\alpha(s) = 0 \quad \text{for} \quad s < 4m^2; \quad \operatorname{Im}\alpha(s) > 0 \quad \text{for} \quad s > 4m^2. \tag{5.3.34}$$

By analogy with potential theory, it is also assumed for bosons that

$\alpha(s)$ is regular in the entire s-plane except for the branch cut on the real axis, so it satisfies a dispersion relation,

$$\alpha(s) = \frac{1}{\pi} \int_{4m^2}^{\infty} dx\, \frac{\mathrm{Im}\,\alpha(x)}{x-s}. \qquad (5.3.35)$$

It is assumed that $\quad \mathrm{Im}\,\alpha(s) \to 0 \quad$ as $\quad s \to \infty, \qquad (5.3.36)$

and that for large enough s,

$$\mathrm{Re}\,\alpha(s) < 0 \quad \text{as} \quad s \to \infty. \qquad (5.3.37)$$

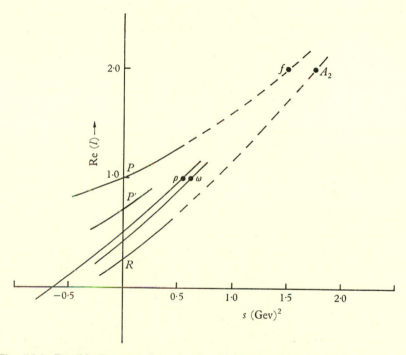

Fig. 5.3.1. Possible Regge trajectories for bosons (see § 9.3 for the methods for determining them). The slopes may be found from experiments for large t and $s < 0$, and/or from observed particles for $s > 0$ and l integer. The intercepts with $s = 0$ may be found from total cross-sections, and from the differences of total cross-sections, as the energy squared s tends to infinity.

In general one would expect a Regge trajectory, $l = \alpha(s)$, to have branch points at each inelastic threshold, but the requirement that they are defined with s nearly real but having a small positive imaginary part makes the trajectories unique. We will see in Chapter 9 that the trajectories are remarkably near to straight lines in the $[\mathrm{Re}\,(l), \mathrm{Re}\,(s)]$ plane, when deduced empirically. This is unlike the situation in potential scattering, but the forces involved are presumed to be much stronger, so there is no obvious contradiction of the simplifying assumptions. However, we will see in §5.6 that, even within the general framework of Regge theory, there are some difficulties in the relativistic theory that are not present in potential theory.

For $s > 0$, the Regge trajectories can sometimes be determined from the requirement that they go through values given by observed particles and resonances, having the same quantum numbers, except for l which changes by 2 units from one physical value to the next. For $s < 0$, the trajectories may be determined in some cases by comparing the predicted cross-sections, for large t in the t channel, with experimentally measured cross-sections (see Fig. 5.3.1). We will discuss these questions in more detail in Chapter 9.

In the above discussion of Regge theory we could equally well have worked with partial waves in the t channel, and from the resulting Regge poles $\alpha(t)$ we would have obtained the asymptotic behaviour of the scattering amplitude $F(s, t)$ in the s channel, as the energy s tends to infinity. In our later use of Regge theory we will use whichever channel is most natural in the context.

5.4   Coupled channels and factorisation

(a) Factorisation of residues

The similarity of Regge poles $l = \alpha(t)$ in the complex angular momentum plane, and of Breit–Wigner resonance poles in the complex energy (t) plane led to the suggestion that residues at Regge poles should be factorisable (Gell-Mann, 1962c; Gribov & Pomeranchuk, 1962a, b). This leads to relations between the asymptotic cross-sections for different processes. We will consider firstly the general argument following Gell-Mann (1962c). Then we will describe the method of Gribov & Pomeranchuk that is based on generalised unitarity.

We will consider the high energy (high s) behaviour for the reaction

$$a + b \rightarrow c + d. \tag{5.4.1}$$

On the assumptions of Regge theory, this will depend on the position of the pole $l = \alpha(t)$ for a partial wave $f(l, t)$ in the t channel; namely for the reaction,

$$\bar{c} + b \rightarrow \bar{a} + d. \tag{5.4.2}$$

Now $\alpha(t)$ will depend only on the quantum numbers of this reaction. If we consider another process having the same conserved quantum numbers as (5.4.2), then $\alpha(t)$ will be the same as before. Let the following reaction have this property,

$$\bar{c}' + b' \to \bar{a}' + d'. \tag{5.4.3}$$

However, although $\alpha(t)$ will be the same for each reaction (5.4.2, 3), the coupling strengths $\beta(t)$ will be different in general. This is analogous to resonance theory where the complex energy remains the same for each reaction in which it occurs but the strength changes for each reaction. Thus (5.4.2) will have a partial wave f,

$$f(l,t) = \frac{\beta(t)}{l - \alpha(t)} + R(l,t), \tag{5.4.4}$$

but (5.4.3) will have a partial wave f',

$$f'(l,t) = \frac{\beta'(t)}{l - \alpha(t)} + R'(l,t). \tag{5.4.5}$$

If we consider a value of t for which l is an integer of appropriate signature, the pole $\alpha(t)$ will correspond to a particle A (bound state or resonance). The residue $\beta(t)$ is then just the product of the coupling parameters to the particle A from $\bar{a}$, d (or a, d) which we write $g(adA)$, and from A to $\bar{c}$, b which we write $g(cbA)$,

$$\frac{\beta(t)}{\pi \operatorname{Re} \alpha'(t)} = g(adA) \, g(cbA). \tag{5.4.6}$$

where α' denotes the derivative of $\alpha(t)$, the Regge pole. Similarly,

$$\frac{\beta'(t)}{\pi \operatorname{Re} \alpha'(t)} = g(a'd'A) \, g(c'b'A). \tag{5.4.7}$$

In the particular case where the trajectory $l = \alpha(t)$ for the Regge particle A is the Pomeranchuk trajectory P, it will dominate total cross-sections. The coupling for $\pi\pi$ scattering will give (see Fig. 5.4.1),

$$\frac{\beta(0)}{\pi \operatorname{Re} \alpha'(0)} = [g(\pi\pi P)]^2. \tag{5.4.8}$$

For πN scattering we will have

$$\frac{\beta'(0)}{\pi \operatorname{Re} \alpha'(0)} = g(\pi\pi P) \, g(NNP), \tag{5.4.9}$$

and for NN scattering,

$$\frac{\beta''(0)}{\pi \operatorname{Re} \alpha'(0)} = [g(NNP)]^2. \tag{5.4.10}$$

Hence as $s \to \infty$, assuming dominance of each total cross-section by the (t channel) Pomeron exchange contribution, we obtain a relation between the total cross-sections,

$$\sigma_t(\pi\pi)\,\sigma_t(NN) = [\sigma_t(\pi N)]^2. \tag{5.4.11}$$

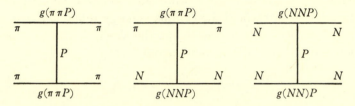

Fig. 5.4.1. Diagrams illustrating the factorisation hypothesis for $\pi\pi$, πN, and NN, scattering at high energy, assuming dominance of Pomeron exchange.

(b) Generalised unitarity and factorisation

The above relation (5.4.11) was observed independently by Gribov & Pomeranchuk (1962 a, b) who used the following argument:

For simplicity, consider reactions involving spinless equal mass particles. Let f_j, g_j, and h_j be the partial wave amplitudes for angular momentum j and for a particular isospin state of the reactions,

$$f_j(t): \quad \pi + \pi \to \pi + \pi, \tag{5.4.12}$$

$$g_j(t): \quad K + \bar{K} \to \pi + \pi, \tag{5.4.13}$$

$$h_j(t): \quad K + \bar{K} \to K + \bar{K}. \tag{5.4.14}$$

The generalised unitarity condition can be derived for potential scattering from the work of Regge (see §5.2). In relativistic theory it follows from our earlier assumptions. With these coupled channels it takes the form (for $4m^2 < t < 16m^2$; $m = m_\pi$),

$$\frac{1}{2i}(f_j - f_{j*}^*) = \left(\frac{k}{\omega}\right) f_j f_{j*}^*, \tag{5.4.15}$$

$$\frac{1}{2i}(g_j - g_{j*}^*) = \left(\frac{k}{\omega}\right) g_j f_{j*}^*, \tag{5.4.16}$$

$$\frac{1}{2i}(h_j - h_{j*}^*) = \left(\frac{k}{\omega}\right) g_j g_{j*}^*, \tag{5.4.17}$$

where k and ω are momentum and energy for the π meson,

$$4\omega^2 = 4(m^2 + k^2) = t. \tag{5.4.18}$$

Each partial wave f_j, g_j, h_j, is a function of j and also of ω. A star denotes complex conjugate, j^* of j, and f^* of f. For j an integer and ω a real value, the above equations give the usual discontinuity relation on the elastic cut. The proof of their uniqueness is similar to that for the uniqueness of extending f_l from integer l to complex l, and is based on Carlson's theorem (see § 5.1).

The generalised unitarity equations can be re-written in the form,

$$f_j = \frac{f_{j*}^*}{1 - 2i\left(\dfrac{k}{\omega}\right)f_{j*}^*}, \tag{5.4.19}$$

$$g_j = \frac{g_{j*}^*}{1 - 2i\left(\dfrac{k}{\omega}\right)f_{j*}^*}, \tag{5.4.20}$$

$$h_j = h_{j*}^* + \frac{2i\left(\dfrac{k}{\omega}\right)(g_{j*}^*)^2}{1 - 2i\left(\dfrac{k}{\omega}\right)f_{j*}^*}. \tag{5.4.21}$$

From these equations it is apparent that in general all the amplitudes will have a pole when

$$1 - 2i\left(\frac{k}{\omega}\right)f_{j*}^*(t) = 0. \tag{5.4.22}$$

For fixed t this gives a value $j = \alpha(t)$, at which each partial wave amplitude has a Regge pole.

When j is near to $\alpha(t)$, we can write approximately,

$$f_{j*}^* = \left(\frac{\omega}{2ik}\right)\left[1 - \frac{(j-\alpha)}{\beta(t)}\right]. \tag{5.4.23}$$

Substituting this approximate form into (5.4.19, 20, 21), we obtain

$$f_j \approx \frac{(\omega/2ik)\,\beta(t)}{j - \alpha(t)}, \tag{5.4.24}$$

$$g_j \approx \frac{\beta g_{\alpha*}^*}{j - \alpha(t)}, \tag{5.4.25}$$

$$h_j \approx \frac{(2ik/\omega)\,\beta(t)\,(g_{\alpha*}^*)^2}{j - \alpha(t)}. \tag{5.4.26}$$

Hence the residues $r_{\pi\pi}(t)$, $r_{\pi K}(t)$ and $r_{KK}(t)$ satisfy

$$(r_{\pi K})^2 = r_{\pi\pi}r_{KK}. \tag{5.4.27}$$

The residues $r_{\pi\pi}(t)$ etc., are assumed to be analytic functions of t, so this factorisation theorem holds for all t.

In particular, when $t = 0$, one obtains relations involving the total cross-sections and the residues for the isospin zero exchange amplitude. Asymptotically one has (Gribov & Pomeranchuk, 1962a),

$$\left.\begin{aligned}
\sigma_t(\pi\pi) &= \frac{12\pi^2}{\mu^2}\, r_{\pi\pi}(0), \\[2mm]
\sigma_t(\pi K) &= \frac{12\pi^2}{m\mu}\, r_{\pi K}(0), \\[2mm]
\sigma_t(KK) &= \frac{12\pi^2}{m^2}\, r_{KK}(0),
\end{aligned}\right\} \tag{5.4.28}$$

giving the analogue of (5.4.11) for these reactions,

$$\sigma_t(\pi\pi)\,\sigma_t(KK) = [\sigma_t(\pi K)]^2. \tag{5.4.29}$$

More generally, for any coupled channel reaction involving channels $1, 2, \ldots, n$, it is assumed that the following result, proved for potential scattering, holds in relativistic theory,

$$\frac{r_{11}}{r_{1a}} = \frac{r_{b1}}{r_{ba}} \quad \text{for} \quad \begin{cases} a = 1, 2, \ldots, n, \\ b = 1, 2, \ldots, n, \end{cases} \tag{5.4.30}$$

where r_{ij} denotes the residue (at a Regge pole) of the reaction amplitude from an initial channel i to a final channel j.

We will consider other experimental consequences of this factorisation of residues at Regge poles in Chapter 9.

5.5   Fermion Regge poles, pion-nucleon scattering

There are two related complications in Regge theory when one or more of the colliding particles has non-zero spin. One is the problem of writing the scattering amplitude in the direct (s) channel in terms of amplitudes that can be analytically continued to the crossed channels (t) or (u). This problem has been considered in §§ 3.6 and 4.9. The second problem arises when, in the relevant crossed channel, there is a fermion rather than a boson; then one requires a treatment of fermion Regge poles. We will illustrate these problems by considering pion-nucleon scattering.

For $\pi^+ p$ scattering there are three channels related by analytic continuation,

$$(s):\quad \pi^+ + p \to \pi^+ + p, \tag{5.5.1}$$

$$(t):\quad p + \overline{p} \to \pi^+ + \pi^-, \tag{5.5.2}$$

$$(u):\quad \pi^- + p \to \pi^- + p. \tag{5.5.3}$$

In the t channel the isospin can be 1 or 0, and in addition to an intermediate particle with the quantum numbers of the vacuum (P or P') we could have an intermediate ρ meson (see § 9.1 for a fuller discussion). These particles can be treated as Regge poles. On the usual assumptions, for fixed t as $s \to \infty$, the Regge poles $\alpha(t)$ with the largest real parts will dominate the scattering amplitudes. In this case, the only complications arise from the crossing conditions considered in § 4.9.

The situation is different in the u channel (5.5.3), where the neutron is a possible intermediate particle. If we are interested in asymptotic behaviour for fixed u as $s \to \infty$, then we must consider Regge behaviour associated with a fermion, in this case the neutron. We will see in this section that this situation arises in π^+p backward scattering, and in § 9.4 we will note that experimental results confirm that the neutron pole plays a significant role.

(a) Fermion Regge poles

These have been considered by Mandelstam (1963), Calogero, Charap & Squires (1963), Gell-Mann et al. (1963). The general method was outlined by Stapp (1962a, b), and the special case of pion-nucleon scattering was considered by Gribov & Pomeranchuk (1962b) and by Singh (1963). In this section we will follow the treatment of Singh, except that we will assume parity and time reversal invariance, which reduces the number of independent amplitudes.

For pion-nucleon scattering (see § 3.6 and § 4.9), the scattering amplitude can be written in the form

$$F(s, u) = A(s, u) - i(\gamma \cdot Q) B(s, u), \tag{5.5.4}$$

where $Q = \frac{1}{2}(q_1 + q_2)$. We begin by developing the Regge theory for s channel partial waves.

Using the conventional amplitudes, f_1 and f_2, (defined in § 3.6),

$$F(s, u) = f_1 + (\boldsymbol{\sigma} \cdot \hat{\mathbf{q}}_2)(\boldsymbol{\sigma} \cdot \hat{\mathbf{q}}_1) f_2. \tag{5.5.5}$$

In terms of the invariant amplitudes A and B,

$$f_1 = \frac{E+m}{8\pi W}\{A + (W-m)B\}, \tag{5.5.6}$$

$$f_2 = \frac{E-m}{8\pi W}\{-A + (W+m)B\}, \tag{5.5.7}$$

where E is the nucleon energy and W the total energy in the centre of mass system,

$$s = W^2, \quad E = (W^2 + m^2 - \mu^2)/2W. \tag{5.5.8}$$

The partial wave expansions are (cf. (3.6.32, 33)),

$$f_1(s, u) = \sum_J [f_{(J-\frac{1}{2})+}(W) P'_{J+\frac{1}{2}} - f_{(J+\frac{1}{2})-}(W) P'_{J-\frac{1}{2}}], \qquad (5.5.9)$$

$$f_2(s, u) = \sum_J [f_{(J+\frac{1}{2})-}(W) P'_{J+\frac{1}{2}} - f_{(J-\frac{1}{2})+}(W) P'_{J-\frac{1}{2}}], \qquad (5.5.10)$$

with $J = \frac{1}{2}, \frac{3}{2}, \frac{5}{2}, \ldots$ in the sums. The argument of the Legendre polynomial derivatives is z, where

$$z = \cos\theta_s = 1 + \frac{t}{2q^2} = 1 + \frac{(2m^2 + 2\mu^2 - s - u)}{2q^2}, \qquad (5.5.11a)$$

$$4q^2 = s - (2m^2 + 2\mu^2) + \frac{(m^2 - \mu^2)^2}{s}. \qquad (5.5.11b)$$

The partial wave projections of $f_{(J+\frac{1}{2})-}$ and $f_{(J-\frac{1}{2})+}$ have been given in (3.6.39). Using the Mandelstam representation, these can be transformed to give a form suitable for analytic continuation in J. We have, in (5.5.6 and 7), expressions for f_1 and f_2 in terms of A and B. The corresponding relations between the partial wave amplitudes are

$$f_{(J\mp\frac{1}{2})\pm} = \frac{E+m}{16\pi W}[A_{J\mp\frac{1}{2}} + (W - m) B_{J\mp\frac{1}{2}}]$$

$$+ \frac{E-m}{16\pi W}[-A_{J\pm\frac{1}{2}} + (W + m) B_{J\pm\frac{1}{2}}]. \qquad (5.5.12)$$

The amplitudes A and B are assumed to satisfy the Mandelstam representation, so their partial wave amplitudes can be transformed by the Froissart–Gribov method that was used for bosons in § 5.3. This gives, for example,

$$A_{J-\frac{1}{2}}(W) = \frac{1}{\pi q^2} \int_{x_0}^{\infty} dx \left[A_1(s, x) + (-1)^{J-\frac{1}{2}} \right.$$

$$\left. \times A_2 \left\{ s, x + \frac{(m^2 - \mu^2)^2}{s} \right\} \right] Q_{J-\frac{1}{2}} \left(1 + \frac{x}{2q^2} \right). \qquad (5.5.13)$$

We can obtain a unique analytic continuation in J by defining even amplitudes $A^e_{J-\frac{1}{2}}$ for $(J - \frac{1}{2})$ an even integer, and odd amplitudes $A^\phi_{J-\frac{1}{2}}$ for $(J - \frac{1}{2})$ an odd integer, (as in § 5.3),

$$A^e_{J-\frac{1}{2}}(W) = \frac{1}{\pi q^2} \int_{x_0}^{\infty} dx [A_1(s, x) + A_2(s, x')] Q_{J-\frac{1}{2}} \left(1 + \frac{x}{2q^2} \right), \qquad (5.5.14)$$

$$A^\phi_{J-\frac{1}{2}}(W) = \frac{1}{\pi q^2} \int_{x_0}^{\infty} dx [A_1(s, x) - A_2(s, x')] Q_{J-\frac{1}{2}} \left(1 + \frac{x}{2q^2} \right), \qquad (5.5.15)$$

where $x' = x + (m^2 - \mu^2)^2/s$.

Even and odd amplitudes $B^e_{J-\frac{1}{2}}$ and $B^\phi_{J-\frac{1}{2}}$ are similarly defined.

These even and odd amplitudes, defined from A and B, can be analytically continued as in §5.3. This leads to an analytic continuation for the corresponding even and odd amplitudes that are defined from (5.5.12). For example,

$$f^e_{(J-\frac{1}{2})+}(W) = \frac{E+m}{16\pi W}[A^e_{J-\frac{1}{2}} + (W-m)B^e_{J-\frac{1}{2}}]$$
$$+ \frac{E-m}{16\pi W}[-A^\phi_{J+\frac{1}{2}} + (W+m)B^\phi_{J+\frac{1}{2}}]. \quad (5.5.16)$$

Another even amplitude $f^e_{(J+\frac{1}{2})-}$ is similarly defined. In addition there are two odd amplitudes $f^\phi_{(J-\frac{1}{2})+}$ and $f^\phi_{(J+\frac{1}{2})-}$, giving four continued partial wave amplitudes in all.

The partial wave series for the amplitudes $f_1(s,t)$ and $f_2(s,t)$ can be re-expressed as a Sommerfeld–Watson transform using the analytic continuations of $f^e_{(J-\frac{1}{2})+}$ etc. to complex J. This gives (Singh 1963),

$$f_{1,2}(s,t) = \pm\frac{i}{4}\int_C \frac{dJ}{\cos\pi J}[f^e_{(J-\frac{1}{2})+}(W)\{P'_{J\pm\frac{1}{2}}(-z) \pm P'_{J\pm\frac{1}{2}}(z)\}$$
$$+ f^\phi_{(J-\frac{1}{2})+}(W)\{P'_{J\pm\frac{1}{2}}(-z) \mp P'_{J\pm\frac{1}{2}}(z)\}$$
$$- f^e_{(J+\frac{1}{2})-}(W)\{P'_{J\mp\frac{1}{2}}(-z) \mp P'_{J\mp\frac{1}{2}}(z)\}$$
$$- f^\phi_{(J+\frac{1}{2})-}(W)\{P'_{J\mp\frac{1}{2}}(-z) \pm P'_{J\mp\frac{1}{2}}(z)\}],$$
$$(5.5.17)$$

where the upper signs refer to f_1, and the lower to f_2, and C is the undistorted contour surrounding the real J axis.

(b) Kinematic singularities

There are inconvenient kinematic singularities in the above partial wave amplitudes at $s = 0$, ($W = \sqrt{s}$), due to the kinematic factors in (5.5.12) which arise from the nucleon spin. It is clearly an advantage to work in the W-plane, and not the s-plane as with bosons. There are additional branch points at $q^2 = 0$ ($q = $ centre of mass momentum) when J is non-physical.

The kinematic singularities can be removed by working with modified partial wave amplitudes,

$$h^{(e,\phi)}_{(J\mp\frac{1}{2})\pm}(W) = \left(\frac{16\pi W}{E\pm m}\right)\frac{1}{(2q^2)^{J-\frac{1}{2}}}[f^{(e,\phi)}_{(J\mp\frac{1}{2})\pm}(W)]. \quad (5.5.18)$$

The analogue for the general spin case has been considered by several authors (Wang, 1966; Hara, 1964; Williams, 1963; Hepp, 1964; and

Fox, 1966). The amplitudes h are assumed to be meromorphic in the complex J-plane.

These amplitudes also have an important symmetry relation, (first noted for physical J by MacDowell, 1960), which follows from,

$$f_1(W) = -f_2(-W).$$ (5.5.19)

This gives (Singh, 1963),

$$h^{(e,\phi)}_{(J-\frac{1}{2})+}(W) = -h^{(e,\phi)}_{(J+\frac{1}{2})-}(-W).$$ (5.5.20)

It follows that, if there is a pole of

$$h^{(e,\phi)}_{(J+\frac{1}{2})-}(W) \quad \text{at} \quad J = \alpha(W),$$ (5.5.21)

then there is a pole of

$$h^{(e,\phi)}_{(J-\frac{1}{2})+}(W) \quad \text{at} \quad J = \alpha(-W),$$ (5.5.22)

and there are relations between the residues at these poles.

Unitarity

There is no coupling between the above amplitudes and elastic unitarity therefore gives

$$\text{Im}\, h^e_{(J\mp\frac{1}{2})\pm} = \frac{q(E \pm m)(2q^2)^{J-\frac{1}{2}}}{16\pi W} |h^e_{(J\mp\frac{1}{2})\pm}|^2,$$ (5.5.23)

with a similar relation for the odd amplitude.

A Regge trajectory can be specified by giving the signature τ or the J parity, the space parity P, and the isospin. Thus there will be a family of particles associated with the nucleon having even J parity, even space parity, and isospin $\frac{1}{2}$. The particles on the nucleon trajectory will have

$$I = \tfrac{1}{2}; \quad \tau = +; \quad J = \tfrac{1}{2}, \tfrac{5}{2}, \tfrac{9}{2}, \dots.$$ (5.5.24)

If the nucleon is interpreted as a Regge particle, it will correspond to a pole in the amplitude,

$$h^e_{(J+\frac{1}{2})-}(W) \quad \text{at} \quad J = \tfrac{1}{2}, \quad W = 0.94 \,\text{Gev}.$$

Similarly, the $N^*(1238)$ resonance has $I = \tfrac{3}{2}$; $P = +$; $\tau = -$; $J = \tfrac{3}{2}$.

(c) Backward scattering $\pi^+ p \to \pi^+ p$

The Regge poles in the u channel are expected to dominate the scattering at fixed u for large s, which is near to backward scattering. We recall that in the u channel, instead of (5.5.11) we have,

$$z_u = \cos\theta_u = 1 + \frac{2m^2 + 2\mu^2 - u - s}{2k^2},$$ (5.5.25a)

where

$$4k^2 = u - (2m^2 + 2\mu^2) + \frac{(m^2 - \mu^2)^2}{u}. \tag{5.5.25b}$$

For fixed u as $s \to \infty$, $\cos \theta_u \to \infty$, so the amplitude should be dominated by the Regge pole with largest $\operatorname{Re} J$. Assume that this is the neutron pole for $\pi^+ p$ backward scattering (note that for $\pi^- p$ backward scattering the u channel pole must have charge 2 so it cannot be a neutron, although it could be $N^*(1238)$).

We can make use of crossing symmetry between u and s, which tells us that the u channel expansions of f_1, f_2 are similar to (5.5.17), except that z_u replaces z and the partial waves are functions of $W_u = \sqrt{u}$ instead of $W = \sqrt{s}$, and of E_u the nucleon energy in the u channel centre of mass system. We require the contribution from the neutron pole

$$J = \alpha = \alpha_N(W_u) \quad \text{in} \quad h^e_{(J+\frac{1}{2})-}(W_u). \tag{5.5.26}$$

Using (5.5.16) with E, q, W replaced by E_u, k, W_u, the resulting contribution to (5.5.17) from the neutron pole, (isospin $\frac{1}{2}$) is

$$[f^{(\frac{1}{2})}_{1,2}(u,s)]_N = \frac{(E_u - m)\,(2k^2)^{\alpha(W_u) - \frac{1}{2}}\,b_N(W_u)}{32 W_u \cos{[\pi\alpha(W_u)]}} [P'_{\alpha \mp \frac{1}{2}}(z_u) \mp P'_{\alpha \mp \frac{1}{2}}(-z_u)]$$

$$- \frac{(E_u + m)\,(2k^2)^{\alpha(-W_u) - \frac{1}{2}}\,b_N(-W_u)}{32 W_u \cos{[\pi\alpha(-W_u)]}} [P'_{\alpha \pm \frac{1}{2}}(z_u) \pm P'_{\alpha \pm \frac{1}{2}}(-z_u)], \tag{5.5.27}$$

where $\alpha = \alpha(W_u)$ in the first term and $\alpha = \alpha(-W_u)$ in the second term; $b_N(W_u)$ is the residue at the pole of h^e in (5.5.26), and

$$E_u = (u + m^2 - \mu^2)/2u^{\frac{1}{2}}.$$

The upper signs in (5.5.27) refer to f_1, the lower to f_2.

When the neutron trajectory satisfies,

$$\alpha(W_u) = -\tfrac{1}{2} \quad \text{so that} \quad \cos \pi\alpha = 0, \tag{5.5.28}$$

it can be argued that the residue $b_N(W_u)$ should also vanish, so as to prevent a pole in the partial wave amplitude $h(W_u)$. The derivatives of Legendre polynomials in (5.5.27) produce a factor $[\alpha(W_u) + \frac{1}{2}]$ in both terms. This suggests that the differential cross-section may have a dip when (5.5.28) is satisfied. Such a dip is observed experimentally (see §9.4 and Stack, 1966). The differential cross-section in the s channel is given by the sum of (3.6.10) and (3.6.11),

$$\frac{d\sigma}{d\Omega} = |f_1 - f_2|^2 + 2 \operatorname{Re}[f_1^* f_2](\cos \theta_s - 1). \tag{5.5.29}$$

In the s channel, f_1 and f_2 are the $\pi^+ p$ scattering amplitudes. Under (s, u) crossing they become the $\pi^- p$ amplitudes, which are then represented by the nucleon Regge pole contribution (5.5.27). The latter is continued analytically to the s channel, and is assumed to be the dominant part of the amplitudes and of (5.5.29) for fixed u as $s \to \infty$. We will discuss in § 9.4 the comparison with experiment of the resulting differential cross-section.

(d) Near forward scattering, $\pi p \to \pi p$

Near to the forward direction we are concerned with fixed t and large s. This will involve Regge poles in the t channel, which in this case describes the process

$$\pi^+ + \pi^- \to \bar{p} + p. \tag{5.5.30}$$

There are two isospin states, $I = 0$ and $I = 1$, in this channel. In order to carry out the analytic continuations in J of the partial amplitudes, one must introduce even J parity and odd J parity amplitudes. In the t channel, crossing symmetry (Bose statistics for the pions) leads to the result that, for isospin $I = 0$, the odd J parity amplitudes are identically zero. Also, for $I = 1$, the even J parity amplitudes must be zero. Thus the ρ meson ($I = 1$, odd J) can occur as an intermediate state; but the ω meson cannot occur since it has $I = 0$ and odd J parity.

We will omit the isospin indices in the following. They can be inserted and crossing relations can be established by the methods given in § 4.9. The total cross-section is given by,

$$\sigma(\pi p, \text{total}) = \frac{4\pi W}{m p_1} \operatorname{Im} (f_1 + f_2)|_{t=0} \tag{5.5.31}$$

where p_1 is the laboratory momentum of the pion. The differential cross-section (5.5.29) can be expressed in terms of the invariant amplitudes A and B (see (3.6.13, 14)). We have

$$\frac{d\sigma}{d\Omega} = |f_1 + f_2|^2 + \frac{t}{q^2} \operatorname{Re} (f_1^* f_2), \tag{5.5.32}$$

giving,

$$\frac{d\sigma}{d\Omega} = \left(\frac{m}{4\pi W} \right)^2 \left[\left(1 - \frac{t}{4m^2} \right) |A'|^2 + \frac{t}{4m^2} \left\{ s - \frac{(m+\nu)^2}{1 - t/4m^2} \right\} |B|^2 \right], \tag{5.5.33}$$

where ν is the laboratory energy of the pion and

$$A' = A + \left(\frac{\nu + t/(4m)}{1 - t/(4m^2)} \right) B. \tag{5.5.34}$$

The total cross-section is given by

$$\sigma(\pi p, \text{total}) = \frac{1}{p_1} \text{Im}\, [A'(s,t)]_{t=0}. \qquad (5.5.35)$$

The amplitude A' can be expressed in terms of even and odd amplitudes, $A'^{(+)}$, $A'^{(-)}$. In the t channel, the even (odd) amplitude $A'^{(+,-)}$ can be expanded in terms of even (odd) partial waves $f_+^{+,J}(\sqrt{t})$, $f_+^{-,J}(\sqrt{t})$, (in the notation of Singh, 1963 and Frazer & Fulco, 1960). Similarly, the even and odd amplitudes $B^{(+)}$ and $B^{(-)}$ can be expressed in terms of even and odd partial waves $f_+^{+,J}$ and $f_-^{-,J}$. After applying the Sommerfeld–Watson transform one obtains, (as $s \to \infty$, for fixed t), terms that correspond to Regge poles in the t channel. These give

$$A'^{(\pm)} \sim C_{\mp}^{\pm}(t) \left(\frac{s}{2m}\right)^{\alpha\pm} \left[\frac{1 \pm \exp\left(-i\pi\alpha^\pm\right)}{\sin\left(\pi\alpha^\pm\right)}\right], \qquad (5.5.36)$$

$$B^{(\pm)} \sim [\alpha^\pm(t)]\, C_{\mp}^{\pm}(t) \left(\frac{s}{2m}\right)^{\alpha\pm-1} \left[\frac{1 \pm \exp\left(-i\pi\alpha^\pm\right)}{\sin\left(\pi\alpha^\pm\right)}\right], \quad (5.5.37)$$

where $\alpha^+(t)$ is the pole having the largest real part in the even J, isospin 0, partial wave. Similarly $\alpha^-(t)$ is the pole with the largest real part in the odd J, isospin 1, partial wave.

The corresponding asymptotic forms of the total cross-sections give

$$\sigma_t(\pi^+ p) + \sigma_t(\pi^- p) \sim C_+^+(0) \left(\frac{s}{2m}\right)^{\alpha^+(0)-1}, \qquad (5.5.38)$$

$$\sigma_t(\pi^+ p) - \sigma_t(\pi^- p) \sim C_+^-(0) \left(\frac{s}{2m}\right)^{\alpha^-(0)-1}. \qquad (5.5.39)$$

It is evident from (5.5.38) that there must be a contribution from the Pomeron if the total cross-sections are asymptotic to a non-zero constant, and that

$$\alpha_P(0) = \alpha^+(0) = 1; \quad \tau = +, \ I = 0. \qquad (5.5.40)$$

The experimental comparison of these results will be considered in Chapter 9.

5.6 Branch cuts and other features

In this section we will briefly consider a few of the complications in Regge theory that have not yet been fully taken into account. These are related to the fundamental problems mentioned earlier, which are: (1) the proof that an analytic continuation of partial wave amplitudes exists (this result has been established only from the Mandelstam representation with certain convergence assumptions related to the number of subtractions), (2) the proof that the analytic partial waves $f(l, s)$, if they exist, have the desired asymptotic properties to permit a Sommerfeld–Watson transform, (3) the proof that the background integral along $\mathrm{Re}\,(l) = -\frac{1}{2}$ contributes less to $F(s, t)$ for large t than pole terms in $\mathrm{Re}\,(l) > -\frac{1}{2}$, (4) the proof that $f(l, s)$ has the desired meromorphy properties in $\mathrm{Re}\,(l) > -\frac{1}{2}$ (or that it can be modified so as to have them).

We will consider difficulties that may arise out of consistency requirements, which suggest that the properties in (1) to (4) above are too simple to be compatible both with each other and with analyticity and unitarity.

(a) Branch cuts for complex l

The possibility of branch points of the partial wave amplitude $f(l, s)$ in $\mathrm{Re}\,(l) > -\frac{1}{2}$, was first indicated from iterations of Feynman diagrams in perturbation theory. In a more sophisticated treatment these branch points appear to follow from Regge pole assumptions, analyticity assumptions and unitarity (Mandelstam, 1963). The iteration of single particle exchange gives a sum of ladder diagrams, Fig. 5.6.1 (a), indicated symbolically by the single ladder (b). This iteration can be solved as a Bethe–Salpeter equation, and shown to have asymptotic properties of Regge theory (Lee & Sawyer, 1962). Alternatively, one can use the perturbation methods noted in §5.2 to prove the same result (Polkinghorne, 1963b; Federbush & Grisaru, 1963; and ELOP, 1966). A pole in $f(l, s)$, from this infinite iteration, produces asymptotic behaviour

$$F(s, t) \sim C(s)\, t^{\alpha(s)}, \tag{5.6.1}$$

where $\alpha(s)$ denotes the pole having the largest real part.

It was first noted by Amati, Fubini & Stangellini (1962) that combinations of Regge poles may produce branch cuts in the complex l-plane. In particular they considered results from the combination shown in Fig. 5.6.1 (c). The branch cuts from this combination were subsequently shown to cancel out (Mandelstam, 1963; Polkinghorne, 1963a, d) at least on the physical sheet. However, Mandelstam (1963) has shown that this cancellation is very unlikely to occur for the combination of Regge poles (the ladders) shown in Fig. 5.6.1 (d).

Although branch cuts introduce a considerable complication, they also bring a simplification of a difficulty that is connected with a possible essential singularity at $l = -1$. It has been suggested by

Gribov & Pomeranchuk (1962*d*) that a paradox, that arises from a pole at $l = -1$, may be resolved by an accumulation of poles giving an essential singularity. The existence of a branch cut at $l = 1$ will serve to 'screen off' this essential singularity, so that it can be ignored for most practical purposes (Mandelstam, 1963). We consider next the less satisfactory consequences of the branch cuts.

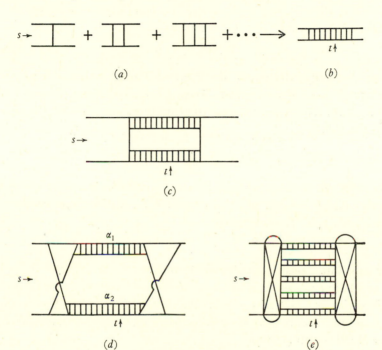

Fig. 5.6.1. (*a*) Ladder diagrams whose sum leads to a Regge pole indicated by diagram (*b*); (*c*) a combination of two Regge poles that does not lead to a branch cut; (*d*) a diagram giving a Mandelstam–Regge branch cut; (*e*) another diagram that is expected to give a branch cut in $f(l, s)$ in the l-plane.

If we denote by W_1 the part of the total energy $W = \sqrt{s}$, carried by the Regge pole α_1 in Fig. 5.6.1 (*d*), then

$$\alpha_1 = \alpha_1(W_1), \quad \alpha_2 = \alpha_2(W - W_1). \tag{5.6.2}$$

The combination of Regge poles in Fig. 5.6.1 (*d*) would produce a pole at

$$\alpha_1 + \alpha_2 - 1, \tag{5.6.3}$$

if W_1 was fixed. However W_1 can vary between certain limits that depend on W and the masses involved. The consequent integration over W_1 converts the pole into a branch cut. The end of the cut, the branch point, lies at β the maximum of (5.6.3),

$$\beta(W) = \text{max.} \, [\alpha_1(W) + \alpha_2(W - W_1) - 1]. \qquad (5.6.4)$$

In general if α_1 carries the required quantum numbers for the s channel, then α_2 can be the Pomeron having $\alpha(0) = 1$. If the value $W_1 = W$ is allowed by the conditions on the range of integration, one might expect to find a branch point located at

$$\beta_1(W) = \alpha_1(W). \qquad (5.6.5)$$

More generally, diagrams like 5.6.1(e) should be considered, with energy W shared between all poles. It can be argued (Mandelstam, 1963) that, when vacuum quantum numbers are allowed, the branch point satisfies

$$\beta(s) = \beta(0) = 1 \quad \text{for} \quad s \leqslant 0, \qquad (5.6.6)$$

$$\beta(s) \sim s^{\frac{1}{2}} \quad \text{as} \quad s \to +\infty. \qquad (5.6.7)$$

The difficulty consequent from this last possibility is that it would lead to an infinite number of subtractions in the t dispersion relation for $F(s, t)$ as $s \to \infty$. It would give

$$F(s, t) \sim t^{\sqrt{s}} \quad \text{as} \quad t \to \infty, \qquad (5.6.8)$$

for any large positive value of s, no matter how large. Thus the Mandelstam representation would not hold because of subtraction difficulties as $s \to +\infty$, $t \to +\infty$. Further aspects of the Mandelstam–Regge branch cuts have been considered by Gribov, Pomeranchuk & Ter Martirosyan (1965).

If there are branch points for $f(l, s)$ in the complex l-plane, they will affect the physical interpretation of asymptotic behaviour. Consider a branch point in Re $(l) > -\frac{1}{2}$ at

$$l = \alpha_c(s). \qquad (5.6.9)$$

If this was a pole, it would lead to a large t behaviour of $F(s, t)$ like t^{α_c}. However, since it is a branch point, there will be an integration, in the l-plane, of the discontinuity across the attached branch cut. This gives for large t,

$$F(s, t) \sim B(s) \, t^{\alpha_c(s)} (\log t)^{-\gamma_c(s)}, \qquad (5.6.10)$$

where $\gamma_c > 0$, and γ_c depends on the discontinuity across the branch cut. If the cut is fixed at $\alpha_c(0)$ when $s < 0$, as has been suggested, we will have for $s < 0$, $t \to \infty$,

$$F(s, t) \sim B(s) \, t^{\alpha_c(0)} (\log t)^{-\gamma_c(s)}. \qquad (5.6.11)$$

In particular, if Pomeron exchange is allowed,

$$\alpha_c(0) = \alpha_P(0) = 1, \tag{5.6.12}$$

and the branch cut contribution will dominate over any pole contribution in $s < 0$ as $t \to \infty$. Only at $s = 0$ would the Pomeron pole dominate over the branch cut.

This consequence of the (possible) existence of branch cuts is not incompatible with experimental evidence, but it places most of the variation of the cross-section with angle in the coefficient $B(s)$ in (5.6.10) for $s < 0$ (note that s is the momentum transfer variable). Regge theory gives little information about the s dependence of $B(s)$, so the theory would become rather unsatisfactory.

The procedure at the present time is to assume that if branch cut contributions exist, they are small in comparison with the dominant pole contributions at the energies of existing experiments. Therefore the experimental comparison of Regge theory considers only contributions from the leading Regge poles in the first instance. In Chapter 9 we will note the success of this procedure but will also note some places where branch cuts may be relevant.

(b) Special features at $s = 0$

For Regge poles in the s channel, there are certain special features that arise for large t, near the forward direction, $\cos \theta_t = 1$. One of these is special to unequal masses of the colliding particles, the other is a consequence of the additional symmetry in the forward direction.

Unequal masses

As an example we will consider pion-nucleon scattering. Then, in the s channel centre of mass system,

$$\cos \theta_s = 1 + \frac{t}{2q^2}, \tag{5.6.13}$$

$$4q^2 = s - 2(m^2 + \mu^2) + \frac{(m^2 - \mu^2)^2}{s}. \tag{5.6.14}$$

For fixed q^2 as $t \to \infty$, we will in general have

$$\cos \theta_s \to \infty \quad \text{as} \quad t \to \infty, \tag{5.6.15}$$

$$F(s, t) \sim C(s) P_\alpha(\cos \theta_s) \sim C(s) \left(\frac{t}{2q^2}\right)^\alpha, \tag{5.6.16}$$

where α is the dominant pole. However, near $s = 0$ there is a difficulty, since if

$$s \sim \frac{(m^2 - \mu^2)^2}{t}, \tag{5.6.17}$$

we find that $\cos\theta_s$ remains finite even when t is large. This difficulty is related to the singularity in the transformation

$$(s, t) \leftrightarrow (s, \cos\theta_s), \qquad (5.6.18)$$

at $s = 0$.

It is evident that special care must be taken in using Regge theory near to a singular point (Frautschi, Gell-Mann & Zachariasen, 1962; and Atkinson & Barger, 1965). Some of the difficulties that arise in the usual formulation of Regge theory near $s = 0$ have been discussed by Goldberger & Jones (1966), Freedman & Wang (1966), and by Leader & Omnes (1966), (see also Low, 1966).

Symmetry conditions

In the forward direction the scattering amplitude must have an additional symmetry. In nucleon-nucleon scattering there are normally five independent invariant scattering amplitudes (Goldberger *et al.* 1960), but in the forward direction there are only three. In particular there is no polarization and the spin correlations take on a simpler form than for non-forward scattering. The consequences for Regge theory of this additional symmetry were noted by Volkov & Gribov (1963) and have been studied further by Gell-Mann & Leader (1966). We will indicate the results of Volkov & Gribov here.

Nucleon-nucleon scattering amplitudes involve five partial wave amplitudes that depend on J and W,

$$f_0(J, W), \quad f_1(J, W), \quad f_{00}(J, W), \quad f_{01}(J, W), \quad f_{11}(J, W), \quad (5.6.19)$$

where f_0, f_1 refer to scattering in singlet states with even and odd parity respectively; f_{00} and f_{11} refer to scattering in triplet states, and f_{01} to the transition between the two triplet states.

The symmetry conditions for forward scattering imply that there are relations between the partial wave amplitudes (5.6.19) for integer J. These generalise to give relations for complex J at $s = 0$; for example

$$f_0(J-1, W) - f_0(J+1, W) - \frac{J-1}{J} f_{11}(J-1, W) + \frac{J+2}{J+1} f_{11}(J+1, W)$$

$$- \frac{2J+1}{J(J+1)} f_1(J, W) = 0. \quad (5.6.20)$$

Each f is a function of $s = W^2$. Relations like (5.6.20) imply that the positions of the singularities (Regge poles) of different states are related at $s = 0$. There is not an obviously unique relation between these Regge poles at $s = 0$, and various possibilities are considered by the authors quoted above. These possibilities are described as 'conspiracy' or 'evasion', but their experimental consequences are not yet fully worked out so the interested reader should refer to the forthcoming papers on these topics (for example, Gell-Mann & Leader, 1966).

CHAPTER 6

ASYMPTOTIC BOUNDS ON THE BEHAVIOUR OF CROSS-SECTIONS

The first derivation of a bound on the asymptotic behaviour of total cross-sections at high energy was given by Froissart (1961 b). He deduced from the Mandelstam representation that $\sigma(\text{total}) < C \log^2 E$ as $E \to \infty$. Later Martin (1966 a) established the Froissart bound using results that have been proved from the axioms of quantum field theory (see §4.8). In this chapter we will describe the principal upper and lower bounds that have been proved for cross-sections or for scattering amplitudes. Many of these results are due to Martin and collaborators and references will be given in context.

6.1 Bounds on total cross-sections

The Froissart bound is

$$\sigma(\text{total}) < C \log^2 (s/s_0) \quad \text{as} \quad s \to \infty, \tag{6.1.1}$$

where s is the energy squared in the centre of mass system and s_0 is an unknown constant. We will frequently take units so that $s_0 = 1$, but this is solely to simplify the notation.

The intuitive justification of the Froissart bound can be seen from the following model: Suppose the target particle has a probability density distribution
$$P(r) = P_0 \exp(-br), \tag{6.1.2}$$

at a distance r from its mean position. Now assume that for an incident particle of energy E, the probability of interaction is bounded by some power E^N, where N is fixed. Then the probability of an interaction at distance r between the incident particle and the target centre will satisfy the inequality
$$P(E, r) < P_0 E^N \exp(-br). \tag{6.1.3}$$

The interaction will be negligible for $r > r_0$, where

$$r_0 = \frac{N}{b} \log E, \tag{6.1.4}$$

giving a bound on the total cross-section

$$\sigma(\text{total}) < r_0^2 < C \log^2 E. \tag{6.1.5}$$

Since E is proportional to s, this indicates the result (6.1.1). The following rigorous derivation is due to Martin (1963a,b, 1965b and 1966a). We will consider the case of equal masses for simplicity, but this is not essential for the validity of the Froissart bound.

The rigorous derivation of the Froissart bound is based on unitarity and the domain of convergence of the partial wave series for the imaginary part of the amplitude. Write

$$8\pi A(s, \cos\theta) = \operatorname{Im} F(s, t). \tag{6.1.6}$$

Then the series

$$A(s, \cos\theta) = \frac{s^{\frac{1}{2}}}{k} \sum_0^\infty (2l+1)\, a_l(s)\, P_l(\cos\theta), \tag{6.1.7}$$

where $a_l(s) = \operatorname{Im} f_l(s)$, will converge for fixed s, when $\cos\theta = z$ is inside the Martin–Lehmann ellipse (see § 4.8). This is an ellipse with foci at -1 and $+1$, which extends to a positive point $\cos\theta = z_0$ that for equal masses is at

$$z_0 = 1 + \frac{2t_0}{s - 4m^2}, \tag{6.1.8}$$

where t_0 is the nearest threshold, $t_0 = 4m^2$. From unitarity $a_l(s)$ is positive, and (see (3.5.18)),

$$0 \leqslant |f_l(s)|^2 \leqslant a_l(s) \leqslant 1. \tag{6.1.9}$$

Since $A(s, \cos\theta)$ is regular for $\cos\theta < z_0$, it is bounded there for each fixed s. But it has also been proved from quantum field theory (see § 4.8) that as $s \to \infty$, $F(s, t)$ is bounded by a polynomial in s, when $t < t_0$. Therefore, putting $z = z_1 < z_0$ in (6.1.7), as $s \to \infty$

$$\frac{s^{\frac{1}{2}}}{k} \sum_{l=0}^\infty (2l+1)\, a_l(s)\, P_l\!\left(1 + \frac{t_1}{2k^2}\right) < s^N, \tag{6.1.10}$$

where N is independent of s.

Since each term in the series (6.1.10) is positive, we can deduce that, as $s \to \infty$,

$$(2l+1)\, a_l(s)\, P_l\!\left(1 + \frac{t_1}{2k^2}\right) < s^N. \tag{6.1.11}$$

For large l and $x > 1$,

$$P_l(x) > \frac{C'}{(2l+1)^{\frac{1}{2}}} [1 + (2x-2)^{\frac{1}{2}}]^l. \tag{6.1.12}$$

Hence for large l,

$$a_l(s) < \frac{C' s^N}{(2l+1)^{\frac{1}{2}}} [1 + (t_1/k^2)^{\frac{1}{2}}]^{-l}, \tag{6.1.13}$$

giving
$$a_l(s) < C'' \exp\left[-2l(t_0/s)^{\frac{1}{2}}(1-\epsilon(s)) + N \log s\right], \qquad (6.1.14)$$

where $\epsilon \to 0$ as $s \to \infty$.

From (6.1.9) and (6.1.14) we see that both the real and the imaginary parts of the partial wave amplitudes $f_l(s)$ decrease exponentially with l, and become negligible when s is large and

$$l > L = Cs^{\frac{1}{2}} \log s. \qquad (6.1.15)$$

For $|t| < \alpha t_0$ where $0 < \alpha < 1$, we can neglect all terms in the series (6.1.7) for $A(s, \cos \theta)$ which have $l > L$. The same conclusion holds for $\cos \theta$ inside the Martin–Lehmann ellipse (at a distance from its boundary not less than c/s, where c is a positive constant). On account of (6.1.9), a similar result applies to the partial wave series for $F(s, t)$,

$$F(s, t) = \frac{8\pi s^{\frac{1}{2}}}{k} \sum_{l=0}^{L} (2l+1) f_l(s) P_l(\cos \theta) + R_L(s, t), \qquad (6.1.16)$$

where
$$|R_L(s, t)| < s^{-M} \quad \text{as} \quad s \to \infty. \qquad (6.1.17)$$

By choice of C in (6.1.15) we can make M arbitrarily large. This result holds, in particular, for the physical region,

$$-1 \leqslant \cos \theta \leqslant 1. \qquad (6.1.18)$$

From the optical theorem, for scattering of equal mass bosons,

$$\sigma(\text{total}) = \frac{4\pi}{ks^{\frac{1}{2}}} A(s, 1) = \frac{1}{2ks^{\frac{1}{2}}} \operatorname{Im} F(s, 0). \qquad (6.1.19)$$

For $t = 0$ ($\cos \theta = 1$), from (6.1.16) using (6.1.9),

$$A(s, 1) < \frac{s^{\frac{1}{2}}}{k} \sum_{0}^{L} (2l+1) \sim 2L^2 \quad \text{as} \quad s \to \infty. \qquad (6.1.20)$$

This gives the Froissart bound,

$$\sigma(\text{total}) < C \log^2 s, \qquad (6.1.21)$$

where C is a constant.

It has been shown by Łukaszuk & Martin (1967) (see also Jin & Martin (1964)) that as $s \to \infty$,

$$C \leqslant \frac{\pi}{\mu^2} \quad (\mu = \text{pion mass}). \qquad (6.1.22)$$

This bound on C applies when s is sufficiently large, except possibly on intervals whose mean length tends to zero as $s \to \infty$. We will see in § 6.7 that the Froissart bound continues to hold when the particles have non-zero spin.

The importance of the Martin extension of the Lehmann ellipse becomes apparent by considering (6.1.14). In the analyticity domain given by the Lehmann ellipse $t_0 \sim c/s$, so that (6.1.15) has $L \sim Cs \log s$. This yields the Greenberg–Low bound (1961),

$$\sigma(\text{total}) < Cs \log^2 s. \tag{6.1.23}$$

The Greenberg–Low bound was improved by four powers of $\log s$, by Martin (1965b) and Eden (1966a), to give

$$\sigma(\text{total}) < \frac{Cs}{\log^2 s}. \tag{6.1.24}$$

This result establishes the validity of forward dispersion relations with only two subtractions. This provides a possible starting point for Martin's extension of the Lehmann ellipse which we describe in the next chapter, and which is necessary for the derivation of the Froissart bound.

The forward scattering amplitude also gives an upper bound for $A \equiv \text{Im} F(s,t)$, since

$$|P_l(\cos\theta)| < 1 \quad \text{for} \quad -1 < \cos\theta < 1. \tag{6.1.25}$$

This shows that, for $4m^2 - s < t < 0$,

$$|\text{Im} F(s,t)| < \text{Im} F(s,0) \quad \text{as} \quad s \to \infty, \tag{6.1.26}$$

and hence $\text{Im} F(s,t)$ is bounded by $Cs \log^2 s$ as $s \to \infty$. Using crossing symmetry, (see §4.3), the amplitude

$$F(-s - i\epsilon, 0) \tag{6.1.27}$$

will denote the physical scattering amplitude for the crossed reaction $s \leftrightarrow u$, in the forward direction. For equal mass scalar bosons, the crossed reaction and the direct reaction are the same. Thus

$$-\text{Im} F(-s + i\epsilon, 0) \tag{6.1.28}$$

is bounded by $Cs \log^2 s$ as $s \to \infty$, using the Froissart bound for the crossed channel.

We therefore see that the Froissart bound plus crossing establishes convergence of the dispersion relation, for $t \leqslant 0$,

$$F(s,t) = C_0 + C_1 s + \frac{s^2}{\pi} \int_{4m^2}^{\infty} dx \frac{\text{Im} F(x,t)}{x^2(x-s)} + \frac{s^2}{\pi} \int_{-\infty}^{-t} dx \frac{\text{Im} F(x,t)}{x^2(x-s)}. \tag{6.1.29}$$

The improved Greenberg–Low bound (6.1.24) is also sufficient to ensure that this dispersion relation converges with only two subtractions. However with the Froissart bound (6.1.21) a stronger result can be obtained. Jin & Martin (1964) have shown, using the Froissart bound, that only two subtractions are required, provided that $t < 4\mu^2$ (μ = pion mass). Thus

$$|F(s,t)| < s^{2-\epsilon} \quad \text{with } \epsilon > 0 \text{ for } t < 4\mu^2. \qquad (6.1.30)$$

This is an important result since it shows that no Regge trajectory can have $l = \alpha(t)$ exceed 2, when t is below the threshold $t = 4\mu^2$.

It should be noted that our assumption of equal masses is not essential for the validity of the Froissart bound. We will also see later that the bound continues to hold when the colliding particles have spin.

6.2   Inequalities involving the forward peak

A number of useful inequalities can be obtained directly from the partial wave series (6.1.7), using unitarity (6.1.9) and the properties of Legendre polynomials. These will be given in outline here (for fuller details see Eden, 1966c and 1967).

For large s, $A = \operatorname{Im} F(s,t)/8\pi$ is given by,

$$A(s,t) \sim 2 \sum_0^L (2l+1)\, a_l(s)\, P_l(1+2t/s), \qquad (6.2.1)$$

where $\qquad L = Cs^{\frac{1}{2}}\log s \quad \text{and} \quad a_l = \operatorname{Im} f_l \geqslant 0.$

Rearrange this series to give

$$A(s,t) \sim 4 \sum_0^L b_n(s)\, \frac{t^n}{(n!)^2}, \qquad (6.2.2)$$

where $b_n(s)$ is defined by this relation. Then

$$b_0(s) \sim \sum_1^L l a_l(s), \qquad (6.2.3)$$

$$b_1(s) \sim \frac{1}{s} \sum_1^L l^2(l-1)\, a_l \sim \frac{1}{s} \sum_1^L l^3 a_l, \qquad (6.2.4)$$

$$b_n(s) \sim \frac{n!}{4} \left(\frac{d}{dt}\right)^n A(s,t)\bigg|_{t=0} \sim \frac{1}{s^n} \sum_1^L l^{2n+1} a_l. \qquad (6.2.5)$$

The last asymptotic equality holds when n is restricted so that

$$n \ll L \sim Cs^{\frac{1}{2}}\log s; \qquad (6.2.6)$$

however, we will later take s and L to be arbitrarily large. The optical theorem leads to the relation,

$$b_0(s) \sim \frac{s\sigma_t}{32\pi}, \tag{6.2.7}$$

where σ_t denotes $\sigma(\text{total})$. Define $g(s)$ as the logarithmic derivative of $A(s, t)$ at $t = 0$, written as

$$g(s) = \frac{d}{dt} \log A(s, 0). \tag{6.2.8}$$

Then

$$b_1(s) = g(s) b_0(s), \left. \phantom{\frac{}{}} \right\} \atop \sim s g \sigma_t / 32\pi. \tag{6.2.9}$$

Since $0 \leqslant a_l \leqslant 1$,

$$\left[2 \sum_0^L l a_l \sum_0^{l-1} m a_m - \sum_0^L l^3 a_l \right] \leqslant 0. \tag{6.2.10}$$

Taking only the leading terms, this gives

$$b_0^2 - s b_1 \leqslant 0, \quad \text{i.e. } b_0^2 - s g b_0 \leqslant 0, \tag{6.2.11}$$

$$\sigma(\text{total}) \leqslant 32\pi g(s). \tag{6.2.12}$$

This inequality between the total cross-section and the logarithmic derivative was observed to follow from Regge theory by Leader (1963) (see also Martin, 1963b). Leader's method makes use of the inequality

$$\sigma(\text{elastic}) \leqslant \sigma(\text{total}), \tag{6.2.13}$$

and the assumption that $F(s, t)$ has the Regge form

$$F(s, t) = F(s, 0) s^{\alpha(t) - \alpha(0)}. \tag{6.2.14}$$

For small t, we can approximate

$$s^{\alpha(t) - \alpha(0)} \sim \exp{(\alpha' t \log s)}, \tag{6.2.15}$$

where α' denotes the derivative of $\alpha(t)$ at $t = 0$. The elastic cross-section is given by,

$$\sigma(\text{elastic}) = \frac{4\pi}{s(s - 4m^2)} \int_{-s + 4m^2}^{0} dt \left| \frac{F(s, t)}{8\pi} \right|^2. \tag{6.2.16}$$

With the approximation (6.2.15)

$$\sigma(\text{elastic}) \sim \left| \frac{F(s, 0)}{8\pi s} \right|^2 \frac{4\pi}{2\alpha' \log s}. \tag{6.2.17}$$

Now

$$\sigma(\text{total}) \sim \frac{\text{Im} \, F(s, 0)}{s} < \frac{|F(s, 0)|}{s}. \tag{6.2.18}$$

Hence, from (6.2.13)

$$\frac{(\sigma_t)^2}{32\pi\alpha'\log s} \leqslant \sigma_t,$$

$$\sigma_t \leqslant 32\pi\alpha'\log s.$$

$$(6.2.19)$$

This is a special case of the general result (6.2.12), since for Regge theory $g(s) = \alpha'\log s$.

Since α' in (6.2.19) is independent of s, we see that the assumption of a Regge behaviour in the forward peak leads to a stronger bound than that of Froissart. This 'improved' bound can be attained if the scattering amplitude has the form

$$F(s,t) = iC(\log s)\, s^{\alpha(t)} \quad \text{with } \alpha(0) = 1. \qquad (6.2.20)$$

Provided $\alpha(t) < 1$ for $t < 0$, and $C < 32\pi\alpha'$, this amplitude satisfies unitarity in the s channel. It can also be given the correct analyticity in t for small t, and can even be made crossing symmetric under $s \leftrightarrow u$, (as in Chapter 7). It is therefore a useful counter-example to any 'improvement' towards a constant cross-section that does not use additional information (like crossing $s \leftrightarrow t$). A more detailed discussion of counter-examples is given by Kinoshita, Loeffel & Martin (1964).

The general case does not require a bound so strong as (6.2.19). Indeed we can have $g(s) \sim \log^2 s$ in (6.2.12) which allows the possibility of saturation of the Froissart bound. The corresponding series for $A(s,t)$ can be studied from (6.2.2–6.2.9) using Cauchy's inequality for real x_n and y_n,

$$\sum_0^N x_n^2 \sum_0^N y_n^2 \geqslant \left[\sum_0^N x_n y_n\right]^2. \qquad (6.2.21)$$

From the series for b_2 and b_0,

$$b_2 b_0 \sim \frac{1}{s^2}\sum_1^L l^5 a_l \sum_1^L l a_l \geqslant \left[\frac{1}{s}\sum_1^L l^3 a_l\right]^2 \sim (b_1)^2 \sim g^2 b_0^2. \qquad (6.2.22)$$

More generally, provided $n \ll L$,

$$b_n \geqslant g^n b_0 \sim g^n s\sigma_t/32\pi. \qquad (6.2.23)$$

From (6.2.21) we have,

$$b_{n+1} b_{n-1} \geqslant b_n^2. \qquad (6.2.24)$$

Writing $b_n = g^n c_n b_0$, this becomes

$$\frac{c_{n+1}}{c_n} \geqslant \frac{c_n}{c_{n-1}}. \qquad (6.2.25)$$

For $t \geqslant 0$,

$$A(s,t) \sim \frac{s\sigma_t}{8\pi}\sum_0^L \frac{c_n(s)}{(n!)^2}[tg(s)]^n. \qquad (6.2.26)$$

By taking $c_n = (n!)^\epsilon$, and $g = (\log s)^{2-\epsilon}$, $\sigma_t = (\log s)^{2-\epsilon}$, $\epsilon > 0$, we obtain a series that nearly saturates the Froissart bound. If we had taken $c_n = 1$, the resulting series would contradict unitarity. However with $\epsilon > 0$, it is probable that 'local' unitarity (in the s channel) will be satisfied.

It is instructive to re-derive the Froissart bound from the series (6.2.26) using the above inequalities and the properties of entire functions (note that as $s \to \infty$ in (6.2.26) $L \to \infty$). For $0 \leqslant t \leqslant 4\mu^2$,

$$A(s,t) > \frac{s\sigma_t}{8\pi} \sum_0^L \frac{(tg)^n}{(n!)^2}. \tag{6.2.27}$$

If $g(s)$ is bounded as $s \to \infty$, we obtain $\sigma_t <$ constant from (6.2.12), so we will assume $g(s) \to \infty$ as $s \to \infty$. Then for fixed $t > 0$, and large s, (6.2.27) is equivalent to

$$A(s,t) > \frac{s\sigma_t}{8\pi} \exp\left[(tg)^{\frac{1}{2}}\right]. \tag{6.2.28}$$

However, $A(s,t)$ must be bounded by a polynomial in s; hence (6.2.28) requires that

$$g(s) \leqslant \log^2 s, \tag{6.2.29}$$

from which the result (6.2.12) gives the Froissart bound.

6.3   Elastic cross-sections

Using the fact, established in § 6.1, that only the first L partial waves contribute significantly to the total and the elastic cross-sections, MacDowell & Martin (1964) have obtained a lower bound on σ(elastic), in terms of σ(total) and the logarithmic derivative $g(s)$ defined in (6.2.8). In terms of partial wave amplitudes f_l, the elastic cross-section is given, in the case of equal masses, by

$$\sigma(\text{elastic}) = \frac{4\pi}{k^2} \sum_0^\infty (2l+1) |f_l(s)|^2, \tag{6.3.1}$$

where $s = 4(m^2 + k^2)$. Write a_l for $\text{Im} f_l$,

$$f_l(s) = c_l + ia_l. \tag{6.3.2}$$

Then $\quad\quad \sigma(\text{elastic}) > \sigma(\text{el. im.}) = \dfrac{4\pi}{k^2} \sum_0^\infty (2l+1)(a_l)^2, \tag{6.3.3}$

where el. im. stands for 'elastic imaginary', and σ(el. im.) is defined by this equation. The total cross-section is

$$\sigma_t = \frac{4\pi}{k^2} \sum_0^L (2l+1) a_l, \tag{6.3.4}$$

with neglect of a term smaller than s^{-M} with $M > 2$ (say). The logarithmic derivative g is given in terms of $b_1(s)$ by (6.2.4), or more accurately,

$$b_1(s) = \frac{1}{4k^2} \sum_1^L (l+\tfrac{1}{2}) l(l+1) a_l = sg\sigma_t/32\pi. \qquad (6.3.5)$$

We will approximate by taking only the leading terms, so that

$$b_1(s) \sim \frac{1}{s} \sum_1^L l^3 a_l, \qquad (6.3.6)$$

$$\sigma(\text{el. im.}) \sim \frac{32\pi}{s} \sum_1^L l a_l^2, \qquad (6.3.7)$$

$$\sigma_t \sim \frac{32\pi}{s} \sum_1^L l a_l. \qquad (6.3.8)$$

Now minimise $b_1(s)$, assuming that $\sigma(\text{el. im.})$ and σ_t are given constants, and that a_l ($l = 1, ..., L$) are variables that are restricted by the constraints (6.3.7), (6.3.8), with the additional unitarity constraint $0 \leqslant a_l \leqslant 1$. With Lagrange multipliers p and q, we obtain

$$l^3 - 64\pi p l a_l - 32\pi q l = 0, \qquad (6.3.9)$$

which can be rewritten with parameters λ, μ, as

$$a_l = \lambda - \mu l^2, \qquad (6.3.10)$$

subject to $0 \leqslant a_l \leqslant 1$.

There are now two possibilities: (a) $\lambda > 1$, and (b) $\lambda < 1$. The solution in case (a) leads to

$$\sigma(\text{el. im}) \geqslant \tfrac{2}{3}\sigma(\text{total}), \qquad (6.3.11)$$

which is unlikely on physical grounds, since it does not agree with experimental results at high energies for $\sigma(\text{elastic})$ and $\sigma(\text{total})$. The solution in case (b) leads to

$$a_l = \lambda - \mu l^2 \quad \text{for} \quad l^2 < \left(\frac{\lambda}{\mu}\right), \qquad (6.3.12)$$

$$a_l = 0 \qquad \text{for} \quad l^2 > \left(\frac{\lambda}{\mu}\right). \qquad (6.3.13)$$

Substitute in (6.3.7, 8) and replace sums by integrals (valid to leading order when s is large), giving

$$\sigma_t = \frac{32\pi}{s} \left(\frac{\lambda^2}{4\mu}\right), \qquad (6.3.14)$$

$$\sigma(\text{el. im.}) = \frac{32\pi}{s} \left(\frac{\lambda^3}{6\mu}\right). \qquad (6.3.15)$$

Bound $b_1(s)$ by its minimum value, obtained by substituting (6.3.12, 13) in (6.3.6) and replacing the sum by an integral. Then use (6.2.9) to give

$$g(s)\,\sigma_t(s) > \frac{32\pi}{s^2}\left(\frac{\lambda^3}{12\mu^2}\right). \tag{6.3.16}$$

Using (6.3.14, 15)

$$g(s) > \frac{\lambda}{3\mu s} = \frac{\sigma_t}{36\pi}\left\{\frac{\sigma_t}{\sigma(\text{el. im.})}\right\}. \tag{6.3.17}$$

Hence as $s \to \infty$,

$$\sigma(\text{elastic}) > \left[\frac{\sigma_t(s)}{36\pi g(s)}\right]\sigma_t(s), \tag{6.3.18}$$

where

$$g(s) = \frac{d}{dt}\log\left(\text{Im}\,F(s,t)\right)\Big|_{t=0}. \tag{6.3.19}$$

In deriving the result (6.3.18) we have neglected terms of order $1/s$ compared with the leading term. If these are included one obtains (MacDowell & Martin, 1964), instead of (6.3.17),

$$g(s) > \frac{\sigma_t^2}{36\pi\sigma(\text{el. im.})} - \frac{1}{9k^2}. \tag{6.3.20}$$

This inequality can be compared directly with experiment, and it is found to be nearly saturated by observed experimental values for g, σ_t and $\sigma(\text{elastic})$.

A somewhat weaker bound on $\sigma(\text{elastic})$ can be obtained from (6.3.18) using (6.2.29), $g \leqslant \log^2 s$,

$$\sigma(\text{elastic}) > C\frac{\sigma_t^2}{\log^2 s}. \tag{6.3.21}$$

This is of course a weaker statement than (6.3.18). For example, if Regge theory is valid we have $g(s) \sim \log s$, so that (6.3.18) becomes

$$\sigma(\text{elastic}) > C\frac{\sigma_t^2}{\log s}. \tag{6.3.22}$$

These inequalities are important experimentally because they establish that elastic scattering will not decrease faster than $(\log s)^{-2}$ if total cross-sections are asymptotically constant. Thus, very high energy experiments will still include a substantial proportion of elastic scattering. It may transpire that this proportion is asymptotically constant but this has not yet been established on theoretical grounds.

6.4 Non-forward scattering

Upper bounds for forward scattering amplitudes and for total cross-sections are approached to within a factor $\log^2 s$ by experimental results. The situation is quite different for non-forward scattering, either at fixed momentum transfer or at a fixed angle. For example, at a fixed angle the experimental differential cross-sections decrease exponentially with exponent $-s^{\frac{1}{2}}$, but the theoretical bounds are no better than an inverse power, s^{-2}. It is evident that there is a complicated cancellation between contributions from different partial waves. It is therefore unlikely that a method based on the partial wave series will give a strong fixed angle bound unless information has first been obtained about the l-dependence of partial wave amplitudes. Although such information can be obtained from specific models, rather little progress has been made in the general case.

In this section I will outline the upper bounds obtained by Martin (1963a) and by Kinoshita, Loeffel & Martin (1964). In the next section we will consider lower bounds.

For $|\cos\theta| < 1$ we have the inequalities,

$$|P_l(\cos\theta)| < 1 \quad \text{and} \quad |P_l(\cos\theta)| < \left(\frac{2}{\pi l \sin\theta}\right)^{\frac{1}{2}}. \qquad (6.4.1)$$

From the partial wave series (6.1.7) for $A(s, \cos\theta)$,

$$A(s, \cos\theta) \sim \frac{s^{\frac{1}{2}}}{k} \sum_0^L (2l+1)\, a_l(s)\, P_l(\cos\theta), \qquad (6.4.2)$$

we obtain for $|\cos\theta| < 1$,

$$|A(s, \cos\theta)| \leqslant 2 \sum_0^L (2l+1) \left(\frac{2}{\pi l \sin\theta}\right)^{\frac{1}{2}}. \qquad (6.4.3)$$

For large s, (large L), we can replace the sum by an integral, giving

$$|A(s, \cos\theta)| \leqslant 4 \left(\frac{2}{\pi}\right)^{\frac{1}{2}} \int_0^L dl\, \frac{l^{\frac{1}{2}}}{(\sin\theta)^{\frac{1}{2}}},$$

$$\leqslant \frac{CL^{\frac{3}{2}}}{(\sin\theta)^{\frac{1}{2}}}. \qquad (6.4.4)$$

Hence, for θ real and $(0 < \theta < \pi)$,

$$|A(s, \cos\theta)| \leqslant \frac{Cs^{\frac{3}{4}}(\log s)^{\frac{3}{2}}}{(\sin\theta)^{\frac{1}{2}}}. \qquad (6.4.5)$$

This is not a very strong bound and from unitarity one can readily see that it can be saturated only at isolated points. If the bound was attained in any interval $\theta_1 < \theta < \theta_2$, there would be a contribution of magnitude $s^{\frac{1}{2}}(\log s)^3$ to the elastic cross-section, since

$$\sigma(\text{elastic}) > \frac{2\pi}{s} \int_{-1}^1 d(\cos\theta)\, |A(s, \cos\theta)|^2. \qquad (6.4.6)$$

This saturation of the bound would cause σ(elastic) to increase faster than σ(total), which would contradict unitarity. It is therefore evident that an improved use of unitarity will give a better bound (Martin, 1963b; and Kinoshita, Loeffel & Martin, 1964). Their results are obtained assuming the Mandelstam representation and give

$$|F(s, \cos\theta)| < \frac{C(\log s)^{\frac{3}{2}}}{\sin^2\theta}, \tag{6.4.7}$$

for fixed θ as $s \to \infty$. Note that for large s,

$$\sin^2\theta \sim -\frac{4t}{s} - \frac{4t^2}{s^2}, \tag{6.4.8}$$

so that (6.4.7) and (6.4.5) have the same form for fixed t and large s.

At fixed momentum transfer, (6.4.7) can be replaced by

$$|F(s, t)| < Cs(\log s)^{\frac{3}{2}} \quad (t < 0). \tag{6.4.9}$$

This result cannot be deduced immediately from (6.4.7), since it was assumed there that $\theta \neq 0$ as $s \to \infty$, but it can be established directly from the partial wave series (Martin, 1963a).

If $t > 0$, (and s is large and positive), the scattering amplitude $F(s, t)$ refers to an unphysical region of the variables. As noted in (6.1.30) Jin & Martin (1964) have shown that

$$|F(s, t)| < s^{2-\epsilon}, \quad 0 < t < 4\mu^2. \tag{6.4.10}$$

The consequences of this result for subtractions in dispersion relations, and for Regge poles $\alpha(t)$ with t below threshold, were discussed in § 6.1.

6.5 Lower bounds for forward scattering amplitudes and total cross-sections

(a) Lower bounds for amplitudes

The following lower bound on the forward amplitude $F(s, 0)$, has been shown to hold as $s \to \infty$,

$$|F(s, 0)| > \frac{1}{s^{2+\epsilon}} \quad \text{with} \quad \epsilon > 0. \tag{6.5.1}$$

This bound holds if $|F|$ does not oscillate. If, however, there are oscillations, then the least upper bound of $|F|$ will satisfy (6.5.1). Although this result for a lower bound is rather weak, it is of considerable theoretical importance that such a bound exists. It has been established under very general conditions of analyticity by Jin &

Martin (1964) using a method based on Herglotz functions, and by Sugawara (1965) using the phase representation of Sugawara & Kanezawa (1961) (see also Sugawara & Tubis, 1963). We will describe the method using Herglotz functions here, and will introduce the phase representation in Chapter 7 in connection with crossing symmetry and phase relations.

A Herglotz function $H(z)$ satisfies the following conditions (see Shohat & Tamarkin, 1943):

(*a*) it is regular in $\operatorname{Im} z > 0$, and such that $\operatorname{Im} H(z) > 0$ when $\operatorname{Im} z > 0$,

(*b*) it has a representation:

$$H(z) = A + Bz + \frac{1}{\pi} \int_{-\infty}^{\infty} \frac{dx \operatorname{Im} H(x)(1 + zx)}{(1 + x^2)(x - z)}, \qquad (6.5.2)$$

with $B \geqslant 0$, $\operatorname{Im} H(x) \geqslant 0$, and

(*c*) $\displaystyle\int_{-\infty}^{\infty} \frac{dx \operatorname{Im} H(x)}{1 + x^2}$ is convergent.   (6.5.3)

In addition $-[H(z)]^{-1}$ is a Herglotz function, which means that

$$\int_{-\infty}^{\infty} \frac{dx \operatorname{Im} H(x)}{|H(x)|^2 (1 + x^2)} \quad \text{is convergent.} \qquad (6.5.4)$$

For z complex, $|z| > 1$ and $\epsilon < \arg z < \pi - \epsilon$, for some C,

$$\frac{C}{|z|} < |H(z)| < C|z|. \qquad (6.5.5)$$

An analogous result is required for z real. This can be obtained from the function

$$G(z) = \frac{1}{z} \int_{0}^{z} d\omega \, H(\omega). \qquad (6.5.6)$$

Using a straight line for the integration path, it can be seen that $G(z)$ is also a Herglotz function. In particular, putting $G(z)$ into (6.5.4), the integral

$$\int_{A}^{\infty} \frac{dx}{x} \left\{ \int_{0}^{x} dy \operatorname{Im} H(y) \right\} \left| \int_{0}^{x} dy \, H(y) \right|^{-2} \qquad (6.5.7)$$

must converge. Since $\displaystyle\int_{0}^{x} dy \operatorname{Im} H(y)$ is non-decreasing, this means that

$$\int_{0}^{x} dy \, |H(y)| > C(\log x)^{\frac{1}{2}}. \qquad (6.5.8)$$

Hence $\displaystyle\lim_{x \to \infty} x \, |H(x)| \, (\log x)^{\frac{1}{2}} = +\infty. \qquad (6.5.9)$

The above results have been extended by Jin & Martin (1964) to functions that are analytic in a cut plane, with $\operatorname{Im} F(x+i0)$ positive on the right-hand cut and negative on the left-hand cut (as holds for a symmetric forward scattering amplitude). The symmetric scattering amplitude $F(z)$ (where $z = s$ the square of the invariant energy) may have complex zeros at z_r, z_r^* and real zeros at $x_1, \ldots, x_{2n+1}$. Write it in the form,

$$F(z) = \Pi(z - z_r)(z - z_r^*) \prod_1^{2n}(z - x_p) H(z). \qquad (6.5.10)$$

Then it can be shown that $H(z)$ is a Herglotz function (ref. Symanzik, 1960). By leaving one zero x_{2n+1} on the real axis between the branch cuts, $\operatorname{Im} H(x+i0)$ becomes positive (or zero) along the entire real axis. This can be extended to include the possibility of poles on the real axis below threshold.

The conditions on $F(x)$ analogous to (6.5.8, 9) can be shown to be (Jin & Martin, 1964),

$$\int_{a_0}^x dy\,|F(y)|\,y > C(\log x)^{\frac{1}{2}} \quad \text{as} \quad x \to \infty, \qquad (6.5.11)$$

$$\lim_{x \to \infty} x^2(\log x)^{\frac{1}{2}}\,|F(x)| = +\infty. \qquad (6.5.12)$$

If $|F(x)|$ oscillates at infinity one uses the least upper bound in (6.5.12).

From the above conditions it follows that $F(x)$ cannot decrease arbitrarily fast. In particular using (6.5.12):

If $\qquad |F(s, 0)| \sim s^\alpha \quad \text{as} \quad s \to \infty \quad \text{then} \quad \alpha \geqslant -2. \qquad (6.5.13)$

For the non-forward scattering amplitude $F(s, t)$ at fixed t, when $0 < t < 4m^2$, we have $\operatorname{Im} F(s, t) > 0$, and we can proceed as before. Thus the bound (6.5.13) will hold also for $t > 0$. However for physical values of t, $(t < 0)$, the problem becomes much more difficult unless special assumptions are made. The difficulty is that for any fixed negative t, it is possible that, as $s \to \infty$, $\operatorname{Im} F(s, t)$ may oscillate an infinite number of times and have an infinity of zeros (see below and Eden & Łukaszuk, 1967). But if, following Jin & Martin (1964), one makes the special assumption

$$F(s, t) \sim s^{\alpha(t)} f(t) \quad \text{as} \quad s \to \infty, \qquad (6.5.14)$$

then there can be only a finite number $n(t)$ of oscillations of $\operatorname{Im} F(s, t)$ as $s \to \infty$. This weakens the result (6.5.11) and gives instead,

$$\int_{a_0}^x dy\,y^{n+1}\,|F(y)| > C(\log x)^{\frac{1}{2}} \quad \text{as} \quad x \to \infty, \qquad (6.5.15)$$

giving in (6.5.14)　　　　$\alpha \geqslant -2 - n(t).$　　　　　(6.5.16)

If assumptions are made about low energy behaviour of $F(s, 0)$ the lower bound (6.5.13) can be improved (Jin & Martin, 1964 and Sugawara, 1965). Similarly if assumptions are made about the number of oscillations of partial wave amplitudes, bounds can be obtained on their behaviour at infinity using similar methods (Jin, 1966 and Kinoshita, 1966). Lower bounds have also been considered by Wit (1965a).

(b) *The problem of oscillations*

If it is assumed that for large enough s the amplitude does not have oscillations as $s \to \infty$ for fixed t, one obtains a considerable simplification in some important aspects of high energy scattering. For $t < 4m^2$, one already has this simplifying feature in Regge theory if dominance as $s \to \infty$ of a single pole $l = \alpha(t)$ is assumed. The asymptotic form is then given by (6.5.14) which does not oscillate. However for $t > 4m^2$, $\alpha(t)$ is complex, $\alpha = \alpha_1 + i\alpha_2$, and, for example,

$$\operatorname{Im} F(s, t) \sim f(t) s^{\alpha_1} \sin [\alpha_2 \log s]. \qquad (6.5.17)$$

This has oscillations of decreasing frequency as $s \to \infty$. Unfortunately there is no evidence that this result would remain true without special assumptions such as those of Regge theory. Alternative criteria for power behaviour of F as $s \to \infty$ have been discussed by Bessis & Kinoshita (1966).

We will see in Chapter 7 that consideration of the phase of scattering amplitudes at high energy is greatly simplified if one assumes a non-oscillatory form (as $s \to \infty$) for $F(s, t)$ with $t < 4m^2$. The assumed absence of oscillations has not given any improvement on the Froissart bound or on the Martin enlargement of the Lehmann ellipse. It may be important for fixed angle scattering where questions of cancellation in the partial wave series are vital.

The nature of the general problem of oscillations at high energy has been considered by Eden & Łukaszuk (1967). It should be noted that the derivation by Martin of the Froissart bound described in §6.1 does not depend on any assumptions about oscillations. The lower bounds described in the present section are too weak to exclude oscillations that take $|F|$ down to arbitrarily small values as the energy tends to infinity. It might be thought that regularity of $F(s, t)$ in the upper half s-plane and polynomial boundedness would restrict oscillations. Such restrictions do exist for entire functions and one can obtain for them a result on oscillations analogous to (6.5.17), where the frequency of oscillations decreases as s tends to infinity. However, for a function having a branch cut along the real axis there can be arbitrary oscillations even though it is polynomial bounded and regular in the upper half plane. The possibility of restricting oscillations requires some information about Im $F(s, t)$ on the real axis that is stronger than the lower bounds so far obtained.

It is possible, however, to obtain restrictions on the number of zeros of $F(s, t)$ for s real and $|t| < t_0$(const.), from the known bounds on the amplitude for large s. The upper bound on $|F(s, t)|$ is s^2 for $|t| < 4m^2$, (see (6.4.10)), and the lower bound for $|F(s, 0)$ is s^{-2}, (6.5.13), provided we are not too close to a zero (in general complex) of $F(s, t)$. Because $F(s, t)$ is analytic for fixed s, as a function of t in $|t| < t_0$ with $F(s, 0) \neq 0$, we can write Jensen's theorem in the form:

$$\int_0^r \frac{dy}{y} N(s, y) = \frac{1}{2\pi} \int_0^{2\pi} d\theta \log \left[\frac{F(s, r \exp i\theta)}{F(s, 0)} \right], \qquad (6.5.18)$$

where $r < t_0$, and $N(s, y)$ is the number of zeros of $F(s, t)$ inside

$$|t| < y. \qquad (6.5.19)$$

We can obtain a bound on the number of zeros inside a circle of radius $b < r$, by taking an upper bound on the integral on the right-hand side of (6.5.18), giving

$$N(s, b) < \frac{4 \log s}{\log (r/b)}. \qquad (6.5.20)$$

The maximum possible number of zeros, for any finite $(r/b) > 1$, is therefore increasing, for large s, at most like $\log s$. The same applies to zeros of Im $F(s, t)$.

It has been shown by Bessis (1966) that the first zero cannot occur nearer to $t = 0$ than $|t| \sim (\log s)^{-2}$. The result (6.5.20) shows that in this case the distance between zeros must increase on average so that no more than N zeros ($N \sim \log s$) lie within $|t| < b < r < t_0$. In general the zeros would occur for complex values of t, when s is real, so they would lead to diffraction minima in the differential cross-section $d\sigma/dt$.

The nature of the singularity of $F(s, t)$ as $s \to \infty$ is of great importance in determining high energy behaviour. The point at infinity is an accumulation point for the normal threshold branch points, and it would be surprising if oscillations were absent for general values of t. However, it is sometimes convenient to assume their absence in order to obtain a possible indication of a general result.

(c) *Lower bounds on total cross-sections*

The lower bound on the modulus of the forward scattering amplitude (6.5.1) can be used to give a bound on the corresponding total cross-section. We recall that in the derivation of this lower bound, it was assumed that we could construct from the amplitude, a function having an imaginary part of definite sign on the right and left-hand cuts. This cannot be done when there is a part of the cut that is unphysical, as with collisions between particles having unequal masses.

Since it is possible that $F(s, 0)$ is purely real in the asymptotic limit in which (6.5.1) holds, we cannot use the optical theorem to obtain a lower bound on the total cross-section. Instead, we must bound it via the elastic cross-section, assuming the narrowest allowed forward peak, which is obtained from $g(s) = \log^2 s$ (see (6.2.8)). Then (6.2.16) gives the bound

$$\sigma(\text{total}) > \sigma(\text{elastic}) > \frac{C}{s^6 \log^2 s}. \qquad (6.5.21)$$

Using special assumptions about $F(s, 0)$ at threshold, Jin & Martin (1964) and Sugawara (1965), have improved this bound to give,

$$\sigma(\text{total}) > \frac{C}{s^2 \log^2 s}. \qquad (6.5.22)$$

6.6 Lower bounds for large angle scattering

An interesting study of large angle scattering has been made by Cerulus & Martin (1964). Their method provides an elegant illustration of the power of analytic function theory. Their work is concerned primarily with a bound for the amplitude $F(s, \cos\theta)$ near $\cos\theta = 0$, but it includes inequalities that hold for any real scattering angle other than zero. As with the other bounds we have described, the method is based on analyticity in $z = \cos\theta$ for large values of s. However, they use a domain of analyticity that goes outside the Martin–Lehmann ellipse. Their domain D is smaller than the cut plane assumed for the Mandelstam representation, but within D they also assume a polynomial bound as $s \to \infty$.

Their assumptions are:

(i) Analyticity of $F(s, z)$ in $z = \cos\theta$ for a domain D whose approximate form is shown in Fig. 6.6.1 (a). The boundary of D meets the real axis at $z = \pm\rho$, given by

$$\rho = 1 + \frac{4b}{s}, \qquad (6.6.1)$$

where $t = 2b$ is the nearest singularity in t. The z-plane is assumed to have branch cuts $(-\infty, -\rho)$ and (ρ, ∞).

(ii) A polynomial bound,

$$|F(s,z)| < s^N \quad \text{as} \quad s \to \infty, \tag{6.6.2}$$

where N is independent of s and z, within D.

(iii) On a segment of the real axis $(-a, a)$ the modulus of F is bounded by a given function,

$$|F(s,z)| < \exp\left[-\phi(s)\right], \quad -a < z < a, \tag{6.6.3}$$

where $0 < a < 1$. For z in $(-a, a)$ the momentum transfer is large, $t \sim -s$.

The objective is to obtain a bound on the behaviour of the function $\phi(s)$ in (6.6.3), by investigating whether it is compatible with known bounds on the forward amplitude $F(s, 1)$. The basic idea is that an analytic function, that has the bounds (6.6.3) near $z = 0$, and (6.6.2) for $|z|$ at a specified value (greater than one) within D, must increase reasonably steadily through intermediate values. This can be put in a precise form through Hadamard's three circles theorem (see, for example, Titchmarsh, 1939). This theorem states:

Let $f(\zeta)$ be an analytic function, regular for $r_1 < \zeta < r_3$. Let $r_1 < r_2 < r_3$ and let M_1, M_2, M_3 be the maxima of $|f(\zeta)|$ on the three circles, $|\zeta| = r_1, r_2, r_3$ respectively. Then

$$M_2^{\log(r_3/r_1)} \leqslant M_1^{\log(r_3/r_2)} M_3^{\log(r_2/r_1)}. \tag{6.6.4}$$

In order to apply this theorem, Cerulus and Martin make a conformal transformation that takes:

(i) the segment of the real axis $-a < z < a$ into the circle $|\zeta| = r_1 = 1$,

(ii) a closed curve Γ_3 through $z = \pm 1$, lying within D into the circle $|\zeta| = r_2 = E$,

(iii) the domain D into the interior of the circle $|\zeta| = r_3 = R$.

A suitable transformation is given by the result of the following two transformations,

$$\omega = \frac{\rho}{z}[\rho - (\rho^2 - z^2)^{\frac{1}{2}}], \tag{6.6.5}$$

$$\zeta = \frac{1}{a_1}[\omega + (\omega^2 - a_1^2)^{\frac{1}{2}}], \tag{6.6.6}$$

where

$$a_1 = \frac{\rho}{a}[\rho - (\rho^2 - a^2)^{\frac{1}{2}}]. \tag{6.6.7}$$

The resulting configuration is shown in Fig. 6.6.1(b). The points labelled $0, 1, E, R$ in the ζ-plane, correspond in the z-plane to $0, a, 1, \rho$. The form of D is specified from the requirement (iii) given above, with R defined as the value

$$R = \frac{1}{a_1} [\rho + (\rho^2 - a_1^2)^{\frac{1}{2}}]. \tag{6.6.8}$$

The value E is defined similarly by the result of the transformation on $z = 1$. Applying Hadamard's theorem to the three circles,

$$r_1 = 1, \quad r_2 = E, \quad r_3 = R, \tag{6.6.9}$$

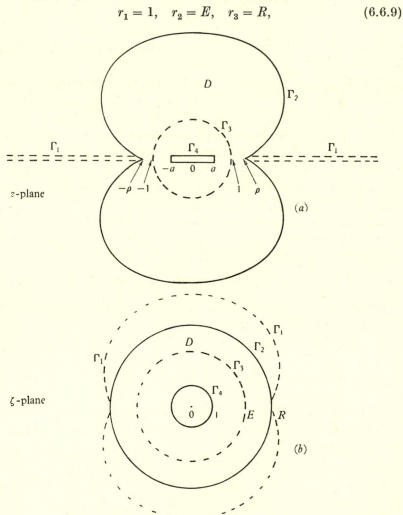

Fig. 6.6.1. (a) The domain D in which analyticity and boundedness is required, shown in the $z = (\cos\theta)$-plane; (b) the transform of D in the ζ-plane, in which Hadamard's three circles theorem can be applied.

one gets,

$$\log R \log M_2 \leqslant [\log R - \log E] \log M_1 + \log E \log M_3. \quad (6.6.10)$$

The bound M_2 is $|F(s, 1)|$; the bound M_1 is given by (6.6.3), and the bound M_3 by (6.6.2). Substituting for R and E, and approximating with s large, one obtains

$$|F(s, 1)| \leqslant \exp\left[-\frac{C\phi(s)}{s^{\frac{1}{2}}} + N \log s \left(1 - \frac{C}{s^{\frac{1}{2}}}\right)\right], \quad (6.6.11)$$

where C can be found in terms of a and b. Cerulus and Martin discuss the consequences of taking the bound (6.6.3) to have $\phi(s)$ of the form,

$$\phi(s) = C_1 s^{\alpha} (\log s)^{\beta}. \quad (6.6.12)$$

If $\alpha > \frac{1}{2}$, then (6.6.11) gives

$$|F(s, 1)| < s^{-M} \quad \text{as} \quad s \to \infty, \quad \text{for any } M. \quad (6.6.13)$$

This bound is inconsistent with the lower bound (6.5.13), so we must have $\alpha \leqslant \frac{1}{2}$.

If $\alpha < \frac{1}{2}$, (6.6.11) is equivalent to the assumption (6.6.2) and no new information is gained. If $\alpha = \frac{1}{2}$, it is necessary to have $\beta \leqslant 1$. Otherwise one would obtain again an inequality (6.6.13) that contradicts the lower bound on $F(s, 1)$. For $\alpha = \frac{1}{2}$, $\beta = 1$, a condition on the constants C_1 and N can be obtained. If $\alpha = \frac{1}{2}$ and $\beta < 1$, (6.6.11) again becomes equivalent to (6.6.2).

It is interesting that the experimental results on large angle scattering have an order of magnitude that indicates

$$F(s, \cos\theta = 0) \sim (\text{Polynomial}) \exp\left[-C\sqrt{s}\right]. \quad (6.6.14)$$

Thus experiment comes quite close to the theoretical minimum obtained by Cerulus and Martin. However one should note that there is some doubt (due to possible Regge branch cuts) about the assumption (6.6.2), which in this section has been used in a region where both s and t are large and positive.

6.7 Bounds when particles have spin

It has been shown that total cross-sections and differential cross-sections have essentially the same high energy bounds when the colliding particles have non-zero spin as when they are spinless bosons. The proofs are given by Yamamoto (1963), Hara (1964) and Cornille (1964).

Yamamoto bases his method on a discussion of the trace of the scattering amplitude. He shows that the quantity

$$\mathrm{Tr}\,[F(s,t)], \tag{6.7.1}$$

where the trace is taken with respect to the spin variables, has the same bound as an elastic scattering amplitude for spinless bosons. This gives, for particles having arbitrary spin,

$$\sigma(\mathrm{total}) < C\log^2 s. \tag{6.7.2}$$

He also considers the trace of the spin eigen-amplitudes,

$$\mathrm{Tr}\,[P_\beta F(s,t)\,P_\alpha], \tag{6.7.3}$$

where P_α picks out the simultaneous eigenstates of $(\boldsymbol{\sigma}_1 . \mathbf{p}_\alpha)$ and $(\boldsymbol{\sigma}_2 . \mathbf{p}_\alpha)$ for initial momentum $\mathbf{p}_\alpha$ in the centre of mass system, and P_β is the corresponding operator for final centre of mass momentum $\mathbf{p}_\beta$. He shows that the trace (6.7.3) satisfies essentially the same conditions as the scattering amplitude for spinless bosons. From these results it follows that the traces have the same asymptotic behaviour as in the scalar case.

The method of Hara is based on helicity amplitudes that are modified to be analytic in an ellipse surrounding the physical region, $(-1 \leqslant \cos\theta \leqslant 1)$. These modified helicity amplitudes have the form

$$(c,d|\,F_1\,|a,b) = (\cos\tfrac{1}{2}\theta)^{-|\lambda+\mu|}(\sin\tfrac{1}{2}\theta)^{-|\lambda-\mu|}\,(c,d|\,F\,|a,b), \tag{6.7.4}$$

where $(c,d|\,F\,|a,b)$ denotes the usual helicity amplitude, a denotes (λ_a, p_a) etc. and $\lambda = \lambda_a - \lambda_b$; $\mu = \lambda_c - \lambda_d$. This amplitude is analytic inside an ellipse in the z-plane ($z = \cos\theta$), with foci at $+1$ and -1 and semi-major axis $(1 + 2t_0/s)$ for large s, where t_0 is the nearest singularity. The helicity partial wave amplitudes f_λ^J can be expressed as integrals, round this ellipse, of an integrand involving the amplitude F_1 and Legendre functions $Q_J(z)$. The amplitude F_1 is assumed to be polynomial bounded. Then the integrand is dominated by the rapid decrease of $Q_J(z)$ on the ellipse, and a bound is obtained on the helicity partial wave f_λ^J that is similar to the bound (6.1.14) in the scalar case. The method of Hara is a development of that used by Greenberg & Low (1961) for scalar particles.

The third method, which was applied by Cornille (1964) to nucleon-nucleon scattering, uses the scalar amplitudes of Goldberger *et al.* (1960). He then applies the methods of Kinoshita, Loeffel & Martin

(1964) to obtain the improved bound (6.4.7) for the non-forward direction. His methods also give the Froissart bound on the total cross-section.

6.8 Summary of bounds

The following tables were prepared by Martin for the Berkeley Conference 1966 and I am indebted to him for allowing me to use them here.

Table 6.8.1. *Upper bounds*

	Assumptions	References (see after Table 6.8.3)
Forward scattering		
$\|F(s, \cos\theta = 1)\| < Cs \log^2 (s/s_0)$, $\sigma(\text{total}) < C' \log^2 (s/s_0)$, arbitrary spin	Axiomatic quantum field theory	(1), (2), (3), (4)
$C' < \pi/\mu^2$ for $\pi\pi$, πK, KK	Axiomatic quantum field theory	(20)
$C' < 12\pi/\mu^2$ for πN	Axiomatic quantum field theory	(5)
Fixed transfer		
$\|F(s, t)\| < s(\log s)^{\frac{3}{2}}$, $t < -\epsilon < 0$,	Axiomatic	(3)
$\|F(s, t)\| < s^{2-\epsilon}$, $0 < t < 4\mu^2$,	Axiomatic	(5)
$\|F(s, t)\| < \text{constant},$ $0 < t < 4\mu^2,$ if $\|F(s, 0)\| < s^{-\epsilon},$ $(\epsilon > 0)$		(6)
Fixed angle		
$\|F(s, \cos\theta)\| < C[s^{\frac{3}{4}}(\log s)^{\frac{3}{2}}]/(\sin\theta)^{\frac{1}{2}}$	Axiomatic	(1), (3)
$\|F(s, \cos\theta)\| < C(\log s)^{\frac{3}{2}}/(\sin^2\theta)$	Mandelstam	(7)
$\dfrac{d\sigma}{d\Omega} < C(\log s)^3/(s \sin^4\theta)$ valid for arbitrary spins	Mandelstam	(8)

Table 6.8.2. *The diffraction peak*

	Assumptions	References (see after Table 6.8.3)
$\dfrac{d}{dt}[\log A_s(s, t)]_{t=0} > \dfrac{1}{9}\left[\dfrac{\sigma_t^2}{4\pi\sigma_{\text{el.}}} - \dfrac{1}{k^2}\right].$	Unitarity and partial wave series	(9)
$\dfrac{d}{dt}[\log A_s(s, t)]_{t=0} < C(\log s)^2.$	Axiomatic	(10)
If $A_s(s, \cos\theta = 1) > C/\log s$, $\left[\dfrac{\displaystyle\int_{-1}^{1} d(\cos\theta)\,\|F(s, \cos\theta)\|^2}{\|F(s, \cos\theta = 1)\|^2}\right] \to 0.$	Axiomatic	(11)

Table 6.8.3. *Lower bounds*

	Assumptions	References (see after this Table)
Elastic cross-sections		
$\sigma(\text{elastic}) > C\,\dfrac{(\sigma_t)^2}{\log^2 s}$	Axiomatic	(12)
Forward amplitude (strict for complex s; average sense for real s)		
(a) $|F(s, t = 0)| > s^{-2}$ (when there is no unphysical cut)	Axiomatic	(13)
$\sigma(\text{total}) > \dfrac{1}{s^6 \log^2 s}$	Axiomatic	(13)
(b) If F(threshold) is sufficiently small, $|F(s, t = 0)| > \text{constant} > 0,$	Axiomatic	(13), (14)
$\sigma(\text{total}) > \dfrac{1}{s^2 \log^2 s}$		
Fixed angle amplitude		
$|F(s, \cos\theta)| > \exp\left[-s^{\frac{3}{4}}\log s . C(\theta)\right]$	Mandelstam	(15)
$C(\theta) \sim \sqrt{\theta}$, for small θ	Mandelstam	(16)
Form factors		
$|F(t)| > \exp\left[-Ct^{\frac{1}{2}}\right], t > 0$	Axiomatic	(17)
Fixed t amplitude (t < 0)		
$|F(s, t)| > s^{-N}$	Axiomatic	(18)
Fixed u amplitude (u < 0)		
If pole at $u = m^2$, $|F(s, u)| > s^{-N}$	Slightly more than axiomatic	(19)

References given in the above Tables

(1) Froissart (1961*b*)
(2) Greenberg & Low (1961)
(3) Martin (1963*a*)
(4) Martin (1966*a*)
(5) Jin & Martin (1964)
(6) Kinoshita (1964*b*)
(7) Kinoshita, Leoffel & Martin (1964)
(8) Yamamoto (1963), Hara (1964), Cornille (1964)
(9) MacDowell & Martin (1964)
(10) Kinoshita (1964*b*)
(11) Jin (1966), Aramaki (1963)
(12) Martin (1963*c*)
(13) Jin & Martin (1964)
(14) Sugawara (1965)
(15) Cerulus & Martin (1963)
(16) Kinoshita (1964*a*)
(17) Martin (1965*a*)
(18) Martin (unpublished)
(19) Gribov & Pomeranchuk (1962*c*)
(20) Łukaszuk & Martin (1967)

CHAPTER 7

CROSSING SYMMETRY AND ANALYTICITY IN THE ENERGY

Crossing relations for scattering amplitudes have been described in Chapter 4. They have been established rigorously from quantum field theory by Bros, Epstein & Glaser (1964, 1965), who show that for scalar bosons a single amplitude $F(s,t)$ can, by analytic continuation, be used to describe scattering in the three channels for which s, t and u respectively correspond to the square of the energy. We will not be using the full content of crossing in this chapter but only the part that relates the s and u channels at fixed t. When t is fixed and real, a dispersion relation holds for $F(s,t)$ if $|t|$ is not too large (Hepp, 1964). This dispersion relation can be used to relate the real and imaginary parts of F. Alternatively one can use more general theorems to obtain such relations. For forward scattering, the imaginary part of F can be found experimentally from the total cross-section, and the real part from Coulomb interference. In this chapter we will consider general theorems that relate these two quantities.

Three types of approach will be described. The first, in §7.1, uses simplifying assumptions about the absence of oscillations at high energies. The results are simply stated and are applied in §7.2 to the real part of the forward amplitude. The second approach does not make the simplifying assumption about oscillations, and its full discussion is rather complicated, but the main results are outlined in §7.2. The third approach uses the combined information about analyticity in the energy s in a cut plane, and analyticity in the momentum transfer t in the Lehmann ellipse, together with unitarity. In §7.3 we consider the method of Martin (1966 a) which uses this third approach to enlarge the Lehmann ellipse. In §7.4 we show how information can be obtained about partial wave amplitudes if some additional assumptions are made.

7.1 Crossing symmetry and phase relations

(a) *Crossing relations—symmetric and antisymmetric amplitudes*

We have noted in § 4.3 that the forward scattering amplitude $F(E)$ for equal mass scalar bosons is analytic in the complex energy plane cut along $(-\infty, -m)$ and (m, ∞), where E denotes the laboratory energy. Crossing symmetry takes the form

$$F(-E-i0) = F(E+i0), \tag{7.1.1}$$

where $E \pm i0$ indicates a limit from the upper (lower) half plane on to the branch cut. Since F is real between the branch cuts, it is hermitean analytic, and (7.1.1) can be written

$$F(-E+i0) = F^*(E+i0), \tag{7.1.2}$$

or, restricting ourselves to the upper half plane,

$$F(E \exp i\pi) = F^*(E), \tag{7.1.3}$$

where $F(x)$ denotes $F(x+i0)$.

This relation holds for a symmetric amplitude (for example, $\pi^0\pi^0$ scattering) but these are comparatively rare for physical scattering processes. For π^+p and π^-p scattering, write F_+, F_-, for the forward amplitudes,

$$F_+(E) = F(\pi^+p \to \pi^+p), \quad F_-(E) = F(\pi^-p \to \pi^-p). \tag{7.1.4}$$

Then crossing gives

$$F_+(E \exp i\pi) = F_-^*(E). \tag{7.1.5}$$

We can form a symmetric amplitude F_S and an antisymmetric amplitude F_A,

$$F_S = \tfrac{1}{2}(F_+ + F_-), \quad F_A = \tfrac{1}{2}(F_+ - F_-). \tag{7.1.6}$$

For these amplitudes, crossing takes the form,

$$F_S(E \exp i\pi) = F_S^*(E), \tag{7.1.7}$$

$$F_A(E \exp i\pi) = -F_A^*(E). \tag{7.1.8}$$

The optical theorem (3.3.4) gives

$$\operatorname{Im} F_S(E) = M(E^2 - m^2)^{\frac{1}{2}} \left[\sigma(\pi^+p \text{ total}) + \sigma(\pi^-p \text{ total})\right], \tag{7.1.9}$$

and a similar expression for $\operatorname{Im} F_A$.

When writing asymptotic relations, constant factors are often omitted, so, as $E \to \infty$, one writes

$$\operatorname{Im} F_S(E) \sim E[\sigma_t(\pi^+p) + \sigma_t(\pi^-p)], \tag{7.1.10}$$

$$\operatorname{Im} F_A(E) \sim E[\sigma_t(\pi^+p) - \sigma_t(\pi^-p)]. \tag{7.1.11}$$

With this convention, the sign $\sim$ is to be interpreted as indicating asymptotic equality to within a constant factor.

(b) Dispersion relations

The forward dispersion relation for F_S will require two subtractions (assuming $\sigma_t \geqslant$ constant for large E),

$$F_S(E) = F_S(0) + \frac{E^2}{\pi} \int_m^\infty \frac{dx \operatorname{Im} F_S(x)}{x^2(x-E)} + \frac{E^2}{\pi} \int_{-\infty}^{-m} \frac{dx \operatorname{Im} F_S(x)}{x^2(x-E)}, \quad (7.1.12)$$

$$= F_S(0) + \frac{2E^2}{\pi} \int_m^\infty \frac{dx \operatorname{Im} F_S(x)}{x(x^2 - E^2)}. \quad (7.1.13)$$

The integration is along the top of the branch cut $(x+i0)$. Normally with two subtractions we would expect two constants, $F_S(0)$ and its derivative $F_S'(0)$, but because we have a symmetric amplitude the second constant is zero.

If the Pomeranchuk hypothesis (§ 8.1) is valid, the total cross-sections for π^+p and π^-p scattering will be asymptotically equal. Experimentally their difference decreases like $E^{-\frac{1}{2}}$, so only one subtraction in the dispersion relation for F_A would seem to be necessary. We must not assume this for the general theory, so for F_A we write in general,

$$F_A(E) = EF_A'(0) + \frac{2E^3}{\pi} \int_m^\infty \frac{dx \operatorname{Im} F_A(x)}{x^2(x^2 - E^2)}. \quad (7.1.14)$$

However, if $E^{-1} \operatorname{Im} F_A$ tends to zero sufficiently fast, we can write

$$F_A(E) = \frac{2E}{\pi} \int_m^\infty \frac{dx \operatorname{Im} F_A(x)}{x^2 - E^2}. \quad (7.1.15)$$

Given a simple asymptotic form for $\operatorname{Im} F_S$, or $\operatorname{Im} F_A$, one can determine the asymptotic form of $\operatorname{Re} F_S$, or $\operatorname{Re} F_A$, from these dispersion relations. For example if

$$\operatorname{Im} F_A(E) \sim C_A E \quad \text{as} \quad E \to \infty, \quad (7.1.16)$$

we find from (7.1.14)

$$\operatorname{Re} F_A(E) \sim -\frac{2E}{\pi} C_A \log E. \quad (7.1.17)$$

Then the asymptotic form of F_A clearly satisfies the crossing relation (7.1.8), since it has the form,

$$F_A(E) \sim -\frac{2}{\pi} C_A E \left(\log E - \frac{i\pi}{2} \right). \quad (7.1.18)$$

(c) Use of the Phragmén–Lindelöff theorem

Instead of using the dispersion relations to determine $\operatorname{Re} F$, or the phase, it is sometimes more convenient to use the asymptotic uniqueness that follows from the Phragmén–Lindelöff theorem. We require this theorem in the following form (adapted from Boas, 1954):

Theorem

Let D be the region in $z = r e^{i\theta}$ between two straight lines, intersecting at the origin at an angle π/α. Let $f(z)$ be an analytic function of z, regular in D and continuous on the lines themselves. Suppose that as $r \to \infty$,

$$f(z) = O[\exp(r^{\beta})] \quad (\beta < \alpha), \tag{7.1.19}$$

in the region D, and that

$$|f(z)| \leqslant M \tag{7.1.20}$$

on the lines, then the inequality (7.1.20) holds throughout the region D.

We also require the following result. Suppose that $g(z)$ is analytic in the region D described above, and $g(z)$ has limiting values L_1 and L_2 when $z \to \infty$ along the two lines bounding the region D. Then either $L_1 = L_2$ or else $g(z)$ is unbounded in the region D.

We will use these theorems in the case when D is the half-plane $\mathrm{Im}\, z > 0$, when we obtain the following corollary. If $f(z)$ is bounded by a polynomial, and $f(z)$ tends to the limits L_1 and L_2 along the rays $z = x + i0$ as $x \to +\infty$ and $-\infty$, then we must have $L_1 = L_2$.

The usefulness of these and similar theorems for determining relations between the modulus and the phase of F was first noted by Meimann (1962) in the present context, and has been developed by Van Hove (1964), Logunov *et al.* (1963) and Wit (1965 *b*).

Assume the following functional form for the symmetric scattering amplitude $F_S(E)$, as $E \to \infty$ along the real axis,

$$\frac{F_S(E)}{E^{\alpha}} \to M \exp i\theta \quad \text{as} \quad E \to \infty, \tag{7.1.21}$$

where α and M are constants. Then to determine θ we use the above corollary which gives

$$\frac{F_S(E \exp i\pi)}{E^{\alpha} \exp i\pi\alpha} \to M \exp i\theta. \tag{7.1.22}$$

We then use the symmetric crossing relation (7.1.7),

$$F_S(E \exp i\pi) = F_S^*(E), \tag{7.1.23}$$

to obtain

$$\exp(2i\theta) = \exp(-i\pi\alpha), \tag{7.1.24}$$

$$\theta = n\pi - \tfrac{1}{2}\alpha\pi. \tag{7.1.25}$$

For large E,

$$F_S(E) \sim \pm M E^{\alpha} \exp(-i\pi\alpha/2). \tag{7.1.26}$$

The $\pm$ can be determined by the requirement that

$$\text{Im } F_S(E) > 0. \tag{7.1.27}$$

Thus if we are given a polynomial form for the total cross-section for a symmetric scattering process (or a symmetric sum), such as

$$\sigma_t(\pi^+ p) + \sigma_t(\pi^- p) \sim 2CE^{\alpha - 1}, \tag{7.1.28}$$

then the optical theorem, combined with the crossing symmetric form (7.1.26), gives

$$F_S(E) \sim CE^\alpha \left[i - \cot \frac{\pi \alpha}{2} \right]. \tag{7.1.29}$$

The main point to note is that a function of $(-iE)$ with real coefficients will be automatically crossing symmetric. Another example is given by

$$F_S(E) \sim \pm C_S(-iE)^\alpha \left[\log E - \frac{i\pi}{2} \right]^\beta, \tag{7.1.30}$$

where α and β are constants.

For the antisymmetric amplitude $F_A(E)$, the crossing relation (7.1.8) leads to an extra factor i. If we are given

$$\sigma_t(\pi^+ p) - \sigma_t(\pi^- p) \sim 2C_A E^{\alpha - 1}, \tag{7.1.31}$$

the optical theorem and (7.1.8) leads to,

$$F_A(E) \sim C_A \frac{iE^\alpha \exp(-i\pi\alpha/2)}{\cos(\pi\alpha/2)} \sim C_A E^\alpha \left[i + \tan\left(\frac{\pi\alpha}{2}\right) \right]. \tag{7.1.32}$$

A more general form analogous to (7.1.30) is,

$$F_A(E) \sim \pm C_A i(-iE)^\alpha \left[\log E - \frac{i\pi}{2} \right]^\beta. \tag{7.1.33}$$

The additional factor i is particularly significant when $\alpha = 1$ since it then leads to $\text{Re } F_A \gg \text{Im } F_A$. We will return to this point in discussing the Pomeranchuk hypothesis in Chapter 8.

(d) Non-forward scattering

If t is non-zero, crossing symmetry involves a less convenient variable than E. We must take F to be a function of W and t where in the equal mass case,

$$W = s + \tfrac{1}{2} t - 2m^2. \tag{7.1.34}$$

The substitution $s = 4m^2 - u - t$ gives

$$-W = u + \tfrac{1}{2} t - 2m^2, \tag{7.1.35}$$

so the substitution of u for s, with t fixed, corresponds to changing the sign of W.

Crossing symmetry between the physical regions where s and u respectively are the squared energies therefore gives

$$F(\pi^-p, -W+i0, t) = F^*(\pi^+p, W+i0, t), \tag{7.1.36}$$

$$F_S(W \exp i\pi, t) = F_S^*(W, t), \tag{7.1.37}$$

$$F_A(W \exp i\pi, t) = -F_A^*(W, t). \tag{7.1.38}$$

We will require these relations when we discuss the Pomeranchuk hypothesis for non-forward scattering in Chapter 8. Our next task is to use the results of the present section to indicate possible characteristics of the forward amplitude and total cross-sections at high energy.

7.2   Real part of the forward amplitude

(a) *Simple examples using non-oscillating functions*

At high energies, experimental total cross-sections appear to be decreasing towards constant values. In particular the π^+p and π^-p total cross-sections appear to be approaching the same constant. We will see how the phase conditions of the previous section are applied in a simple model. Assume that as $E \to \infty$,

$$\sigma(\pi^+p) \sim C + 2C_1 E^{\alpha-1}, \tag{7.2.1}$$

$$\sigma(\pi^-p) \sim C + 2C_2 E^{\beta-1}, \tag{7.2.2}$$

where C, C_1, C_2 are positive, and $C_2 > C_1$. Then if

$$0 < \alpha \leqslant \beta < 1, \tag{7.2.3}$$

we will have $\sigma(\pi^-p) > \sigma(\pi^+p)$, as is observed experimentally. But, if $0 < \beta < \alpha < 1$, the cross-sections may cross over as E increases; this possibility is not excluded by present experimental evidence. However, for simplicity we will take

$$C_2 > C_1 \quad \text{and} \quad \alpha = \beta. \tag{7.2.4}$$

The symmetric amplitude is found from (7.2.1, 2) using the optical theorem and (7.1.29) applied to terms in each power of E separately. It has the asymptotic form,

$$F_S(E) \sim iEC + (C_1 + C_2) E^\alpha [i - \cot(\pi\alpha/2)]. \tag{7.2.5}$$

The corresponding antisymmetric amplitude gives,

$$F_A(E) \sim ED + (C_1 - C_2) E^\alpha [i + \tan(\pi\alpha/2)], \tag{7.2.6}$$

where D is a constant that is not determined by the total cross-sections. The ratios of real to imaginary parts of the physical amplitudes for this model, with E large, will be given by

$$\frac{\operatorname{Re} F(\pi^+ p)}{\operatorname{Im} F(\pi^+ p)} \sim \frac{ED - E^\alpha[(C_1 + C_2)\cot \tfrac{1}{2}\pi\alpha + (C_2 - C_1)\tan \tfrac{1}{2}\pi\alpha]}{EC}, \qquad (7.2.7)$$

$$\frac{\operatorname{Re} F(\pi^- p)}{\operatorname{Im} F(\pi^- p)} \sim \frac{-ED - E^\alpha[(C_1 + C_2)\cot \tfrac{1}{2}\pi\alpha - (C_2 - C_1)\tan \tfrac{1}{2}\pi\alpha]}{EC}. \qquad (7.2.8)$$

The experimental evidence (see Chapter 2) suggests that both these ratios tend to zero as $E \to \infty$, so we expect that the constant D should be zero. Then, with this model, putting $D = 0$, we should have (since $C_2 > C_1$; $\alpha \approx \tfrac{1}{2}$),

$$\frac{\operatorname{Re} F(\pi^+ p)}{\operatorname{Im} F(\pi^+ p)} < \frac{\operatorname{Re} F(\pi^- p)}{\operatorname{Im} F(\pi^- p)} < 0. \qquad (7.2.9)$$

Note that the general theory that we have used so far does not indicate that D is zero, nor does it limit the types of term in F_S and F_A except as noted in §7.1, in (7.1.29) and (7.1.32), (but see Chapter 8, particularly §8.3 where it is suggested that D must be zero for pion-nucleon scattering).

The amplitudes for proton-proton scattering and antiproton-proton scattering are related by crossing symmetry in the same way as those for $\pi^+ p$ and $\pi^- p$. Thus on a similar model

$$\frac{\operatorname{Re} F(pp)}{\operatorname{Im} F(pp)} \sim \frac{-D' - E^{\alpha'-1}C_3}{C'}, \qquad (7.2.10)$$

and

$$\frac{\operatorname{Re} F(\bar{p}p)}{\operatorname{Im} F(\bar{p}p)} \sim \frac{+D' - E^{\alpha'-1}C_4}{C'}. \qquad (7.2.11)$$

Experimental results given in §2.1 allow the possibility that D' may be non-zero, and that $(-D'/C') \approx -0.2$. If this is the case, we would expect the ratios (7.2.10, 11) to have the same magnitude and opposite signs at high energy. Since, if it is non-zero, D' will dominate the numerators of (7.2.10, 11), this result is of a more general nature than our use of a special model might suggest. However, if $D' = 0$, no general conclusions can be drawn from this model. We will consider this question more generally in §8.3.

(b) General theorems using univalent functions

Although the preceding results using simple functions are of considerable heuristic value and provide a convenient generalisation of

Regge theory, it is important also to know what can be rigorously stated about the phase and rate of growth of a scattering amplitude. This general problem including the possibility of oscillations has been considered by two methods, which will be outlined here. The first to be described will be that of Khuri & Kinoshita (1965c) based on the theory of univalent functions, which is a development from their earlier work (Khuri & Kinoshita, 1965a, b). The second method, to be described later in this section (see (c) below), is due to Jin & MacDowell (1965) and is based on the phase representation developed by Sugawara & Tubis (1963).

We will consider the high energy behaviour of a symmetric forward scattering amplitude,

$$F(E) = F_S(E) = \tfrac{1}{2}[F(\pi^+p) + F(\pi^-p)]. \tag{7.2.12}$$

We have noted in § 6.5 that very little is known about the possibility of oscillations of $\mathrm{Im}\,F$ or of $|F|$ as $E \to \infty$. It is not evident, even on intuitive grounds, that we can neglect possible oscillations, since as $E \to \infty$ there is a condensation of the branch points that occur at normal thresholds. For equal masses these are at,

$$E = (N^2 - 2)\,m/2, \quad N = \text{integer} \geqslant 2. \tag{7.2.13}$$

The work of Khuri & Kinoshita (1965c) (denoted by KK in this discussion) establishes bounds on the behaviour of $|F(E)|$ for given bounds on the behaviour of the ratio of its real to imaginary part. There would be ambiguity in the relation between these bounds if one did not use univalent functions.

A function $f(z)$ is univalent (schlicht, or simple) in a region D if it is regular, one-valued and does not take any value more than once in D. The essential feature of a univalent function that we require here is that it has a simple relation between bounds on its phase and bounds on its modulus. The amplitude itself may not be univalent but from it a univalent function can be constructed using the following properties:

(i) $F(E)$ is analytic in E, regular in the entire plane cut along the real axis on $(-\infty, -m)$ and (m, ∞).

(ii) $F(E)$ is hermitean, in particular for E real,

$$F(E - i\epsilon) = F^*(E + i\epsilon).$$

(iii) $F(E)$ is crossing symmetric

$$F(-E - i\epsilon) = F(E + i\epsilon).$$

(iv) $F(E)$ satisfies the Froissart bound,

$$|F(E)| < E \log^2 E \quad \text{as} \quad E \to \infty.$$

From these conditions, using $F'(0) = 0$, KK show that the function $G(E)$ is univalent in $\operatorname{Im} E > 0$, where G is defined by,

$$G(E) = \int_0^E dz \, \frac{F(z) - F(0)}{z^2}, \tag{7.2.14}$$

and z is restricted to $\operatorname{Im} z > 0$.

The properties of $F(E)$ ensure that as $|E| \to \infty$,

$$\left| \frac{G(E)}{E} \right| < \text{constant}. \tag{7.2.15}$$

Using the univalent function $G(E)$, a wide range of theorems can be established that relate its rate of growth to its phase. Some of these theorems are listed by KK, of which the following are examples.

KK *Theorem* 1. If for E real and greater than E_0,

$$\tan \pi\beta \leqslant \frac{\operatorname{Re} G(E)}{\operatorname{Im} G(E)} \leqslant \tan \pi\gamma, \tag{7.2.16}$$

with $0 < \beta \leqslant \gamma < \frac{1}{2}$, then for $E > E_0$,

$$C \left(\frac{E}{E_0} \right)^{2\beta} \leqslant |G(E)| \leqslant C' \left(\frac{E}{E_0} \right)^{2\gamma}. \tag{7.2.17}$$

In the special case for which

$$F(E) \sim E^\alpha \exp(i\theta), \tag{7.2.18}$$

we obtain $\beta = \gamma = (\alpha - 1)/2$, which agrees with our special result (7.1.26).

Note that this theorem of KK will give information about G even if F has severe oscillations. For example if $|F|$ is much smaller than $|E|^\alpha$ only for decreasingly small intervals as $|E| \to \infty$, then $G(E)$ defined by (7.2.14) will be dominated by the larger values of $F(E)$, so that the above theorem could still apply. Conversely it should also be noted that restrictions on the phase (or rate of growth) of $G(E)$ do not of necessity correspond to bounds on the phase (or the rate of growth) of $F(E)$, except in some 'average' sense. Our consideration of oscillations of $F(E)$ in § 6.5 suggests that one may have to be content with information about its 'average' behaviour as $E \to \infty$, for example as defined through the univalent function $G(E)$.

Further theorems by KK consider other possible rates of growth of $G(E)$.

KK Theorem 2. If for E real and greater than E_0,

$$\frac{\operatorname{Re} G(E)}{\operatorname{Im} G(E)} \leqslant \frac{C}{(\log E)^\alpha} \quad (\alpha \geqslant 1), \tag{7.2.19}$$

then for $E > E_0,\ \alpha > 1$,

$$|G(E)| \leqslant \text{constant}. \tag{7.2.20}$$

If $\alpha = 1$, then

$$|G(E)| \leqslant C''(\log E)^{C'}, \tag{7.2.21}$$

where C' can be found from E_0, C and α, and $C' \to 2(2C)^{\frac{1}{2}}$ as $E_0 \to \infty$.

KK Theorem 3. If, for E real,

$$\operatorname{Re} G(E) \geqslant b \quad \text{for} \quad E > E_0, \tag{7.2.22}$$

then, for $E > E_0$,

$$|G(E)| \geqslant \frac{2b}{\pi} \log\left(\frac{E}{E_0}\right) + \text{constant}. \tag{7.2.23}$$

KK Theorem 4. If, for E real,

$$0 \leqslant \operatorname{Re} G(E) \leqslant b' \quad \text{for} \quad E > E_0, \tag{7.2.24}$$

then, for $E > E_0$,

$$|G(E)| \leqslant \frac{2b'}{\pi} \log\left(\frac{E}{E_0}\right) + \text{constant}. \tag{7.2.25}$$

An example of the situation envisaged in Theorems 3 and 4 was considered in §7.1. Taking for large E

$$F(E) \sim iE(\log E - i\pi/2)^\beta, \tag{7.2.26}$$

we have

$$\frac{d\,G(E)}{dE} \sim \frac{i(\log E - i\pi/2)^\beta}{E}, \tag{7.2.27}$$

giving

$$G(E) \sim \frac{i(\log E - i\pi/2)^{\beta+1}}{(\beta+1)}. \tag{7.2.28}$$

Hence

$$\operatorname{Re} G(E) \sim \tfrac{1}{2}\pi(\log E)^\beta. \tag{7.2.29}$$

For $\beta > 0$, the conditions of Theorem 3 are satisfied, and for $\beta < 0$, we have the conditions of Theorem 4.

Reference should be made to the paper by Khuri & Kinoshita (1965c) for other results and for proofs of the above theorems. Their work can of course be extended to give analogous theorems for anti-symmetric amplitudes.

(c) *The phase representation*

The phase representation for a scattering amplitude $F(s, t)$ was developed by Sugawara & Tubis (1963). It provides a useful method for considering lower bounds, alternative to that of Herglotz functions (§ 6.5), and for considering asymptotic relations between the phase and the modulus of a scattering amplitude. We will outline the development and application of the phase representation to the latter problem, which was presented by Jin & MacDowell (1965).

Let $F(z)$ be the symmetric forward scattering amplitude, considered as a function of z, which is fined for πN scattering by

$$z = \frac{(s - m^2 - M^2)^2}{4m^2 M^2}. \tag{7.2.30}$$

Then $F(z)$ has the following properties:

(i)   it is analytic in the z-plane cut along $(1, \infty)$,

(ii) $|F| < |z|$ as $|z| \to \infty$,

(iii) $F(z^*) = (F(z))^*$.

In addition it is assumed that

(iv) For x real > 1, $\operatorname{Im} F(x)$ is continuous and has only a finite number of zeros (alternatively one can assume that $F(z)$ itself has only a finite number of zeros in the cut plane).

Such a function has only a finite number of zeros (Jin & Martin, 1964), say at $z_1, z_2, ..., z_n$. Then the function $\log G(z)$ is analytic in the cut z-plane, where

$$G(z) = \frac{F(z)}{\prod\limits_{1}^{n} (z - z_r)}. \tag{7.2.31}$$

Its discontinuity across the cut gives $\delta(x)$ the phase of $F(x)$,

$$\delta(x) = \tan^{-1}\left[\frac{\operatorname{Im} F(x)}{\operatorname{Re} F(x)}\right] = \frac{1}{2i} \log\left[\frac{F(x + i\epsilon)}{F(x - i\epsilon)}\right], \tag{7.2.32}$$

$$= \frac{1}{2i} \left[\log G(x + i\epsilon) - \log G(x - i\epsilon)\right]. \tag{7.2.33}$$

From assumption (iv) above, $\delta(x)$ is bounded, and the convention $\delta(1) = 0$ is adopted. Using this condition and properties (i), (ii) and (iii) the phase representation can be established. The phase representation for the amplitude $F(z)$ is given by

$$F(z) = A \prod_{1}^{n} (z - z_i) \exp\left[\frac{z}{\pi} \int_{1}^{\infty} \frac{\delta(x)\, dx}{x(x - z)}\right]. \tag{7.2.34}$$

When there is a pole at z_0, this expression gets multiplied by $(z - z_0)^{-1}$.

To illustrate the method we will assume: (a) that $F(z)$ has no poles; (b) that it has no zeros in the physical sheet; and (c) that $F(1) > 0$;

(more general possibilities are considered by Jin & MacDowell, 1965, denoted JM). It is also assumed that

(v) For $x > x_0$, $\left| \dfrac{\mathrm{Re}\, F(x)}{\mathrm{Im}\, F(x)} \right| <$ constant.   (7.2.35)

Then the phase of F must be bounded,

$$0 < 2\epsilon < \pi\delta_1 < \delta(x) < \pi\delta_2 < \pi - 2\epsilon. \tag{7.2.36}$$

From (7.2.34) with $n = 0$, and restricted by (7.2.36), one obtains the following bounds for $z \to \infty$ in any complex direction ($\theta \neq 0$),

$$C_2 |z|^{-\delta_2} (\sin \tfrac{1}{2}\theta)^{(\delta_2 - \delta_1)} < |F(z)| < C_1 |z|^{-\delta_1} (\sin \tfrac{1}{2}\theta)^{(\delta_1 - \delta_2)}. \tag{7.2.37}$$

Along the real axis ($\theta = 0$) the discussion is more complicated and we will just quote some of the results of JM.

JM Theorem 1. For $\theta \neq 0$ the bounds (7.2.37) are valid. Along the real axis ($\theta = 0$), ($0 < \delta_1 < \delta_2 < \tfrac{1}{2}$),

$$\frac{F(x)}{x^p} \text{ is integrable } L(1, \infty) \quad \text{for } p > 1 - \delta_1, \tag{7.2.38}$$

$$\frac{x^p}{F(x)} \text{ is integrable } L(1, \infty) \quad \text{for } p < -1 - \delta_2, \tag{7.2.39}$$

where $L(1, \infty)$ means that the integral from 1 to ∞ is finite (the above theorem becomes modified when there are zeros or poles).

JM Theorem 2. If for x real $> x_0$,

$$\beta_1 < [\delta(x) - \tfrac{1}{2}\pi] \log x < \beta_2, \tag{7.2.40}$$

then $\dfrac{F(x)}{x^{\frac{1}{2}}(\log x)^p}$ is integrable $L(1, \infty)$ for $p > 1 - \dfrac{2\beta_1}{\pi}$,   (7.2.41)

$$\frac{x^{-\frac{1}{2}}(\log x)^p}{F(x)} \text{ is integrable } L(1, \infty) \quad \text{for } p < -1 - \frac{2\beta_2}{\pi}. \tag{7.2.42}$$

If $n_0 =$ (number of zeros) minus (number of poles) of $F(z)$, the power of x in (7.2.41) is $-(n_0 + \tfrac{1}{2})$ instead of $-\tfrac{1}{2}$, and in (7.2.42) is $+(n_0 - \tfrac{3}{2})$. Thus if the Froissart bound holds ($F \leqslant x^{\frac{1}{2}} \log^2 x$), we must have either $n_0 = 0$, or $n_0 = 1$ and $\beta_2 > -\pi$.

JM Theorem 3. If for $x > x_0$, σ(total) (for a symmetric amplitude) decreases steadily to $\sigma(\infty) =$ constant, but not faster than $C/(x^{\frac{1}{2}} \log x)$, then $\mathrm{Re}\, F(x)$ becomes negative for sufficiently large x and tends to minus infinity.

The proof of this result is given by JM. Its plausibility is indicated by the following example chosen to satisfy the crossing conditions of § 7.1 for a symmetric amplitude.

$$F(E) \sim iE - C(-iE)^{1-\alpha}\left(\log E - \frac{i\pi}{2}\right)^{-\beta}. \qquad (7.2.43)$$

Note that $E = x^{\frac{1}{2}}$, where x is the variable used above and by JM. Then

$$\sigma(\text{total}) \sim \frac{\text{Im}\,F}{E} \sim 1 + \frac{C}{E^\alpha}\left[\sin\frac{(1-\alpha)\,\pi}{2}\right](\log E)^{-\beta}, \qquad (7.2.44)$$

and

$$\text{Re}\,F(E) \sim -C\left[\cos\frac{(1-\alpha)\,\pi}{2}\right]E^{1-\alpha}(\log E)^{-\beta}. \qquad (7.2.45)$$

Thus to satisfy the conditions of JM Theorem 3,

$$C > 0, \quad 0 < \alpha \leqslant 1, \quad 0 < \beta < 1. \qquad (7.2.46)$$

Then

$$\text{Re}\,F(E) \to -\infty \quad \text{as} \quad E \to \infty. \qquad (7.2.47)$$

It should be noted that in this example, although (7.2.47) holds, we have

$$\frac{\text{Re}\,F(E)}{\text{Im}\,F(E)} \to 0 \quad \text{as} \quad E \to \infty, \qquad (7.2.48)$$

where the limit is approached through negative values.

7.3   Martin's extension of the Lehmann ellipse

In this section we will follow the main features of the paper by Martin (1966a), but for illustration will include some results based on non-oscillating functions (Eden, 1967). We will begin with a simplified form of Martin's proof that neglects the left-hand cut in the energy plane, and will later indicate what modifications are needed to include it. We will assume the improved Greenberg–Low bound (proved by Martin, 1965b and Eden 1966a)

$$\text{Im}\,F(E, 0) < E^2(\log E)^{-2}. \qquad (7.3.1)$$

Note that the proof of the Froissart bound in § 6.1 depends on Martin's extension of the Lehmann ellipse. The result (7.3.1) permits a dispersion relation with only two subtractions,

$$F(E, 0) = A + BE + \frac{E^2}{\pi}\int_m^\infty \frac{dx\,\text{Im}\,F(x, 0)}{x^2(x - E)} + \frac{E^2}{\pi}\int_{-\infty}^{-m} \frac{dx\,\text{Im}\,F(x, 0)}{x^2(x - E)}. \qquad (7.3.2)$$

with $A = F(0, 0)$, $B = (d/dE)\,F(E, 0)$ evaluated at $E = 0$. The first step is to establish that a similar dispersion relation exists for the derivatives of $F(E, t)$ with respect to t at $t = 0$. This step is achieved from (7.3.2) using unitarity and the properties of Legendre polynomials. We consider equal mass particles, for which $s = 2m(E + m)$. Then

$$\text{Im } F(E, t) = \frac{8\pi s^{\frac{1}{2}}}{k} \sum_0^\infty (2l + 1)\, a_l(E)\, P_l(\cos\theta). \qquad (7.3.3)$$

From unitarity, for real positive E,

$$0 \leqslant a_l(E) \leqslant 1, \qquad (7.3.4)$$

and for real $t < 0$, $|\cos\theta| < 1$, so that

$$|P_l(\cos\theta)| < 1. \qquad (7.3.5)$$

Hence for physical values of E, t, $(t < 0)$,

$$|\text{Im } F(E, t)| < \text{Im } F(E, 0). \qquad (7.3.6)$$

Also, for $|\cos\theta| < 1$,

$$\left| \left(\frac{d}{d\cos\theta} \right)^n P_l(\cos\theta) \right| < \left(\frac{d}{d\cos\theta} \right)^n P_l(1). \qquad (7.3.7)$$

The right-hand expression denotes the nth derivative of $P_l(\cos\theta)$ evaluated at $\cos\theta = 1$.

(a) A simplified method

From the inequality (7.3.6), it follows that the dispersion relation (with two subtractions) for $F(E, t)$ converges for $t \leqslant 0$. Neglecting the left-hand cut and the subtraction terms for the present, we have for $E = E_0$,

$$F(E_0, t) = \frac{E_0^2}{\pi} \int_m^\infty \frac{dx\, \text{Im } F(x, t)}{x^2(x - E_0)}. \qquad (7.3.8)$$

It is convenient to choose E_0 in the range $(-m < E_0 < m)$. In this range $F(E_0, t)$ is regular for some finite domain in t (see references given in § 4.8),

$$|t| < R \quad (R > 0). \qquad (7.3.9)$$

Hence $|F|$ is bounded by some constant M on $|t| = R$, and

$$\left| \left(\frac{d}{dt} \right)^n F(E_0, 0) \right| = \left| \frac{n!}{2\pi i} \oint_{|t|=R} \frac{dt\, F(E_0, t)}{t^{n+1}} \right| < \frac{Mn!}{R^n}, \qquad (7.3.10)$$

where the left-hand expression denotes the nth derivative of $F(E, t)$ at $t = 0$.

For $-m < E_0 < m$, the denominator in (7.3.8) is positive. The inequality (7.3.10) with $n = 1$, can be used with (7.3.3) to (7.3.9) to prove that

$$\frac{d}{dt} F(E_0, 0) = \frac{E_0^2}{\pi} \int_m^\infty dx \, \frac{\frac{d}{dt} \operatorname{Im} F(x, t = 0)}{x^2 (x - E)}. \tag{7.3.11}$$

The inequality (7.3.7) is of particular importance with (7.3.4) in proving that the dispersion relation (7.3.11) converges. Similarly, Martin (1966a) shows that for any n, the nth derivative at $t = 0$ satisfies a convergent dispersion relation,

$$\left(\frac{d}{dt}\right)^n F(E_0, 0) = \frac{E_0^2}{\pi} \int_m^\infty dx \, \frac{\left(\frac{d}{dt}\right)^n \operatorname{Im} F(x, 0)}{x^2 (x - E_0)}. \tag{7.3.12}$$

Using (7.3.10), $\quad \dfrac{M n!}{R^n} > \dfrac{E_0^2}{\pi} \displaystyle\int_m^\infty dx \, \dfrac{\left(\frac{d}{dt}\right)^n \operatorname{Im} F(x, 0)}{x^2 (x - E_0)}.$ (7.3.13)

(b) *The Froissart bound*

Before following Martin's method of using this inequality to enlarge the Lehmann ellipse, we will derive the Froissart bound assuming that the total cross-section does not oscillate. From the inequalities obtained in § 6.2 (see also Eden, 1967), we have for energy x,

$$(n!) \left(\frac{d}{dt}\right)^n \operatorname{Im} F(x, 0) > x [\sigma_{\text{total}}(x)]^{n+1}. \tag{7.3.14}$$

If for large x, $\quad\quad\quad \sigma_{\text{total}}(x) > x^\alpha,$ (7.3.15)

then from (7.3.13), we obtain a contradiction if $\alpha > 0$, because for n sufficiently large the integral would not be convergent. Hence $\alpha \leqslant 0$. If for large x

$$\sigma_{\text{total}}(x) > (\log x)^\beta, \tag{7.3.16}$$

then from (7.3.14), for consistency with (7.3.13) we must have,

$$\frac{M n!}{R^n} > \frac{E_0^2}{\pi} \int_m^\infty \frac{dx \, (\log x)^{(n+1)\beta}}{x^2 (x - E_0) \, (n!)}. \tag{7.3.17}$$

The integral is dominated by the value of the integrand near

$$\log x \approx n, \tag{7.3.18}$$

when n is large. It follows that, in order not to contradict the inequality, (7.3.17),

$$\beta \leqslant 2, \tag{7.3.19}$$

so that for large E we have the Froissart bound,

$$\sigma_{\text{total}}(E) \leqslant (\log E)^2. \tag{7.3.20}$$

(c) *The Martin–Lehmann ellipse*

We return now to the method of Martin (1966a). The relation (7.3.12) holds for any E_0 in $(-m, m)$. It can therefore be continued analytically from E_0 to complex E, as long as the integral converges. The integral converges also for arbitrary complex E_1 because, for $x > m$,

$$\left| \frac{\left(\frac{d}{dt} \right)^n \operatorname{Im} F(x, 0)}{x - E_1} \right| \leqslant \left[\frac{\left(\frac{d}{dt} \right)^n \operatorname{Im} F(x, 0)}{x - E_0} \right] B(E_1, E_0), \qquad (7.3.21)$$

where

$$B(E_1, E_0) = \max_{x > m} \left| \frac{x - E_0}{x - E_1} \right|. \qquad (7.3.22)$$

Hence,

$$\left| \left(\frac{d}{dt} \right)^n F(E_1, 0) \right| < \left(\frac{E_1}{E_0} \right)^2 B(E_1, E_0) \frac{Mn!}{R^n}. \qquad (7.3.23)$$

This proves that the series

$$F(E_1, t) = \sum_0^\infty c_n(E_1) \, t^n, \qquad (7.3.24)$$

converges for

$$|t| < R, \qquad (7.3.25)$$

where R does not depend on E_1. This is Martin's main result. It should be compared with the shrinkage C/E of the analytic neighbourhood of $t = 0$ that corresponds to the Lehmann ellipse (see §4.8).

From (7.3.22) we see that,

$$\text{for} \quad E_1 = E + i\delta, \quad B \sim E/\delta. \qquad (7.3.26)$$

So for large E, and $|t| < R$,

$$|F(E + i\delta, t)| < \frac{E^3}{\delta E_0^2} |F(E_0, t)|. \qquad (7.3.27)$$

Thus $F(E, t)$ is polynomial bounded, provided $|t| < R$.

The extension of the domain (7.3.25) to an ellipse follows from the partial wave series and properties of Legendre polynomials. Consider the series (7.3.3) for $\operatorname{Im} F(s, t)$, for values of t such that $z = \cos \theta$ lies on a certain ellipse with foci at $z = \pm 1$. Each term of the series (7.3.3) is a maximum when t is real and positive (i.e. when z is real and greater than one). The series converges for this real positive value of t if it lies inside $|t| < R$. Hence it converges inside the corresponding ellipse.

The domain of analyticity has been further extended by Martin (1966a). His work includes a proof that $R = 4\mu^2$, where μ is the pion mass. By crossing symmetry and analytic completion it is possible to obtain further extensions of the domain of analyticity of $F(s, t)$ as a function of the two complex variables (see § 4.8).

(d) The left-hand cut

We return now to the terms omitted from the dispersion integral (7.3.8) in the simplified method described above. The amplitude $F(E_0, t)$ is known to be regular in $|t| < R$, and $F(E, t)$ is regular near $E = 0$. The subtraction terms are simply $F(0, t)$ and the derivative $(d/dE) F(0, t)$. Both of them are regular near $t = 0$. Hence the difference of F and the subtraction terms is regular in $|t| < R$, and one can proceed as before.

The integral over the left-hand cut causes considerable complication (Martin, 1966a) but purely of a technical nature. The difficulty is that the left-hand integral has the form for $t \neq 0$

$$\frac{E_0^2}{\pi} \int_{-\infty}^{L(t)} dx \, \frac{\operatorname{Im} F(x, t)}{x^2(x - E_0)}, \tag{7.3.28}$$

where $L(t) = -(t - 2m^2)/2m$. When we differentiate with respect to t, there are contributions from the integrand that are positive at $t = 0$, but in addition there are contributions from the t-dependent limit of integration that are not positive. This prevents the use of (7.3.21) to prove convergence of the new form of (7.3.12) when E_0 is replaced by E_1. This difficulty can be overcome by introducing a new function related to $F(E, t)$ whose derivatives do have the desired properties. The reader is referred to the original paper by Martin (1966a) for further details.

7.4 Partial wave amplitudes at high energy

The domain of analyticity of $F(s, t)$ established by Martin (1966b) leads to a corresponding domain of analyticity of the partial wave amplitude $f_l(s)$ for integer l in the complex s-plane. For the $\pi\pi$ partial wave amplitude, this domain extends from $s = 78m^2$ to $s = -28m^2$, but is cut on the real axis below $s = 0$ and above $s = 4m^2$, where m is the pion mass. Its dimensions are of similar magnitude in the direction of imaginary s, and there are no other branch cuts within this (finite) domain of analyticity.

With additional assumptions further results can be established. Some of the methods used will be outlined in this section. They all assume that $f_l(s)$ is analytic in the s-plane, cut along part of the real axis (e.g. along $(-\infty, 0)$ and $(4m^2, \infty)$). This result has been proved for every term in perturbation theory in the equal mass case (Eden, Landshoff, Olive & Polkinghorne, 1966), but it has not been proved in general. The result follows from the cut-plane analyticity of $F(s, t)$ in s and t, that is assumed for the Mandelstam representation, but it would also hold with weaker analyticity assumptions about $F(s, t)$.

If we assume cut-plane analyticity of $f_l(s)$, then we can discuss the problem of its behaviour as s tends to infinity. We will do this in two stages, (a) by relating the behaviour of $f_l(s)$ to specific simple models of $F(s, t)$, and (b) by discussing the number of subtractions required in a dispersion relation for $f_l(s)$ when specific assumptions are made about the oscillations of $\mathrm{Im}\, f_l(s)$ for $s < 0$, as $s \to -\infty$.

(a) Partial wave amplitudes for special models

A model giving partial wave amplitudes has been described by Van Hove (1963, 1964); another model is provided by Regge theory (see Chapter 5); both will be used in part, for this discussion. Intuitively it is clear that a large number of partial waves should contribute near to the forward direction at high energies. Since $\mathrm{Im}\, f_l \leqslant 1$, at least $L \sim \sqrt{s}$ partial waves must contribute to forward scattering, if the total cross-section is constant. At $t = 4m^2$, the partial wave series must diverge, so using the result (6.1.14), we can deduce that

$$\mathrm{Im}\, f_l(s) \sim C \exp\left[-l(t_0/s)^{\frac{1}{2}}\right] \quad \text{as } l \to \infty, \tag{7.4.1}$$

where t_0 is constant, and l has real integer values. There are, of course, many functions with this behaviour for large l. For smaller values of l, in the neighbourhood of $L \sim (s/t_0)^{\frac{1}{2}}$ for large s, $\mathrm{Im}\, f_l(s)$ appears to have a somewhat different form if one assumes, in agreement with experimental results, that

$$\mathrm{Im}\, F(s, t) \sim F_1(s) \exp\left(-As\theta^2\right), \tag{7.4.2}$$

where θ is the scattering angle in the centre of mass system. Then using $(As) \gg 1$, and

$$\lim_{l \to \infty} P_l(1 - x^2/2l^2) = J_0(x), \tag{7.4.3}$$

$$\int_0^\infty x\, dx \exp\left(-\tfrac{1}{2}x^2\right) J_0(x\xi) = \exp\left(-\tfrac{1}{2}\xi^2\right), \tag{7.4.4}$$

one finds

$$\mathrm{Im}\, f_0(s) \sim \frac{F_1(s)}{64\pi As}, \tag{7.4.5}$$

$$\mathrm{Im}\, f_l(s) \sim \mathrm{Im}\, f_0(s) \exp\left[-l^2/4As\right]. \tag{7.4.6}$$

Unitarity requires $\mathrm{Im} f_0 \leqslant 1$, and an asymptotically constant total cross-section requires that $F_1 \sim C_1 s$, as $s \to \infty$.

For large momentum transfer, one might expect considerably fewer partial waves to contribute to the differential cross-section

$$\frac{d\sigma}{dt} \sim \left| \frac{F(s,t)}{s} \right|^2. \tag{7.4.7}$$

However, taking only a few partial waves will not be sufficient to make $d\sigma/dt$ decrease like $\exp[-Ct^{\frac{1}{2}}]$, as observed experimentally for large t and large s (or $\exp[-Cs^{\frac{1}{2}}]$ at fixed angle, (Orear, 1964)). There must in addition be a rather intricate cancellation. However the cancellation between partial waves must not be so good as with the Gaussian form (7.4.6) since this gives $F(s,t) \sim \exp[-Ct]$ for large t, which is too small. It has been noted (Hagedorn, 1965) that a statistical model for partial waves would give behaviour having a magnitude like

$$F(s,t) \sim \exp[-Ct^{\frac{1}{2}}] \tag{7.4.8}$$

for large t. However this would also produce fluctuations in $d\sigma/dt$ as t varies (Ericson, 1964), which are not observed (Van Hove, 1966). It seems probable that a mixture of the above possibilities could fit all the data, but it is difficult to make this very qualitative discussion more precise, neither is it necessarily reasonable to assume that partial waves give a useful basis for formulating a model for high energy scattering.

We conclude these remarks on special models with a note about the partial wave amplitude using a Regge model. Assume a single pole model and consider a crossing symmetric process. Then

$$F(s,t) \sim -C(-is)^{\alpha(t)}. \tag{7.4.9}$$

One should use $W = (s + \frac{1}{2}t - 2m^2)$ instead of s, for exact crossing symmetry, but the correction term is small as $s \to \infty$. Assuming $\alpha(0) = 1$, the partial wave amplitude will be, for $l \ll L \sim s^{\frac{1}{2}}$,

$$f_l(s) \sim f_0(s) \sim \frac{iC'}{\log s}. \tag{7.4.10}$$

Thus this crossing symmetric Regge model gives $f_l(s) \to 0$, as $s \to \infty$. For a model with a non-shrinking forward peak one would get $f_l(s) \to iC''$, as $s \to \infty$. Unitarity would require $0 \leqslant C'' \leqslant 1$. It should be noted that results of this kind are model-dependent.

(b) Subtractions in partial wave dispersion relations

The previous discussion has been for real $s \to \infty$. The problem of subtractions requires information about $f_l(s)$ for fixed integer l as $|s| \to \infty$, with s complex. This has been considered by Kinoshita (1966), who proves under certain assumptions that the dispersion relation for $f_l(s)$ requires no more than one subtraction for any l. His assumptions are that $f_l(s)$ has the following properties:

(1) It is regular in the s-plane cut along $(-\infty, s_0)$ and $(4\mu^2, \infty)$; real in $(s_0, 4\mu^2)$; continuous on the cuts; and has no essential singularity at any finite point on the cuts.

(2) It has threshold behaviour on $s + i0$,

$$f_l(s) = \frac{(s - 4\mu^2)^{l+\frac{1}{2}}}{s^{\frac{1}{2}}} F_l(s), \qquad (7.4.11)$$

where $F_l(s)$ remains finite as $s \to 4\mu^2$.

(3) On the right-hand cut $f_l(s)$ is unitary,

$$0 \leqslant |f_l(s)|^2 \leqslant \mathrm{Im} f_l(s) \leqslant 1. \qquad (7.4.12)$$

(Note that the normalisation of $f_l(s)$ used here differs from that used by Kinoshita.)

(4) For sufficiently large s, $f_l(s)$ satisfies,

$$|f_l(s)| < \exp[C(\log|s|)^{2-\epsilon}] \quad \text{where} \quad \epsilon > 0. \qquad (7.4.13)$$

(5) The number of times that the discontinuity $\mathrm{Im} f_l(s + i0)$ changes its sign in the interval $(s, 0)$ does not exceed $C'[\log|s|]^{1-\epsilon}$ as $s \to \infty$ along the negative real axis, with $\epsilon > 0$.

The proof that not more than one subtraction is required in the dispersion relation for $f_l(s)$ is based on the method of Herglotz functions, which was described in § 7.2. The details are given by Kinoshita (1966). He has noted the rather unsatisfactory feature that the assumptions (4) and (5) have not been derived from more basic axioms. The assumption (4) is not obviously compatible with the asymptotic behaviour deduced from Mandelstam–Regge branch cuts (Mandelstam, 1963). The assumption (5) suggests there may be some relation between this problem and the problem of establishing limits on possible oscillations of the forward amplitude which we discussed in § 6.5. The use of partial wave dispersion relations is crucially dependent on establishing a polynomial bound on the high energy behaviour of partial wave amplitudes. To do this rigorously from quantum field theory, or S-matrix theory, one requires a further study of the foundation of the above assumptions.

CHAPTER 8

ASYMPTOTIC RELATIONS BETWEEN CROSS-SECTIONS

It was pointed out by Pomeranchuk (1956), and by Okun & Pomeranchuk (1956), that the vanishing of charge exchange cross-sections at high energy, together with isospin conservation, leads to asymptotic relations between cross-sections for different processes. In 1958 Pomeranchuk proved from plausible assumptions that total cross-sections for particle-particle scattering and for antiparticle-particle scattering become asymptotically equal. In this chapter we will consider these results and their generalisations. Amongst these is the asymptotic equality of differential cross-sections for particle-target and antiparticle-target collisions at fixed momentum transfer.

The asymptotic equality of particle and antiparticle total cross-sections will be referred to as the Pomeranchuk theorem. The assumption that charge exchange cross-sections tend to zero at high energies will be called the Pomeranchuk–Okun rule. An important converse of the latter noted by Foldy & Peierls (1963), is that if one exchange process dominates over all others at high energy, that process can involve only the exchange of vacuum quantum numbers. This does not exclude the possibility of some inelastic processes having constant cross-sections at high energy, as in diffraction dissociation (see § 8.4, Good & Walker, 1960; Morrison, 1966).

We begin (§ 8.1) by considering the specific consequences of crossing symmetry. Next we examine (§ 8.2) the combination of crossing symmetry and isospin invariance, and use the latter to illustrate the results on vacuum exchange dominance. In § 8.3 we summarise some consequences of more general symmetries, and in § 8.4 we consider some further examples to illustrate general results on exchange amplitudes.

8.1 The Pomeranchuk theorems on total and differential cross-sections

There are useful reviews of these theorems by Greenberg (1964) and Van Hove (1964) and to some extent I will follow the latter's methods which are based on the work of Meimann (1962). They involve essentially the same mathematics as we used in Chapter 7, when discussing phase relations and high energy behaviour.

(a) Differential cross-sections

Define the 'crossing' energy variable W, by

$$W = s + \tfrac{1}{2}t - (m^2 + M^2), \qquad (8.1.1)$$

where m and M are the masses of the colliding particles. Crossing symmetry relates the processes,

$$\text{I} \qquad A + B \to A + B \quad \text{energy } s,$$

$$\text{II} \qquad A + \bar{B} \to A + \bar{B} \quad \text{energy } u,$$

$$\text{III} \qquad A + \bar{A} \to B + \bar{B} \quad \text{energy } t.$$

As in Chapter 7, we obtain

$$F_{\text{I}}(-W + i0, t) = F_{\text{II}}^*(W + i0, t). \qquad (8.1.2)$$

Following Meimann (1962) and Logunov *et al.* (1963), we assume that,

(i) for fixed real t, $F_{\text{I}}(W, t)$ and $F_{\text{II}}(W, t)$ are regular in the region $\text{Im } W > 0$, and continuous on $\text{Im } W = 0$,

(ii) for fixed real t, F_{I} and F_{II} are polynomial bounded in W, as $|W| \to \infty$,

(iii) for fixed real t, as $W \to \infty$ through real values,

$$\frac{F_{\text{I}}(W + i0, t)}{W^\alpha (\log W)^\beta} \to C_1(t), \qquad (8.1.3)$$

$$\frac{F_{\text{II}}(W + i0, t)}{W^\alpha (\log W)^\beta} \to C_2(t). \qquad (8.1.4)$$

From the Phragmén–Lindelöff theorem (see § 7.1), the functions in (8.1.3, 4) must tend to the same limits, C_1 and C_2 respectively, as $W \to -\infty$. Using (8.1.2) this gives

$$C_1(t) \sim \frac{F_{\text{I}}(-W + i0, t)}{(-W)^\alpha [\log(-W)]^\beta} \to C_2^*(t) \exp(-i\pi\alpha), \qquad (8.1.5)$$

where $-W$ denotes $W \exp(i\pi)$. Therefore,

$$C_1(t) = C_2^*(t) \exp(-i\pi\alpha). \qquad (8.1.6)$$

The asymptotic equality of the differential cross-sections for processes I and II follows from (8.1.6) since, as $W \to \infty$,

$$\frac{(d\sigma_{\text{I}}/dt)}{(d\sigma_{\text{II}}/dt)} = \left| \frac{F_{\text{I}}(W + i0, t)}{F_{\text{II}}(W + i0, t)} \right|^2 \to \left| \frac{C_1(t)}{C_2(t)} \right|^2 = 1. \qquad (8.1.7)$$

This does *not* establish the asymptotic equality of total cross-sections, since these are related to $\mathrm{Im}\,F(W, 0)$ which depends on the phase. Note also (*a*) that this result holds only for fixed t and does not give any asymptotic relation at fixed angle, and (*b*) that we have assumed the absence of oscillations for $t \leqslant 0$; although this may hold asymptotically, it has not been proved from the axioms of field theory.

An interesting consequence of (8.1.7) is the asymptotic equality of total elastic cross-sections, assuming that they are dominated by the forward peak. For example, as the energy tends to infinity,

$$\frac{\sigma(\text{elastic}, \pi^-p)}{\sigma(\text{elastic}, \pi^+p)} \to 1. \tag{8.1.8}$$

(b) *Total cross-sections*

The Pomeranchuk theorem (1958) asserts that total cross-sections for particle-target collisions and for antiparticle-target collisions become asymptotically equal at high energy. The proof of this theorem requires assumptions that are additional to results that have been established from quantum field theory. These assumptions involve (A) some form of restriction on oscillations at high energy, and (B) a requirement that the real part of the forward amplitude is not too large compared with the imaginary part. The precise ingredients for these assumptions can be varied. We will outline a proof of Pomeranchuk's result and will indicate some possible variations in the assumptions (A) and (B).

We make use of the following properties of the forward scattering amplitude $F(E, 0)$,

(i) $F(E, 0)$ is analytic in E, and regular in the E-plane, cut along $(-\infty, -m)$ and (m, ∞), where m is the mass of the incident particle (pion), for F_{I} and F_{II},

(ii) $|F(E, 0)| < |E \log^2 E|$, as $|E| \to \infty$, for F_{I} and F_{II}. Our additional assumption (A) is that $\sigma_{\mathrm{I}}(\text{total})$ and $\sigma_{\mathrm{II}}(\text{total})$ satisfy the asymptotic relations,

$$(A) \quad \sigma_{\mathrm{I}}(\text{total}) \to C_{\mathrm{I}}, \quad \sigma_{\mathrm{II}}(\text{total}) \to C_{\mathrm{II}}, \quad \text{as } E \to \infty, \tag{8.1.9}$$
$$C_A = C_{\mathrm{I}} - C_{\mathrm{II}}.$$

Our additional assumption (B) is

$$(B) \quad \frac{\mathrm{Re}\,F(E, 0)}{\log E \,\mathrm{Im}\,F(E, 0)} \to 0 \quad \text{for both } F_{\mathrm{I}} \text{ and } F_{\mathrm{II}}, \text{ as } E \to \infty. \tag{8.1.10}$$

Define the antisymmetric amplitude F_A, by

$$F_A(E, 0) = F_{\mathrm{I}}(E, 0) - F_{\mathrm{II}}(E, 0). \tag{8.1.11}$$

It was shown in § 7.1 that F_A satisfies a dispersion relation which, from (ii) above, will have no more than two subtractions,

$$F_A(E, 0) = EF'_A(E = 0, 0) + \frac{2E^3}{\pi} \int_m^\infty \frac{dx \operatorname{Im} F_A(x, 0)}{x^2(x^2 - E^2)}. \quad (8.1.12)$$

The even subtraction terms in F_A are zero, due to its antisymmetry, and its reality near $E = 0$. From the optical theorem, for large E,

$$\operatorname{Im} F_A(E, 0) \sim C_A E, \quad (8.1.13)$$

where C_A is the difference (8.1.9). Substituting this into the dispersion relation we obtain,

$$\operatorname{Re} F_A(E, 0) \sim -\frac{2}{\pi} C_A E \log E. \quad (8.1.14)$$

For the symmetric amplitude, $\operatorname{Re} F_S$ is small compared with $\operatorname{Im} F_S$ (see § 7.1), hence from (8.1.14) using (8.1.11),

$$\frac{\operatorname{Re} F_{\mathrm{I}}(E, 0)}{\operatorname{Im} F_{\mathrm{I}}(E, 0)} \sim \frac{-(2/\pi) C_A \log E}{\sigma_{\mathrm{I}}(\text{total})}. \quad (8.1.15)$$

The assumption (8.1.10) applied to (8.1.15) gives

$$C_A = 0, \quad (8.1.16)$$

and hence, $\qquad \sigma_{\mathrm{I}}(\text{total}) - \sigma_{\mathrm{II}}(\text{total}) \to 0 \quad \text{as } E \to \infty. \quad (8.1.17)$

This proves the Pomeranchuk theorem on the asymptotic equality of total cross-sections.

It is unfortunate that assumption (B), (8.1.10), is so directly related to the result and is not obvious on physical grounds. At one time it was thought that $\operatorname{Re} F/\operatorname{Im} F$ should tend to zero if total cross-sections are to be constant, but we have seen in § 7.2 that this need not be true, and in Chapter 2 that experiments on proton-proton scattering may give $(\operatorname{Re} F/\operatorname{Im} F) \sim$ constant at high energies. Also, there is no rigorous proof that $(\operatorname{Re} F/\operatorname{Im} F)$ must not increase like $\log E$. However this seems unlikely, as we can see from the following example.

Assume that the total cross-sections σ_{I} and σ_{II} are asymptotically constant, so (8.1.14) holds. Then, from the inequalities obtained in § 6.3, for each process I and II,

$$\sigma(\text{elastic}) > \frac{C|F(E, 0)|^2}{E^2 g(E)} \quad \text{for large } E, \quad (8.1.18)$$

where $g(E) \leqslant \log^2 E$, is the value of the inverse half-width of the forward peak (in deriving (8.1.18) one must also use the methods of

Chapter 7 to show that $\text{Re } F$ and $\text{Im } F$ vary in a similar way near $t = 0$). Substituting from (8.1.14) into (8.1.18), we obtain

$$\sigma_{\text{I}}(\text{total}) \geqslant \sigma_{\text{I}}(\text{elastic}) > \frac{C_A^2 \log^2 E}{g(E)}. \qquad (8.1.19)$$

This is consistent with $\sigma(\text{total}) \sim$ constant and C_A non-zero, only if $g(E)$ attains its bound $\log^2 E$, so that there is the maximum rate of shrinkage of the forward peak. In particular if Regge theory holds, $g(E) \sim \log E$, and the Pomeranchuk theorem follows, since C_A must be zero (from (8.1.19)).

The theorem can also be proved from unitarity (Eden, 1966b), if the total cross-sections increase so that

$$[\sigma_{\text{I}} - \sigma_{\text{II}}] \sim C_A (\log E)^m \quad 0 < m \leqslant 2. \qquad (8.1.20)$$

Then (8.1.14) is replaced by

$$\text{Re } F_A(E, 0) \sim -\frac{2C_A E}{\pi(m+1)} (\log E)^{m+1}. \qquad (8.1.21)$$

Now (8.1.19) becomes, for process I (or for II),

$$(\log E)^m > \frac{C_A^2 (\log E)^{2m+2}}{g(E)} \quad \text{for large } E. \qquad (8.1.22)$$

Even if $g(E)$ takes its maximum value $\log^2 E$, the result (8.1.22) is inconsistent for $m > 0$ unless $C_A = 0$. This proves that the Pomeranchuk theorem follows from unitarity if total cross-sections increase as fast as $(\log E)^\epsilon$ for any positive ϵ. In this case the theorem asserts that the ratio of particle-target and antiparticle-target cross-sections tends to unity. The most likely direction of extending this result to the physical case, where cross-sections appear to be asymptotically constant, is to obtain an improvement on the bound $g(E) \leqslant \log^2 E$. From the results of §6.2, any improved bound on $g(E)$ would also improve the Froissart bound. The counter example given in §6.2 show that this probably cannot be done trivially.

(c) Possible oscillations

If the difference between the cross-sections σ_{I} and σ_{II} oscillates infinitely many times as $E \to \infty$, very little can be established from crossing symmetry. There may, for example, be infinitely many limit points. To illustrate the problem we conclude this section by quoting the following theorem (see Meimann, 1962).

Let $F(E)$ be regular in $\text{Im } E > 0$, continuous on $\text{Im } E = 0$. Denote by M_1 the manifold of limit points of $F(E + i0)$ as $E \to +\infty$ along the real axis, and by M_2 the manifold of limit points as $E \to -\infty$. Each of these manifolds consists of either one point or a continuum. The theorem asserts that if $|F(E)|$ is bounded in Im $E \geqslant 0$, then *either* the manifolds M_1 and M_2 have a point in common, *or* one manifold surrounds the other.

For a further discussion of the problem of oscillations, see Bessis & Kinoshita (1966), and Eden & Łukaszuk (1967).

8.2   Consequences of isospin invariance

In this section we take isospin invariance as an example of a symmetry principle that leads to certain relations between cross-sections at high energy. Some generalisations to other symmetry principles are considered in § 8.3.

(a) Charge exchange scattering; *consequence of the Pomeranchuk–Okun rule*

Following Pomeranchuk (1956), and Okun & Pomeranchuk (1956), we make the hypothesis that charge exchange cross-sections become vanishingly small compared with the corresponding elastic cross-sections at high energy. We will refer to this hypothesis as the 'Pomeranchuk–Okun rule'; it is supported by experimental results. We begin by considering two examples that illustrate the general features of the problem. For antinucleon-nucleon scattering there are three antiproton reactions that do not involve particle production; denote the cross-sections by σ_1, σ_2, σ_3,

$$\sigma_1: \quad \bar{p}+p \to \bar{p}+p, \tag{8.2.1}$$

$$\sigma_2: \quad \bar{p}+n \to \bar{p}+n, \tag{8.2.2}$$

$$\sigma_3: \quad \bar{p}+p \to \bar{n}+n. \tag{8.2.3}$$

There are also three antineutron reactions, which are equivalent to the antiproton reactions by charge symmetry.

Assuming isospin conservation, these reactions can be described by two amplitudes $f_0(s,t)$ and $f_1(s,t)$ that correspond to isospin 0 and 1 respectively. Then,

$$\left(\frac{k^2}{\pi}\right)\frac{d\sigma_1}{dt} = \tfrac{1}{4}|f_0+f_1|^2, \tag{8.2.4}$$

$$\left(\frac{k^2}{\pi}\right)\frac{d\sigma_2}{dt} = |f_1|^2, \tag{8.2.5}$$

$$\left(\frac{k^2}{\pi}\right)\frac{d\sigma_3}{dt} = \tfrac{1}{4}|f_0-f_1|^2, \tag{8.2.6}$$

where k is the three-momentum in the centre of mass system. If we assume that the charge exchange cross-section σ_3 becomes small compared with σ_1 and σ_2, this implies

$$f_1(s,t) \sim f_0(s,t) \quad \text{as } s \to \infty, \tag{8.2.7}$$

giving
$$\frac{d\sigma_1}{dt} \sim \frac{d\sigma_2}{dt} \quad \text{as } s \to \infty. \tag{8.2.8}$$

Similar results hold for nucleon-nucleon scattering.

For pion-proton scattering, we have (Okun & Pomeranchuk, 1956),

$$\sigma_1: \quad \pi^+ + p \to \pi^+ + p, \tag{8.2.9}$$

$$\sigma_2: \quad \pi^0 + p \to \pi^0 + p, \tag{8.2.10}$$

$$\sigma_3: \quad \pi^0 + p \to \pi^+ + n, \tag{8.2.11}$$

$$\sigma_4: \quad \pi^- + p \to \pi^- + p, \tag{8.2.12}$$

$$\sigma_5: \quad \pi^- + p \to \pi^0 + n. \tag{8.2.13}$$

There are five similar relations involving target neutrons. All these reactions can be expressed in terms of isospin amplitudes $f_{\frac{3}{2}}(s,t)$ and $f_{\frac{1}{2}}(s,t)$, corresponding to isospin $\frac{3}{2}$ and $\frac{1}{2}$ respectively. We have

$$\left(\frac{k^2}{\pi}\right)\frac{d\sigma_1}{dt} = |f_{\frac{3}{2}}|^2; \quad \left(\frac{k^2}{\pi}\right)\frac{d\sigma_2}{dt} = \tfrac{1}{9}|2f_{\frac{3}{2}} + f_{\frac{1}{2}}|^2; \quad \left(\frac{k^2}{\pi}\right)\frac{d\sigma_4}{dt} = \tfrac{1}{9}|f_{\frac{3}{2}} + 2f_{\frac{1}{2}}|^2, \tag{8.2.14}$$

$$\left(\frac{k^2}{\pi}\right)\frac{d\sigma_3}{dt} = \left(\frac{k^2}{\pi}\right)\frac{d\sigma_5}{dt} = \tfrac{2}{9}|f_{\frac{3}{2}} - f_{\frac{1}{2}}|^2. \tag{8.2.15}$$

If we assume that the charge exchange cross-sections σ_3 and σ_5 become small compared with those for elastic scattering, then

$$f_{\frac{3}{2}}(s,t) \sim f_{\frac{1}{2}}(s,t) \quad \text{as} \quad s \to \infty, \tag{8.2.16}$$

$$\frac{d\sigma_1}{dt} \sim \frac{d\sigma_2}{dt} \sim \frac{d\sigma_4}{dt} \quad \text{as} \quad s \to \infty. \tag{8.2.17}$$

(b) Total cross-sections

Consider now the implications of the above results for total cross-sections. Taking $t = 0$ and using the optical theorem, we deduce from (8.2.7) that, for antiproton-nucleon scattering,

$$\sigma\,(\text{total } \bar{p}, p) \sim \sigma\,(\text{total } \bar{p}, n) \quad \text{as} \quad s \to \infty. \tag{8.2.18}$$

By charge symmetry, these are also equal to the antineutron-proton and antineutron-neutron cross-sections. Similarly, for proton-nucleon scattering,

$$\sigma\,(\text{total } p, p) \sim \sigma\,(\text{total } p, n) \quad \text{as} \quad s \to \infty, \tag{8.2.19}$$

and by charge symmetry these are also equal to the neutron-neutron total cross-section. However we do *not* obtain any relation between particle-target and antiparticle-target differential cross-sections or total cross-sections. Thus the result of the Pomeranchuk theorem on total cross-sections discussed in §8.1 is independent of the present result in this case, where particle and antiparticle are not in the same isomultiplet. Combining the two results we obtain asymptotic equality of total cross-sections for collision of any pair of nucleons and/or antinucleons.

For pion-nucleon scattering, the situation is somewhat different (Greenberg, 1964). From the assumption (8.2.16) we can deduce the Pomeranchuk theorem (§8.1) for differential cross-sections for $\pi^+ p$ and $\pi^- p$ elastic scattering; for example (8.2.17) includes the result

$$\frac{d\sigma_1}{dt} \sim \frac{d\sigma_4}{dt} \quad \text{as} \quad s \to \infty. \tag{8.2.20}$$

But for pion-nucleon collisions the Pomeranchuk–Okun rule also gives a stronger result. From (8.2.16)

$$F(\pi^+ p \to \pi^+ p) \sim F(\pi^- p \to \pi^- p) \quad \text{as} \quad s \to \infty. \tag{8.2.21}$$

Taking the amplitudes in the forward direction we deduce, using the optical theorem, that

$$\sigma\,(\text{total}, \pi^+ p) \sim \sigma\,(\text{total}, \pi^- p) \quad \text{as} \quad s \to \infty. \tag{8.2.22}$$

Thus the Pomeranchuk theorem on total cross-sections follows from the Pomeranchuk–Okun rule on charge exchange cross-sections and isospin invariance, when particle and antiparticle belong to the same isospin multiplet.

(c) *Real to imaginary ratios of forward amplitudes*

We can also obtain information about the asymptotic behaviour of the ratios,

$$\alpha(\pi^+ p) = \frac{\operatorname{Re} F(\pi^+ p)}{\operatorname{Im} F(\pi^+ p)}; \quad \alpha(\pi^- p) = \frac{\operatorname{Re} F(\pi^- p)}{\operatorname{Im} F(\pi^- p)}. \tag{8.2.23}$$

We assume the Pomeranchuk–Okun rule, and also assume that total cross-sections are asymptotically constant. Since we can deduce, as above, the equality of total cross-sections, (8.2.22) holds (and also 8.2.21). As in §7.1, form the forward amplitudes that are symmetric and antisymmetric under the substitution

$$E + i0 \to -E - i0. \tag{8.2.24}$$

These are
$$F_S(E) = \tfrac{1}{2}[F(\pi^+p) + F(\pi^-p)], \tag{8.2.25}$$

$$F_A(E) = \tfrac{1}{2}[F(\pi^+p) - F(\pi^-p)]. \tag{8.2.26}$$

Since the total cross-sections are assumed to be asymptotically constant, we can deduce (see § 7.1) that, as $E \to \infty$,

$$\operatorname{Im} F_S \sim CE; \quad F_S \sim iCE; \quad \frac{\operatorname{Re} F_S}{\operatorname{Im} F_S} \to 0. \tag{8.2.27}$$

From (8.2.21)
$$\left| \frac{F_A}{F_S} \right| \to 0 \quad \text{as} \quad E \to \infty. \tag{8.2.28}$$

Since F_S is predominantly pure imaginary, we deduce that the ratios (8.2.23) must tend to zero as $E \to \infty$,

$$\left[\frac{\operatorname{Re} F(\pi^+p)}{\operatorname{Im} F(\pi^+p)} \right] = \alpha(\pi^+p) \to 0; \quad \alpha(\pi^-p) \to 0. \tag{8.2.29}$$

This result was first obtained by Olesen (1965) using pion-nucleon dispersion relations.

There is a different situation for proton and antiproton scattering on protons, because they do not belong to the same isospin multiplet. We cannot deduce the Pomeranchuk theorem on total cross-sections from the vanishing of charge exchange cross-sections. Neither can we proceed in the converse direction. However, from the Pomeranchuk theorem alone, we can make some deductions about the ratios of the real to imaginary parts of the forward scattering amplitudes,

$$\alpha(\bar{p}, p) \quad \text{and} \quad \alpha(p, p). \tag{8.2.30}$$

For simplicity, we will assume that cross-sections tend to constants. Then (8.2.27) holds, with F_S, F_A now defined by

$$F_S = \tfrac{1}{2}[F(\bar{p}, p) + F(p, p)], \tag{8.2.31}$$

$$F_A = \tfrac{1}{2}[F(\bar{p}, p) - F(p, p)]. \tag{8.2.32}$$

Now all that we know about F_A, from the Pomeranchuk theorem, is that
$$\frac{\operatorname{Im} F_A}{\operatorname{Im} F_S} \to 0 \quad \text{as} \quad E \to \infty. \tag{8.2.33}$$

We could, for example, have
$$F_A \sim DE \quad \text{as} \quad E \to \infty, \tag{8.2.34}$$

with $D \neq 0$. In this case, from (8.2.27) and (8.2.31, 32), we would have

$$\alpha(\bar{p}, p) \sim -\alpha(p, p) \to \frac{D}{C} \quad \text{as} \quad E \to \infty. \tag{8.2.35}$$

There are many possible forms of F_A, compatible with (8.2.33) and with crossing symmetry, but more general than (8.2.34). We could have

$$F_A \sim DE(\log E - \tfrac{1}{2}i\pi)^\beta \quad (\beta < 1). \tag{8.2.36}$$

In (8.2.35) we would now have a factor $(\log E)^\beta$. Thus the ratios $\alpha(\overline{p}, p)$ and $\alpha(p, p)$ could tend to infinity, but they would still have opposite signs if $\beta > 0$. However if β were negative, or if

$$F_A \sim DE[E \exp(-\tfrac{1}{2}i\pi)]^{-\gamma} \quad (\gamma > 0), \tag{8.2.37}$$

we could make no deduction about whether $\alpha(\overline{p}, p)$ and $\alpha(p, p)$ have the opposite or the same signs as they tend to zero. This is because, in general, F_S may also include a term like (8.2.37). Only if special models are assumed (like the Regge model), can one predict how any such ratio α tends to zero.

For proton-proton scattering, the experimental results quoted in § 2.1 allow the possibility that $\alpha(p, p) \sim -0.2$ for large energy. If this is the case, the results of this section imply that $\alpha(\overline{p}, p)$ should have asymptotically a value equal in magnitude but opposite in sign to $\alpha(p, p)$. It is possible that the energies used in these measurements (*circa* 20 Gev) are not asymptotic, so one cannot necessarily deduce that $\alpha(\overline{p}, p) \approx 0.2$ at energies now attainable. However, if it is found that $\alpha(\overline{p}, p)/\alpha(p, p)$ differs substantially from -1, then this could be regarded as good evidence that both $\alpha(p, p)$ and $\alpha(\overline{p}, p)$ tend to zero as E tends to infinity.

8.3   Summary of general results

The above results were obtained for two special examples using isospin invariance. We summarise now the general results that they illustrate. These results hold for any internal symmetry group, when the symmetry is exact. However, for SU 3 one must allow for the possibility that symmetry breaking may affect some conclusions.

(a)   Total cross-sections

The Pomeranchuk theorem on total cross-sections (§ 8.1) states

$$\sigma\,(\text{total}, a, b) \sim \sigma\,(\text{total}, \overline{a}, b) \quad \text{as} \quad E \to \infty, \tag{8.3.1}$$

where $\overline{a}$ denotes the antiparticle of a.

(b) Differential cross-sections

The Pomeranchuk theorem on differential cross-sections (§ 8.1) states that, at fixed t,

$$\frac{d\sigma}{dt}(a, b \to a, b) \sim \frac{d\sigma}{dt}(\bar{a}, b \to \bar{a}, b) \quad \text{as} \quad E \to \infty. \quad (8.3.2)$$

(c) Exchange cross-sections

The Pomeranchuk–Okun rule on charge exchange cross-sections (§ 8.2), generalises to give, for fixed t,

$$\frac{\dfrac{d\sigma}{dt}(a, b \to c, d)}{\dfrac{d\sigma}{dt}(a, b \to a, b)} \to 0 \quad \text{as} \quad E \to \infty, \quad (8.3.3)$$

where the reaction $a, b \to c, d$ involves the exchange of discrete quantum numbers that cannot be carried by the vacuum.

If a, c belong to one multiplet associated with a symmetry principle (isospin invariance, or SU3 invariance), and b, d belong to one multiplet, and if (8.3.3) holds for all non-diagonal pairs, we have

$$(a, b| F' |c, d) = \sum_{\text{I}} (a, b| I) F'_{\text{I}}(I|c, d) \to 0, \quad (8.3.4)$$

where F' denotes the amplitude divided by E, and F'_{I} denotes the diagonal amplitude (e.g. the isospin amplitude) for the symmetry group. Then one can show (Greenberg, 1964; Yang, 1963; Foldy & Peierls, 1963) that unitarity implies that $F'_{\text{I}} \to F'$ for all I, and leads to the following result (d).

(d) Elastic amplitudes

All elastic amplitudes are asymptotically equal, for the collisions of any particle in the multiplet (A) and any particle in the multiplet (B). When particles a and c are in multiplet (A) and particles b and d are in multiplet (B) we obtain,

$$(a, b| F' |c, d) \to \delta_{ac} \delta_{bd} F'_{AB} \quad \text{as} \quad E \to \infty, \quad (8.3.5)$$

where a, c and b, d denote the quantum numbers of the particles.

(e) Particle and antiparticle in the same multiplet

From the Pomeranchuk–Okun rule (8.3.3), we can deduce (8.3.5) as noted above. Taking $c = \bar{a}$, the antiparticle of a, we can deduce also the Pomeranchuk theorem for differential cross-sections (8.3.2). From the asymptotic equality of the imaginary parts of the forward amplitudes, we obtain the Pomeranchuk theorem on total cross-sections (8.3.1). The converse deduction is not possible in general.

(*f*) *Ratio of real to imaginary parts* (*particle and antiparticle in the same multiplet*)

When particle and antiparticle belong to the same multiplet, we see from (8.3.5) that

$$(\bar{a}, b| \, F \, |\bar{a}, b) \sim (a, b| \, F \, |a, b) \quad \text{as} \quad E \to \infty. \tag{8.3.6}$$

Using this result when F is the forward amplitude, we deduce that both the amplitudes, $F(\bar{a}, b)$ and $F(a, b)$ in (8.3.6), are asymptotically symmetric under the substitution,

$$E + i0 \to -E - i0. \tag{8.3.7}$$

We will assume that total cross-sections are asymptotically constant. Then it follows from the discussion in Chapter 7 that

$$\alpha(\bar{a}, b) = \frac{\operatorname{Re} F(\bar{a}, b)}{\operatorname{Im} F(\bar{a}, b)} \to 0, \quad \text{as} \quad E \to \infty, \tag{8.3.8}$$

and

$$\alpha(a, b) \to 0 \quad \text{as} \quad E \to \infty. \tag{8.3.9}$$

(*g*) *Ratio of real to imaginary parts* (*different multiplets*)

If particle a, and its antiparticle $\bar{a}$, belong to different multiplets we see from (8.3.1) and (8.3.2) that

$$\operatorname{Im} F(a, b) \sim \operatorname{Im} F(\bar{a}, b), \tag{8.3.10}$$

$$|F(a, b)| \sim |F(\bar{a}, b)|. \tag{8.3.11}$$

It follows that, as $E \to \infty$, *either*

$$\alpha(\bar{a}, b) = \frac{\operatorname{Re} F(\bar{a}, b)}{\operatorname{Im} F(\bar{a}, b)} \sim -\frac{\operatorname{Re} F(a, b)}{\operatorname{Im} F(a, b)} = -\alpha(a, b), \tag{8.3.12a}$$

with non-zero (possibly infinite) limits, *or*

$$\alpha(\bar{a}, b) \to 0 \quad \text{and} \quad \alpha(a, b) \to 0. \tag{8.3.12b}$$

In the latter case we cannot say in general whether $\alpha(\bar{a}, b)$ and $\alpha(a, b)$ have the same or different signs as they tend to zero.

(h) Total cross-sections for particle and antiparticle multiplets

If particle a, and its antiparticle $\bar{a}$ belong to different multiplets, we can generalise (8.3.5) by combining it with (8.3.1). From (8.3.5),

$$\sigma\,(\text{total}, a, b) \sim \sigma\,(\text{total}, c, d), \qquad (8.3.13)$$

for any a, c of one multiplet, and b, d of another multiplet (the multiplets may be the same). From (8.3.1),

$$\sigma\,(\text{total}, \bar{a}, b) \sim \sigma\,(\text{total}, a, b), \qquad (8.3.14)$$

$$\sigma\,(\text{total}, a, \bar{b}) \sim \sigma\,(\text{total}, a, b), \qquad (8.3.15)$$

$$\sigma\,(\text{total}, a, \bar{b}) \sim \sigma\,(\text{total}, \bar{a}, \bar{b}). \qquad (8.3.16)$$

This generalises (8.3.13) to hold when a, c is any pair chosen from a given multiplet and the corresponding multiplet of antiparticles, and b, d is any pair chosen from a given multiplet and its antiparticle multiplet.

(i) Predictions from SU 3

If the strong interactions between hadrons exactly satisfied the symmetry of SU 3, there would be a large number of relations between scattering amplitudes and between two-body reaction amplitudes. These are analogous to those given by the SU 2 symmetry associated with isospin invariance, as in (3.6.4) for example, Due to symmetry breaking in SU 3 the relations between amplitudes are not exact, and there is some difficulty over the question of what energies to use for comparison between particles in the same SU 3 multiplets but having different masses.

At high energies it is intuitively reasonable to suppose that symmetry breaking is not important in general, but there may be special situations where it is relevant (see (10.5.5) for example). If SU 3 becomes exact in the asymptotic limit, then the exchange amplitudes will all tend to zero at high energy (relative to the elastic amplitudes), and the scattering amplitudes, between any particle in multiplet a and any particle in multiplet b, will all become asymptotically equal. Thus, for the elastic amplitudes, we will have

$$F(K, N) \sim F(\pi, N) \sim F(\pi, \Sigma) \sim F(a, b) \quad \text{as} \quad E \to \infty, \qquad (8.3.17)$$

where a denotes a particle in the pseudoscalar meson octet and b

denotes a particle in the spin-half baryon octet. In this case the Pomeranchuk–Okun rule, which leads to (8.3.17), takes the form

$$\frac{F(a_1 b_1 \to a_2 b_2)}{F(a_1 b_1 \to a_1 b_1)} \to 0 \quad \text{as} \quad E \to \infty. \tag{8.3.18}$$

As a consequence of (8.3.17), we will have asymptotic equality between the total cross-sections,

$$\sigma_t(K, N) \sim \sigma_t(\pi, N) \sim \sigma_t(\pi, \Sigma) \sim \sigma_t(a, b). \tag{8.3.19}$$

Most of the results on symmetries that are discussed in this chapter apply to any internal symmetry group, although we have generally used the isospin group for illustration. For the technical aspects of SU3 invariance, particularly the evaluation of Clebsch–Gordan coefficients, recoupling coefficients and crossing matrices, the reader is referred to the article by De Swart (1963), and the books by Gell-Mann & Ne'eman (1964), Lipkin (1965) and Carruthers (1966).

Exact SU3 symmetry leads also to relations between exchange amplitudes. From these one can obtain equalities between certain exchange cross-sections, and also equalities and inequalities involving three or more cross-sections. For example, Meshkov, Levinson & Lipkin (1963) derive relations for a pseudoscalar meson plus a baryon leading to a meson (pseudoscalar or vector) plus a baryon resonance in the spin-$\frac{3}{2}$ decimet. One of these relations is

$$\frac{F(\pi^-, p \to \rho^+, N^{*-})}{F(\pi^-, p \to K^{*+}, Y_1^{*-})} = -(3)^{\frac{1}{2}}. \tag{8.3.20}$$

Comparison with experiment (Meshkov, Snow & Yodh (1964) and references quoted above) gives generally good agreement with SU3 predictions, particularly when allowance is made for the uncertainties associated with symmetry breaking. The possibility that higher symmetries may lead to further relations between cross-sections is still in a state of fluctuation; we will consider certain aspects of SU6 related to the quark model in §10.5. For an extensive discussion of work on higher symmetries see Salam (1966).

8.4 Exchange amplitudes at high energy

In this section we will be concerned with high energy behaviour in the s channel, and we will use isospin invariance for illustration. The amplitude can be expressed in terms of isospin amplitudes in the t channel. These are related to the amplitudes in the s channel by the usual crossing matrices (see §4.9, and Carruthers & Krisch 1965 for crossing matrices with arbitrary isospin). It is of great importance to know which isospin amplitudes in the t channel can dominate the scattering in the s channel at high energy. The Pomeranchuk–Okun rule (§§ 8.2 and 8.3) *assumes* that the exchange of vacuum quantum numbers must dominate. This has not yet been proved, but it has been shown by Foldy & Peierls (1963) that if one exchange amplitude does dominate over all others at high energies in the s channel, this amplitude must have isospin zero. This result has been generalised to any internal symmetry group by Amati, Foldy, Stanghellini & Van Hove (1964). In this section we consider this result for elastic scattering, and its generalisation to inelastic scattering in diffraction dissociation.

(a) Crossing matrices for pion-nucleon scattering

Consider the following reaction in the s channel,

$$\pi^+ + p \to \pi^+ + p, \tag{8.4.1}$$

and the two reactions for $\pi^- + p$. In terms of the isospin amplitudes F^s in the s channel we get the usual results,

$$(\pi^+ p| \, F \, |\pi^+ p) = F^s(\tfrac{3}{2}), \tag{8.4.2}$$

$$(\pi^- p| \, F \, |\pi^- p) = \tfrac{1}{3}F^s(\tfrac{3}{2}) + \tfrac{2}{3}F^s(\tfrac{1}{2}), \tag{8.4.3}$$

$$(\pi^- p| \, F \, |\pi^0 n) = \frac{\sqrt{2}}{3}\{F^s(\tfrac{3}{2}) - F^s(\tfrac{1}{2})\}. \tag{8.4.4}$$

The exchange amplitudes, (i.e. the crossed amplitudes in the t channel), describe the reactions,

$$\pi^+ + \pi^- \to \bar{p} + p, \tag{8.4.5}$$

$$\pi^+ + \pi^0 \to p + \bar{n}. \tag{8.4.6}$$

We can also express these amplitudes in terms of t channel isospin amplitudes F^t. However, by analytic continuation to the s channel, the amplitude describing (8.4.5) is the same as (8.4.2), and (8.4.6) is the same as (8.4.4). Hence, from the inverse to (4.9.24),

$$(\pi^+ p| \, F \, |\pi^+ p) = \frac{1}{\sqrt{6}} F^t(0) - \tfrac{1}{2}F^t(1), \tag{8.4.7}$$

$$(\pi^- p| \, F \, |\pi^- p) = \frac{1}{\sqrt{6}} F^t(0) + \tfrac{1}{2}F^t(1), \tag{8.4.8}$$

$$(\pi^- p| \, F \, |\pi^0 n) = -\frac{1}{\sqrt{2}} F^t(1), \tag{8.4.9}$$

where 0 and 1 denote the isospin in the t channel.

Dominance of the isospin-zero exchange amplitude

It is evident from (8.4.7, 8, 9) that if, as $s \to \infty$,

$$\frac{F^l(1)}{F^l(0)} \to 0, \qquad (8.4.10)$$

then (using also (8.4.2, 3, 4)),

$$F^s(\tfrac{3}{2}) \sim F^s(\tfrac{1}{2}). \qquad (8.4.11)$$

As s becomes large the exchange amplitude (8.4.9) will become small relative to the two elastic scattering amplitudes, which will become equal. This result is made more precise in the following theorem (Foldy & Peierls, 1963; Van Hove, 1965).

Theorem. If one exchange amplitude dominates over the others at high energy (s), and if it is not purely real at $t = 0$, then the dominating amplitude corresponds to the quantum numbers of the vacuum. (The proof of this theorem requires a special assumption, which is stated below.)

For pion-nucleon scattering the theorem states that if $F^l(1)$ dominates as $s \to \infty$, then it must be purely real at $t = 0$. To prove this, write the forward amplitude as,

$$(\pi^\pm, p | F | \pi^\pm, p) \to a_\pm + ib_\pm \quad \text{as} \quad s \to \infty. \qquad (8.4.12)$$

By the optical theorem, $\quad b_+ > 0, \quad b_- > 0. \qquad (8.4.13)$

If $F^l(1)$ dominates in (8.4.7) and (8.4.8), as $s \to \infty$

$$\frac{a_+ + ib_+}{a_- + ib_-} \to -1. \qquad (8.4.14)$$

This can be satisfied (Van Hove, 1965) only if

$$\frac{b_+}{a_+} \to 0, \quad \frac{b_-}{a_-} \to 0 \quad \text{and} \quad \frac{a^+}{a_-} \to -1. \qquad (8.4.15)$$

From this result we see that $F^l(1)$ at $t = 0$, and hence the elastic (π^+, p) and (π^-, p) forward amplitudes, would become asymptotically real. We have seen (compare with (8.3.12a)) that this possibility cannot be ruled out on theoretical grounds, even if we assume the Pomeranchuk theorems (8.3.1) and (8.3.2). However, experimental results for pion-nucleon scattering suggest that these amplitudes do in fact become pure imaginary. The theorem above can be proved from the above argument if we also make the *assumption* that,

$$\left| \alpha(\pi^{\pm}, p) \right| = \left| \frac{\operatorname{Re} F(\pi^{\pm}, p)}{\operatorname{Im} F(\pi^{\pm}, p)} \right| < M \quad \text{as} \quad s \to \infty, \qquad (8.4.16)$$

where M is a constant. In this case, if one exchange amplitude dominates at high energy, it must be $F^{l}(0)$.

If the assumption (8.4.16) is strengthened to

$$\alpha(\pi^{+}p) \to 0 \quad \text{and} \quad \alpha(\pi^{-}p) \to 0 \quad \text{as} \quad s \to \infty, \qquad (8.4.17)$$

and if we assume also the Pomeranchuk theorems (8.3.1) and (8.3.2), we can deduce that

$$\left| \frac{F^{l}(1)}{F^{l}(0)} \right| \to 0 \quad \text{as} \quad s \to \infty, \qquad (8.4.18)$$

which gives the Pomeranchuk–Okun rule. This is the converse of the result obtained earlier (8.3.8) and (8.3.9).

The theorem stated above can be generalised to apply to any internal symmetry group (Amati, Foldy, Stanghellini & Van Hove, 1964, and Yang, 1963).

(b) Exchange dominance for inelastic processes

It is possible for a particle a_1 to go into an excited state a_2 in a collision in which only the quantum numbers of the vacuum are exchanged, thus

$$a_1 + b \to a_2 + b. \qquad (8.4.19)$$

We will describe such a collision as quasi-elastic. There will be a corresponding scattering matrix, that refers purely to quasi-elastic collisions, which we take to be

$$\begin{pmatrix} (a_1, b| \, F \, |a_1, b), & (a_1, b| \, F \, |a_2, b) \\ (a_2, b| \, F \, |a_1, b), & (a_2, b| \, F \, |a_2, b) \end{pmatrix}. \qquad (8.4.20)$$

Since any one of the reactions associated with (8.4.20) can proceed, with exchange only of vacuum quantum numbers, our previous discussion does *not* require that their cross-sections tend to zero at high energy, relative to elastic processes. This means that there is no reason to suppose that the matrix (8.4.20) will become diagonal at high energy. A qualitative interpretation indicating why it does not in general become diagonal has been given by Morrison (1966) to explain the following experimental results.

Quasi-elastic processes (diffraction dissociation)

Experimental situations are observed in which total reaction cross-sections for quasi two-body processes become constant (and non-zero) at high energies (or nearly constant—the experimental errors do not allow one to assert that an asymptotic cross-section is constant). In particular (see § 2.7 and Anderson *et al.* 1966), the following reactions have nearly constant cross-sections, measured between 10 and 30 Gev.

$$p + p \to p_r^* + p \quad (r = 1, 2, 3), \tag{8.4.21}$$

$$\left.\begin{array}{l} p_1^* = N^*(1\cdot52\,\text{Gev}), \quad I = \tfrac{1}{2}, \\ p_2^* = N^*(1\cdot69\,\text{Gev}), \quad I = \tfrac{1}{2}, \\ p_3^* = N^*(2\cdot19\,\text{Gev}), \quad I = \tfrac{1}{2}. \end{array}\right\} \tag{8.4.22}$$

In each of these inelastic processes there is no exchange of quantum numbers except for angular momentum (the latter can be carried by the vacuum). These processes are sometimes called diffraction dissociation by analogy with the low energy phenomenon (Pomeranchuk & Feinberg, 1956; Good & Walker, 1960). The term quasi-elastic is also used. The latter name indicates, that the N^* products must each have the same isospin and charge as the proton.

On a Regge model, these phenomena imply that the Pomeron P is coupled to the p, p_r^* vertex through a coupling 'constant'

$$g(p, p_r^*, P; t), \tag{8.4.23}$$

where the value at $t = 0$ dominates high energy behaviour. The values for $t < 0$ may be expected to depend on the properties of p_r^*. However, for each of the reactions (8.4.21) the slope of the forward peak is approximately the same, so the dependence of g on p_r^* in these examples is not strong. An intuitive explanation by Morrison (1966) considers the emission and re-absorption of a virtual pion,

$$p \to (p + \pi) \to N^*. \tag{8.4.24}$$

At high energies the virtual pion may be regarded as being nearly on the mass shell. Before re-absorption it is elastically scattered by the other proton (with constant elastic cross-section). Thus the total reaction cross-section in this (approximate) model will be proportional to the product squared of the coupling constants, for emission and absorption of the pion, multiplied by the elastic cross-section. More exactly, assuming Regge theory, the asymptotic total reaction cross-section will be given by the square of the coupling constant (8.4.23), times the square of the proton-Pomeron coupling constant. This leads to the relation

$$\sigma\,(\text{total},\,pp \to pp_r^*) \sim \left[\frac{g(p, p_r^*, P)}{g(p, p, P)}\right]^2 \sigma\,(\text{elastic},\,pp) \quad \text{as} \quad E \to \infty.$$

$$(8.4.25)$$

CHAPTER 9

THE EXPERIMENTAL STUDY
OF REGGE THEORY

It was noted in Chapter 5 that many of the important results of Regge theory depend on assumptions that have been proved only for non-relativistic potential scattering. Even with the aid of these assumptions there are other major difficulties in using the theory to study high energy collisions. These are of two kinds, namely, difficulties of principle like the existence and location of branch cuts in the angular momentum plane, and difficulties of approximation. The latter question arises most critically in determining at what energy one can assume dominance of leading terms in the Regge expansion. The energy at which asymptotic approximations become valid will obviously depend on the experimental process under consideration and also on the particular quantities that are being measured.

The uncertainties about some of the assumptions and approximations in Regge theory necessitate detailed experimental study of the predictions that result from various approximations. It seems unlikely that relativistic theory will by itself provide a rigorous justification of the wide range of approximation schemes within Regge theory that are required for experimental comparisons. The corollary to this theoretical impasse is that 'predictions' of Regge theory are likely to be of a heuristic nature for some time to come. Consequently, experimental results should be interpreted as providing information about the scattering amplitude and the complex angular momentum plane, and should lead to a gradual support (and possibly an understanding) of a framework for theoretical approximations.

In this chapter we will begin with a summary of the main features that can be considered as experimental consequences of Regge theory at high energy. By this we mean the theory taken within the framework of the general assumptions and approximations, some of which are open to modification as theory and experiment develop. We next consider in §§ 9.2 and 9.3 specific approximations, that are quite certainly of heuristic character, made for purposes of simplicity. The objectives are to consider what reactions are most favourable for testing specific approximations and to find what measurements in each reaction are most relevant. In §9.2 we concentrate primarily on one pole approximations, and in §9.3 on experiments that require the use of several Regge poles. When studying backward scattering it is appropriate to consider exchanged Regge poles in the u channel, and for some processes the quantum numbers severely restrict the number of allowed poles; examples are described in §9.4. Two classes of experiment that are difficult, but may be of great importance in testing Regge theory, are discussed briefly in §9.5. These involve detailed aspects of differential cross-sections, and the measurement of polarisation and spin parameters. In §9.6 we describe the use of sum rules in Regge theory. Certain aspects of sum rules would also apply in a more general theory, but they have particularly important consequences in the experimental study of Regge theory. Finally, in §9.7 we consider some open questions including possible tests for Regge branch cuts. The experimental status of Regge theory has been reviewed by Leader (1966), Phillips (1966) and Van Hove (1966); their articles provide detailed references for much of the material discussed in this chapter.

9.1 Properties of Regge theory at high energy

In this section we recapitulate some of the properties considered in Chapter 5, with particular reference to their experimental implications. The following topics will be considered: (a) general assumptions, (b) asymptotic total cross-sections, (c) Regge poles and trajectories, particles and ghosts, (d) factorisation, (e) effects of particle spin, (f) leading poles in forward scattering, (g) fermion Regge poles, (h) backward scattering and fermion exchange.

(a) General assumptions

The general assumptions of Regge theory, as distinct from the asymptotic approximations, are indicated in (9.1.1) to (9.1.5):

(i) There are unique analytic extensions of the partial wave amplitudes for even and odd integers such that

$$f^+(l,t) = f_l(t) \quad (l = \text{even integer}), \tag{9.1.1}$$

$$f^-(l,t) = f_l(t) \quad (l = \text{odd integer}), \tag{9.1.2}$$

where t denotes the invariant energy squared. When the colliding particles have non-zero spin, these amplitudes should be replaced by the helicity partial wave amplitudes. The uniqueness of the analytic continuation is ensured by requiring that

$$|f(l,t)| < \exp\left[(\pi - \epsilon)|l|\right] \quad \text{as} \quad |l| \to \infty \quad \text{in} \quad \text{Re}(l) > -\tfrac{1}{2}. \tag{9.1.3}$$

Although the above assumption (i) has not been proved in relativistic theory, it is a much weaker assumption than that required for the Mandelstam representation (from which it can be proved). It could hold for example, if one had less than complete cut-plane analyticity in s and t. It could also hold if the Mandelstam representation failed due to an infinite number of subtraction terms as s and t become infinite (provided the number does not increase too fast compared with s and t).

(ii) The Sommerfeld–Watson transform for the amplitude in the t channel (see §5.1) is assumed to be valid, giving for equal masses as $s \to \infty$,

$$F(s,t) = \sum_n c_n^+(t) P_{\alpha_n}\left(-1 - \frac{s}{2k^2}\right) + \sum_m c_m^-(t) P_{\alpha_m}\left(-1 - \frac{s}{2k^2}\right) + R(s,t),$$

(9.1.4)

where $t = +4(m^2 + k^2)$, $s = -2k^2(1 - \cos\theta)$. The sums over $n(m)$ are taken over poles of the even (odd) partial wave amplitude in $\mathrm{Re}\,(l) > -\frac{1}{2}$, and for finite t it is assumed that

$$R(s,t) < \left(\frac{s}{s_0}\right)^{-\frac{1}{2}} \quad \text{as} \quad s \to \infty.$$

The assumption (ii) includes the requirement that there is no contribution from the deformation at ∞ in the l-plane, that arises when the contour along the positive real l axis is deformed so as to lie along the line $\mathrm{Re}\,(l) = -\frac{1}{2}$ and to surround the Regge poles in $\mathrm{Re}\,(l) > -\frac{1}{2}$ (see Fig. 5.1.1).

The assumption (9.1.4) can be relaxed so as to include the effects of branch cuts in the l-plane (see §5.6). For example, a fixed branch point at $l = 1$, with an attached branch cut in $\mathrm{Re}\,(l) \leqslant 1$, will give an additional term in (9.1.4). As $s \to \infty$, this term is expected to have the form

$$\int_{-\infty}^{1} dl\{\text{disc.}\,[r(l,t)]\} P_l\left(-1 - \frac{s}{2k^2}\right) \sim C(t)\,s(\log s)^{-\beta} \quad (\beta > 0). \quad (9.1.5)$$

We will usually assume that the discontinuity across such a branch cut is small, at least in the range, $-\frac{1}{2} < l \leqslant 1$, so that $C(t)$ will be small compared with the contributions from the pole terms shown in the sums in (9.1.4). However, this assumption is entirely heuristic: it may be found that experimental results require significant contributions to the asymptotic behaviour of collision amplitudes from Regge branch cuts, and we will note later some indications of this possibility. However through most of our discussion branch cuts in the l-plane will be neglected.

(b) *Asympototic total cross-sections*

If total cross-sections are asymptotically constant, it is necessary that at least one Regge pole has the value 1 at $t = 0$. Let the contribution to $F(s, t)$ from this pole be

$$F(s, t) \sim c_1(t) \left(\frac{s}{s_0} \right)^{\alpha_1(t)} \quad \text{as} \quad s \to \infty, \tag{9.1.6}$$

with $\alpha_1(0) = 1$ (cf. (5.3.28, 30)).

This amplitude will be asymptotically symmetric (or anti-symmetric) in the forward direction, if this term comes from the partial wave of even (or odd) signature. This is because (§5.3) the even signature partial wave in the t channel was defined from an amplitude that is symmetric, for large s, under the substitution $s + i0 \to -s - i0$. Hence, the first sum in (9.1.4) will be symmetric in s, and the second sum will be antisymmetric in s (this discussion is more complicated if the colliding particles have general masses but the result is also assumed to hold then). Hence, if the leading pole has $\alpha_1^+(0) = 1$, and comes from $f^+(l, t)$, the amplitude (9.1.6) will have the form (see §7.1),

$$F \sim F_S(s, 0) \sim i |c_1^+(0)| \left(\frac{s}{s_0} \right)^{\alpha_1^+(0)}. \tag{9.1.7}$$

Then the total cross-section will be asymptotically constant,

$$\frac{\operatorname{Im} F(s, 0)}{s} \sim \sigma(\text{total}) \sim \text{constant} \, |c_1^+(0)|. \tag{9.1.8}$$

However if the leading pole came from the partial waves of odd signature, and $\alpha_1^-(0) = 1$, we would have

$$F \sim F_A(s, 0) \sim |c_1^-(0)| \left(\frac{s}{s_0} \right)^{\alpha_1^-(0)}. \tag{9.1.9}$$

Since this amplitude is asymptotically real, its contribution to σ (total) will be zero, so it cannot lead to a constant total cross-section (see also our discussion in §5.3). This establishes the need for the 'Pomeron' Regge pole having the following properties:

The Pomeron

From the above discussion we see that, if total cross-sections are asymptotically constant, there must be a Regge pole of *even* signature ($\tau = +$) having,

$$l = \alpha_1^+(0) = 1. \tag{9.1.10}$$

This is the largest real part of the exponent of s in (9.1.7) at $t = 0$, that is allowed by unitarity (§ 6.2). The corresponding trajectory is called the Pomeranchuk trajectory. The object exchanged at $t = 0$ is called the 'Pomeron'. It is not a physical particle, since it corresponds to a pole at $l = 1$ in the *even* signature partial wave amplitude. However, if, for $t > 0$, the trajectory goes through $l = 2$ (or near $l = 2$ if it is complex), it would then correspond to a physical particle (see Fig. 5.3.1), provided $\mathrm{Re}\,(d/ds)\,\alpha_1^+(s) > 0$ at $l = 2$ (see (5.1.18)).

We have seen in § 8.4 that, on reasonable assumptions (already implied here by (9.1.4)), only the exchange amplitude having the quantum numbers of the vacuum can dominate over all other exchange amplitudes at high energy. This means that the Pomeron must have the quantum numbers of the vacuum.

Phase energy relations

It is important to note that although the phases determined for the even signature term (9.1.7), and the odd signature term (9.1.9), follow from the general arguments given in § 7.1, they are a consequence also of the special simplicity due to the power law dependence on s. This power law is special to Regge theory. It excludes, for example, the appearance of a term involving $(\log E - i\pi/2)^\beta$ as a factor, either in the even or the odd contributions, *unless* the effects of branch cuts are included.

A particular consequence of this power law dependence is that, even if $\alpha_1^-(0) = 1$ (in addition to $\alpha_1^+(0) = 1$), we can deduce from (9.1.7, 9) with $F = F_S + F_A$, for some C,

$$\frac{\mathrm{Re}\,F(s, 0)}{\mathrm{Im}\,F(s, 0)} < C = \text{constant} \quad \text{as} \quad s \to \infty.$$

This result in Regge theory supplements our general result (§ 8.3), when a particle and its antiparticle are in different multiplets. Recall that in § 8.3, the possibility of logarithmic factors allowed that the ratio $\mathrm{Re}\,F/\mathrm{Im}\,F$ might become infinite as $s \to \infty$. When particle and antiparticle are in the same multiplet, our general result, that this ratio tends to zero, holds also for Regge theory.

(c) Regge poles and trajectories, particles and ghosts

For even signature ($\tau = +$), the trajectory $l = \alpha^+(t)$ can correspond to physical particles if it goes near $l = 0, 2, 4, \ldots$. The isospin of such particles will be the same as the isospin of the partial wave amplitude

$f^+(l, t)$, and their (mass)^2 will be given by those positive values of t for which $\alpha^+(t)$ takes on even integer values. Above $t = 4m^2$ (i.e. above the lowest physical threshold in the t channel), $\alpha^+(t)$ will be complex, and the corresponding particles will be unstable. Similar properties hold for $\alpha^-(t)$ (odd signature $\tau = -$), which could correspond to physical particles at $1, 3, 5, \ldots$, if it goes near these values. Note that the bounds given in Table 6.8.1 require that no trajectory can have $\alpha(t)$ greater than 2, if $t < 4m^2$.

If $\alpha^+(t) = 0$ for $t < 0$, a 'ghost' state appears, which would seem to represent a particle in the t channel with a pure imaginary mass (t = mass squared). This is not acceptable and it is assumed that the coefficient of the corresponding term ($c^+(t)$ in (9.1.4)) must vanish at this value of t. This assumption 'kills the ghost'. It has been suggested (Arbab & Chiu, 1966) that, regardless of signature, $c_n^+(t)$ and $c_m^-(t)$ should vanish whenever $\alpha_n^+(t)$ or $\alpha_m^-(t)$, respectively, take on negative integer values or zero. The vanishing of these coefficients, at particular values of $t < 0$, leads to the prediction of dips in the differential cross-sections in the s channel which will be noted later.

Regge recurrences

Successive physical states on a Regge trajectory, $\alpha_n^+(t)$ (or $\alpha_m^-(t)$) at $0, 2, 4, \ldots$ (or $1, 3, 5, \ldots$), are called Regge recurrences. These recurrences on a given trajectory will have the same strangeness, isospin, and baryon number, and only their spin will change (by two units) from one recurrence to the next. An SU3 multiplet with broken symmetry (unequal masses) should lead to similar broken multiplets near the Regge recurrences. Alternatively, one could interpret the associated particles as lying on a multiplet of neighbouring trajectories.

An indication of some possible Regge recurrences is shown in Fig. 9.1.1. When two neighbouring resonances have the same quantum numbers but different masses, it is not possible to be sure which one should be interpreted as belonging to a particular trajectory. It is reasonable to expect that trajectories should be nearly parallel Chew (1962), since trajectory slopes depend on the strength and range of the very strong interactions that determine the masses of particles and resonances, and one might reasonably expect these to be fairly similar for all particles. However, even with this simplication the identification of particles (resonances) and trajectories is still somewhat uncertain.

Trajectories

It is often assumed that Regge trajectories, $l = \alpha(t)$, have only a right-hand cut, and that they satisfy a dispersion relation with just one subtraction. Then

$$\alpha(t) - \alpha(0) = \frac{t}{\pi} \int_{4m^2}^{\infty} \frac{dt' \, \mathrm{Im}\, \alpha(t')}{t'(t'-t)}, \qquad (9.1.11)$$

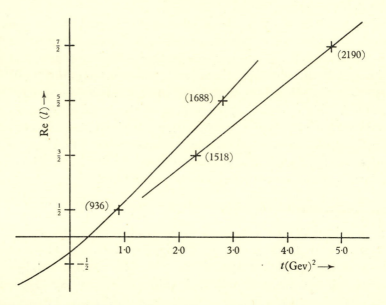

Fig. 9.1.1. Possible Regge trajectories in $t > 0$, showing Regge recurrences for N_α ($I = \frac{1}{2}$, $Y = 1$, $P = +$, $\tau = +$) and N_γ ($I = \frac{1}{2}$, $Y = 1$, $P = -$, $\tau = -$).

and $\mathrm{Im}\, \alpha(t)$ is assumed to be positive or zero. This assumption leads to the result that

$$\left[\left(\frac{d}{dt} \right)^n \alpha(t) \right] > 0, \quad t < 4m^2 \quad (n = 1, 2, \ldots). \qquad (9.1.12)$$

In practice, Regge trajectories $l = \alpha(t)$, identified by the Regge recurrences in $t > 0$, seem to be remarkably like straight lines when projected on to the $(\mathrm{Re}\, l, \mathrm{Re}\, t)$ plane. Also $\mathrm{Im}\, \alpha(t)$ is small, so that if the assumption (9.1.11) is correct, the derivatives (9.1.12) may be quite small. However there is no certainty that one can assume the absence of a left-hand cut even for bosons (Warburton, 1964).

(*d*) *Factorisation*

We have considered the justification of the factorisation assumption in § 5.4. It asserts that, if there are several coupled channels ($\pi\pi, \pi K, KK$) that involve the same Regge pole,

$$f_{ij}(l, t) = \frac{r_{ij}(t)}{l - \alpha(t)} + ..., \tag{9.1.13}$$

where i and j label the channels, then

$$r_{ij}(t) = r_i(t)\, r_j(t). \tag{9.1.14}$$

This leads to relations like (5.4.29) between total cross-sections,

$$\sigma_t(\pi\pi)\, \sigma_t(KK) = [\sigma_t(K\pi)]^2. \tag{9.1.15}$$

These relations are not at all easy to test experimentally, but more detailed consequences hold when the particles have non-zero spin.

(*e*) *Effects of particle spin*

Assuming the factorisation hypothesis, and dominance by a set of poles, the asymptotic amplitude for spinless particles will have the form

$$F(s, t) \sim \sum_j \zeta_j(2s_0) \left(\frac{s}{s_{0j}}\right)^{\alpha_j} \eta^j \eta^{j'}, \tag{9.1.16}$$

where η^j, $\eta^{j'}$ denote the couplings of the exchanged pole to the particles in the collision, and where ζ_j contains the phase factor appropriate to the signature τ_j, (see (5.3.25 and 26)),

$$-\zeta_j = \frac{1 + \tau_j \exp(-i\pi\alpha)}{2 \sin \pi\alpha} \quad (\tau_j = \pm 1). \tag{9.1.17}$$

The sum over j in (9.1.4) is over the exchanged Regge poles for the reaction that is being considered. The modifications when the colliding particles have non-zero spin have been elegantly summarised by Wagner (1963), whose method is followed here (see also Gell-Mann, 1962*b*). For spinless particles, say $\pi\pi$ scattering, the total cross-section (Udgaonkar, 1962) will be given by,

$$\sigma_{\pi\pi}(\text{total}) = \frac{\text{Im}\, F(s, t)}{2ks^{\frac{1}{2}}} \sim \sum_j \tau_j \left(\frac{s}{s_{0j}}\right)^{\alpha_j(0)-1} (\eta_\pi^j)^2. \tag{9.1.18}$$

The elastic differential cross-section will be given by,

$$\frac{d\sigma(\pi\pi \to \pi\pi)}{dt} \sim \frac{1}{4\pi} \sum_{ij} \text{Re}\left[\zeta_i^*\zeta_j\right]\left(\frac{s}{s_{0i}}\right)^{\alpha_i(t)-1}\left(\frac{s}{s_{0j}}\right)^{\alpha_j(t)-1}(\eta_\pi^i)^2(\eta_\pi^j)^2. \quad (9.1.19)$$

Appropriate isospin factors should be inserted to obtain cross-sections for specific charge states of the pions (see below (9.1.23)).

The total cross-section for pion-nucleon scattering, involves only the coupling η_N for no helicity flip, since by the optical theorem it is related to Im F with no change of state. Thus

$$\sigma_{\pi N}(\text{total}) \sim \sum_j \tau_j\left(\frac{s}{s_{0j}}\right)^{\alpha_j(0)-1}(\eta_\pi^i)(\eta_N^j). \quad (9.1.20)$$

However, for the differential cross-section in pion-nucleon scattering, one must include the helicity flip terms ϕ_N, giving (for high energy),

$$\frac{d\sigma(\pi N \to \pi N)}{dt} \sim \frac{1}{4\pi} \sum_{ij} \text{Re}\left[\zeta_i^*\zeta_j\right]\left(\frac{s}{s_{0i}}\right)^{\alpha_i-1}\left(\frac{s}{s_{0j}}\right)^{\alpha_j-1}\eta_\pi^i\eta_\pi^j(\eta_N^i\eta_N^j + \phi_N^i\phi_N^j).$$
$$(9.1.21)$$

Polarisation

Polarisation is produced only when the helicity flip and helicity non-flip amplitudes are out of phase. It is assumed that the couplings η and ϕ are real (since t is less than the lowest threshold of the t channel); hence polarisation depends on the interference between *two or more* Regge poles. For large s, the polarisation P is given by,

$$P_{\pi N}\frac{d\sigma(\pi N \to \pi N)}{dt} \sim \frac{1}{2\pi}\sum_{ij}\text{Im}\left[\zeta_i^*\zeta_j\right]\left(\frac{s}{s_{0i}}\right)^{\alpha_i-1}\left(\frac{s}{s_{0j}}\right)^{\alpha_j-1}\eta_\pi^i\eta_\pi^j\phi_N^i\phi_N^j. \quad (9.1.22)$$

There are analogous formulae for nucleon-nucleon scattering (Wagner 1963; Wagner & Sharp, 1962; see also Leader & Slansky, 1966; and § 9.3).

The helicity-flip coupling ϕ_N^i is related to the couplings used by Gell–Mann (1962b) by

$$\phi_N^i = (-t/4m_N^2)^{\frac{1}{2}}(\eta_2 - \eta_1). \quad (9.1.22a)$$

Thus, as expected, ϕ_N^i is zero at $t = 0$. It is also expected to contain a factor $\alpha_i(t)$ (see § 9.2).

(f) The leading poles in forward scattering

For $\pi\pi$, πN and NN scattering the leading Regge poles near $t = 0$ are assumed to be on trajectories associated with the Pomeron P, a similar object P' (but having a smaller value $\alpha(0)$), the ρ meson, the ω meson, and an R meson. Their quantum numbers are as follows:

	P	C	G	I	τ
P, P'	$+$	$+$	$+$	0	$+$
ρ	$-$	$-$	$+$	1	$-$
ω	$-$	$-$	$-$	0	$-$
R	$+$	$+$	$-$	1	$+$

$$(9.1.23)$$

The R meson is probably the A_2 at $1\cdot3$ Gev with $J^P = 2^+$ (see Fig. 5.3.1). From these quantum numbers one can determine the signs of the couplings (Wagner, 1963; Ahmedzadeh & Leader 1964). The couplings η_π^j, η_N^j and ϕ_N^j are defined to be those involved in coupling a neutral Regge pole to a particle having positive charge. For Regge poles with $I = 0$ and $C = +$ (like P and P') all couplings are the same. For the ω pole,

$$\eta^\omega(p, p) = \eta^\omega(n, n) = -\eta^\omega(\overline{p}, \overline{p}) = -\eta^\omega(\overline{n}, \overline{n}) = \eta_N^\omega, \quad (9.1.24)$$

$$\eta^\omega(\pi, \pi) = \eta_\pi^\omega = 0. \quad (9.1.25)$$

For the ρ pole,

$$\eta^{\rho^0}(\pi^+, \pi^+) = -\eta^{\rho^0}(\pi^-, \pi^-) = -\eta^{\rho^+}(\pi^+, \pi^0)$$

$$= \eta^{\rho^-}(\pi^-, \pi^0) = \eta^{\rho^+}(\pi^0, \pi^-) = -\eta^{\rho^-}(\pi^0, \pi^+) = \eta_\pi^\rho, \quad (9.1.26)$$

$$\eta^{\rho^0}(p, p) = -\eta^{\rho^0}(n, n) = -\eta^{\rho^0}(\overline{p}, \overline{p}) = \eta^{\rho^0}(\overline{n}, \overline{n}) = 2^{-\frac{1}{2}}\eta^{\rho^+}(p, n)$$

$$= 2^{-\frac{1}{2}}\eta^{\rho^-}(n, p) = -2^{-\frac{1}{2}}\eta^{\rho^-}(\overline{p}, \overline{n}) = -2^{-\frac{1}{2}}\eta^{\rho^+}(\overline{n}, \overline{p}) = \eta_N^\rho. \quad (9.1.27)$$

The helicity-flip couplings ϕ satisfy similar relations, and there are analogous relations involving the R pole.

The analogue of (9.1.18) for the nucleon-nucleon total cross-section is

$$\sigma_{NN}(\text{total}) \sim \sum_j \tau_j \left(\frac{s}{s_{0j}}\right)^{\alpha_j(0)-1} \eta_N^j \eta_N^j. \quad (9.1.28)$$

Inserting the contributions appropriate to the five Regge poles (9.1.23), one obtains

$$\sigma(\text{total}, pp) \sim P + P' - \rho - \omega + R, \quad (9.1.29)$$

$$\sigma(\text{total}, \overline{p}p) \sim P + P' + \rho + \omega + R, \quad (9.1.30)$$

$$\sigma(\text{total}, pn) \sim P + P' + \rho - \omega - R, \quad (9.1.31)$$

where the symbol denotes the *positive* contribution appropriate to each pole, P, P', ρ, ω, R,

$$P = \left| \tau_P \left(\frac{s}{s_{0P}} \right)^{\alpha_P - 1} \eta_N^P \eta_N^P \right| = (\eta_N^P)^2, \qquad (9.1.32)$$

$$P' = \left(\frac{s}{s_{0P'}} \right)^{\alpha_{P'} - 1} (\eta_N^{P'})^2, \qquad (9.1.33)$$

$$\rho = \left(\frac{s}{s_{0\rho}} \right)^{\alpha_\rho - 1} (\eta_N^\rho)^2, \qquad (9.1.34)$$

$$\omega = \left(\frac{s}{s_{0\omega}} \right)^{\alpha_\omega - 1} (\eta_N^\omega)^2, \qquad (9.1.35)$$

$$R = \left(\frac{s}{s_{0R}} \right)^{\alpha_R - 1} (\eta_N^R)^2. \qquad (9.1.36)$$

Experimental values of the exponents of s will be discussed later (see (9.3.1)).

Line reversal

The above formulae illustrate the property described as line reversal (Wagner & Sharp, 1962). In general this property can be stated as follows: The two processes

$$a + b \to c + d; \quad a + \bar{c} \to \bar{b} + d, \qquad (9.1.37)$$

have the same quantum numbers in the crossed (t) channel so they can exchange the same Regge poles. These poles therefore contribute the same amounts to each process, except for a possible change of sign. The change of sign is just the signature τ when reversing a spinless boson line in (9.1.37). The change when reversing a nucleon line is a factor C, the charge conjugation quantum number of the t channel meson.

(g) Fermion Regge poles

Fermion Regge poles have been discussed in §5.5, and are treated in detail by Singh (1963), Gribov (1963), Amati *et al.* (1963) and Gell-Mann *et al.* (1963). Exchange of baryons is particularly important for backward scattering, for example, in $\pi^+ p$ backward scattering the trajectory that includes the neutron appears to be dominant. With fermion exchange there are a number of differences from elastic scattering with boson exchange, which include,

   (i)   The mass difference prevents $\cos \theta_u \to \infty$, as $s \to \infty$ near $\theta_s = \pi$.

However it is assumed that Regge asymptotics can still be used (see § 5.6).

(ii) The convenient variable is $W = (u)^{\frac{1}{2}}$ rather than u. This leads to branch points at $u = 0$ in some amplitudes.

(iii) There are two partial wave amplitudes $f_J^+(W)$ and $f_J^-(W)$ in the u channel, for given angular momentum J, having parity $(-1)^{J+\frac{1}{2}}$ and $(-1)^{J-\frac{1}{2}}$ respectively. These amplitudes have Regge poles that coincide at $u = 0$, and are complex conjugate for $u < 0$.

(iv) The asymptotic form of the amplitude, for fixed u and large s, can be associated with a pole in the u channel and gives, for a single pole,

$$\frac{d\sigma}{du} \sim F(u) \left(\frac{s}{s_0}\right)^{2\alpha(u)-3}. \tag{9.1.38}$$

Thus $\mathrm{Re}\,[\alpha(u) - \frac{1}{2}]$ for fermions corresponds to $\mathrm{Re}\,[\alpha(u)]$ for bosons.

(v) Since there are pairs of poles for fermions, there will in general be no results involving fermion exchange that are as simple as those for the one pole approximation for boson exchange.

(h) Backward scattering and fermion Regge exchange

There are at least two aspects of backward scattering where Regge theory gives an interesting comparison with experiment. One is the behaviour of the differential cross-section, near the backward direction of the s channel, when quantum numbers of the u channel restrict exchange to one or two dominant Regge poles. This happens for backward scattering in the (s channel) process

$$\pi^+ + p \to \pi^+ + p. \tag{9.1.39}$$

The u channel corresponds to the process

$$\pi^- + p \to \pi^- + p, \tag{9.1.40}$$

for which the quantum numbers of a neutron are allowed for an intermediate state. Thus exchange of the neutron trajectory will contribute to (9.1.39) in the backward direction. Since the corresponding Regge term contains a factor $[2\alpha_N(t) + 1]$, there should be a dip in the differential cross-section when $\alpha_N(t) = -\frac{1}{2}$. We will discuss this effect in more detail in § 9.4.

Another interesting aspect of backward scattering is related to the remarkable interference effects observed as the energy varies (see § 2.5). The experimental results suggest dominance of direct channel resonances at lower energies, whilst at higher energies the exchange trajectories (i.e. resonances in the u channel) should dominate. A method based on the superposition of s channel and u channel Regge trajectories, proposed by Barger & Cline (1966), is described in § 9.4.

9.2   Experimental tests of one pole approximations

(a) *One pole approximation for elastic scattering*

The asymptotic constancy of total cross-sections implies that, for $t = 0$, the Pomeron will dominate over all other exchange poles when the energy is sufficiently high. However, present energies do not appear to be high enough for this dominance to occur, and one must use the many pole approximation for total cross-sections and for forward amplitudes (§ 9.3).

For non-forward scattering near $t = 0$, a single dominant pole would give a differential cross-section

$$\frac{d\sigma}{dt} \sim |F(t)|^2 \exp\{2(\alpha(t) - 1)\log(s/s_0)\} \tag{9.2.1}$$

$$\sim |F(0)|^2 \exp\{2\alpha' t \log(s/s_0)\}. \tag{9.2.2}$$

This approximation would predict that there is a forward peak at high energy, whose width shrinks with increasing energy like

$$\Delta t \sim \left[\frac{d\alpha(0)}{dt}\log s\right]^{-1}. \tag{9.2.3}$$

Such shrinkage is not observed in general. There is shrinkage for pp and pn scattering, but there is none for πp scattering, and the peak for $\bar{p}p$ scattering expands with increasing energy. This confirms that present experimental energies, around 20 Gev, are not large enough for single pole dominance in elastic scattering. This conclusion may be reinforced if the slope $(d\alpha/dt)$ of the Pomeron trajectory is small at $t = 0$; then the variation of the exponential in (9.2.2), for varying t within the peak, would be small compared with the variation of $|F(t)|^2$. For the process $a + b \to a + b$, the factor $F(t)$ can be written as a product (in the one pole approximation),

$$F(t) \approx \eta_a^P(t)\,\eta_b^P(t). \tag{9.2.4}$$

There are no reliable techniques for estimating the variation with t of the Pomeron vertex function $\eta^P(t)$, although it is possible that the quark model may lead to some estimate when its approximations are better understood (§ 10.5). Alternatively, the peripheral model (§ 10.1), combined with Regge theory, may lead to information about this t dependence.

Another important possibility is that branch cuts, particularly the branch cut ending at $l = 1$, could become dominant for $t < 0$. This would produce a contribution for $t < 0$, (see (9.1.5)),

$$\frac{d\sigma}{dt} \sim |F_1(t)|^2 \exp\left[-2\beta(t)\log\log(s/s_0)\right]. \tag{9.2.5}$$

Such a slow rate of shrinkage as $[\log\log(s/s_0)]^{-1}$ would not be observed in existing experiments, and might also be masked by the variation of $|F_1(t)|^2$ in the t direction. It is of considerable importance to obtain some parametrisation of the effects of branch cuts, so as to assess their magnitude relative to the Regge pole terms.

(b) *One pole approximation for exchange reactions*

For the s channel reaction

$$\pi^- + p \to \pi^0 + n, \tag{9.2.6}$$

the corresponding t channel involves only states with isospin 1,

$$\pi^- + \pi^0 \to \bar{p} + n. \tag{9.2.7}$$

The only known particle in this channel, that has $I = 1$ and negative signature, is the ρ meson. We will consider the contributions to the $\pi^- p$ charge exchange reaction that come from the ρ trajectory, firstly using helicity amplitudes and then in more detail using invariant amplitudes.

Using helicity amplitudes

From (9.1.21) the ρ contribution to the reaction (9.2.6), for large s, will give a differential cross-section,

$$\frac{d\sigma(\pi^- p \to \pi^0 n)}{dt} \sim \left(\frac{s}{s_{0\rho}}\right)^{2\alpha_\rho - 2} \sec^2(\pi\alpha_\rho/2)(\eta^\rho_\pi)^2[(\eta^\rho_\pi)^2 + (\phi^\rho_N)^2]. \tag{9.2.8}$$

Experimental results, on the differential cross-section, show a dip in the forward direction, $t = 0$, and a dip at $t \approx -0.5\,(\mathrm{Gev}/c)^2$. This could be interpreted theoretically as evidence that the spin-flip term ϕ is relatively large for $t \neq 0$ (it vanishes for $t = 0$), and that η and ϕ both vanish when $\alpha_\rho(t) = 0$.

The experimental data for $\pi^- p$ charge exchange scattering is indicated in Fig. 9.2.1(a). It has been analysed by Höhler (1966), for several fixed values of the energy up to 18 Gev, and varying t in $-1 < t < 0$. This data was used to determine the trajectory $\alpha_\rho(t)$ for $t < 0$, and the result is shown in Fig. 9.2.1 (b). The trajectory is found to extrapolate, in $t > 0$, very near to the physical ρ particle, for which

$$\alpha_\rho(t) = 1 \quad \text{at} \quad t = 0.56 \,(\text{Gev})^2. \tag{9.2.9}$$

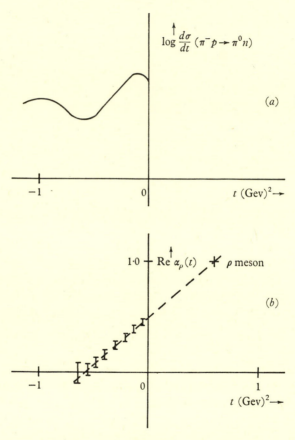

Fig. 9.2.1. (a) The form of the experimental curve for $\pi^- p \to \pi^0 n$ near the forward direction; (b) the ρ-trajectory compared with experiment for $\pi^- p \to \pi^0 n$ scattering (Ter Martirosyan, 1966, who also gives references for the experimental data).

The best fit to the experimental data gives

$$\alpha_\rho(0) = 0.57. \tag{9.2.10}$$

For πp elastic scattering, only P, P' and ρ contribute out of the five poles (9.1.23). One obtains from the optical theorem (compare with (9.1.29)),

$$\sigma(\text{total}, \pi^- p) \sim P + P' + \rho, \tag{9.2.11}$$

$$\sigma(\text{total}, \pi^+ p) \sim P + P' - \rho. \tag{9.2.12}$$

Hence, the difference between these cross-sections involves only the contribution from ρ exchange,

$$\sigma(\text{total}, \pi^- p) - \sigma(\text{total}, \pi^+ p) \sim \left(\frac{s}{s_{0\rho}}\right)^{\alpha_\rho(0)-1} \eta_\pi^\rho(0)\, \eta_N^\rho(0). \tag{9.2.13}$$

The magnitude and energy dependence of this difference of cross-sections is in good agreement with the parameters obtained from (9.2.8) for varying t, namely $\alpha_\rho(0) = 0\cdot57$.

Polarisation measurements create a difficulty for this single pole approximation. According to (9.1.22), the polarisation should be zero if a single Regge pole is dominant. Alternatively, if there is one dominant pole, and a second pole X giving a significant contribution but having a smaller value of $\alpha_X(t)$ than $\alpha_\rho(t)$, the polarisation will be given by

$$P_{\pi N} \frac{d\sigma(\pi^- p \to \pi^0 n)}{dt} \sim \frac{1}{2\pi} \text{Im}\, [\zeta_\rho^* \zeta_X] \left(\frac{s}{s_{0\rho}}\right)^{\alpha_\rho-1} \left(\frac{s}{s_{0X}}\right)^{\alpha_X-1} \eta_\pi^\rho \eta_\pi^X \phi_N^\rho \eta_N^X \tag{9.2.14}$$

(we have approximated by assuming α_ρ and α_X to be real). In general for $\alpha_\rho > \alpha_X$, the phases will be different, so the polarisation should be non-zero. However, comparing (9.2.14) with (9.2.8) we see that the polarisation $P_{\pi N}$ should decrease with energy s. Experimentally this result seems doubtful (Van Hove, 1966), although it is not yet certainly contradicted.

If the polarisation (9.2.14) continues to disagree with further experiments, a possible explanation could be that the branch cut associated with the ρ pole makes a significant contribution to the background. The branch point would be at

$$l = \alpha_\rho(0) + \alpha_P(0) - 1 = \alpha_\rho(0). \tag{9.2.15}$$

The attached branch cut would give a contribution to the differential cross-section that is proportional to

$$\left(\frac{s}{s_{0\rho}}\right)^{\alpha_\rho(0)-1} (\log s)^{-\beta(t)} \quad \text{for} \quad t < 0. \tag{9.2.16}$$

The phase of a term arising from a branch cut is not readily determined. This term is not in general expected to be purely symmetric or purely antisymmetric. There will therefore, in general, be interference between the Regge branch cut term (9.2.16) and the Regge pole term (9.2.8). At present there is no theoretical method of estimating the magnitude and phase of the branch cut term, so the polarisation that would result from the interference is also unknown theoretically.

Using invariant amplitudes

An alternative analysis of pion-nucleon reactions has been presented in Chapter 5 using invariant amplitudes. This can also be applied to the above problem of ρ exchange. We have

$$F(s,t) = -A + \tfrac{1}{2}i\gamma.(q_1+q_2)B, \qquad (9.2.17)$$

where q_1 and q_2 denote the initial and final four-momenta of the pion in the centre of mass system. Write E for the pion laboratory energy and define

$$A' = A + \left(\frac{E+t/4M}{1-t/4M^2}\right)B \qquad (9.2.18)$$

where M is the nucleon mass. Then (Phillips, 1966; Singh, 1963),

$$\frac{d\sigma}{dt} = \frac{1}{\pi s}\left(\frac{M}{4k}\right)^2\left[(1-t/4M^2)\,|A'|^2 + \frac{t}{4M^2}\left(\frac{E^2-m^2-st/4M^2}{1-t/4M^2}\right)|B|^2\right]. \qquad (9.2.19)$$

The total cross-section is

$$\sigma(\text{total}) = \frac{1}{(E^2-mE)^{\frac{1}{2}}}\,\text{Im}\,A'(t=0).$$

The polarisation P relative to $\mathbf{q}_1 \wedge \mathbf{q}_2$ is given by

$$P\frac{d\sigma}{dt} = -\frac{\sin\theta}{16\pi s^{\frac{1}{2}}}\,\text{Im}\,(A'B^*). \qquad (9.2.20)$$

The contribution of the ρ trajectory, (odd signature $\tau = -1$), to A' and B, is given by

$$A'(s,t) = C(t)\,[\tan\tfrac{1}{2}\pi\alpha + i]\left(\frac{s}{s_{0\rho}}\right)^{\alpha_\rho(t)}, \qquad (9.2.21)$$

$$B(s,t) = D(t)\,[\tan\tfrac{1}{2}\pi\alpha + i]\,\alpha(t)\left(\frac{s}{s_{0\rho}}\right)^{\alpha_\rho(t)-1}. \qquad (9.2.22)$$

C and D denote the residues at the ρ pole; using the factorisation hypothesis they are products of the $\pi\rho$ and $N\rho$ coupling constants.

Note the extra factor in (9.2.19), that multiplies $|B|^2$ and compensates for the lower power of s in (9.2.22). The extra $\alpha(t)$ term in (9.2.22) comes from a term involving $(d/d\cos\theta)\,P_\alpha(-\cos\theta)$, (Frazer &

Fulco, 1960; and see §§ 3.6 and 4.9). Since A and B have the same phase, there will be no asymptotic polarisation from the ρ trajectory by itself.

Zeros

At $\alpha_\rho(t) = 0$ in $t < 0$, $B(s, t)$ vanishes. This agrees with the observed dip in the differential cross-section for the charge exchange reaction $(\pi^- p \to \pi^0 n)$. When $\alpha_\rho(t) = -(2N + 1)$, where N is an integer we must also have both $C(t)$ and $D(t)$ equal to zero, if we are to avoid a singularity of the differential cross-section (such a singularity would represent a ghost state; thus the vanishing of C and D is simply the condition that the ghosts must be killed).

The differential cross-section (9.2.19) from the ρ contribution (9.2.21, 22) for large s, is

$$\frac{d\sigma(\pi^- p \to \pi^0 n)}{dt}$$

$$\sim \left(\frac{s}{s_{0\rho}}\right)^{2\alpha_\rho - 2} \sec^2 (\pi\alpha_\rho/2) \left[|C(t)|^2 + \{\alpha_\rho(t)\}^2 \left(\frac{t}{4M^2}\right)^2 |D(t)|^2\right]. \quad (9.2.23)$$

This shows the vanishing of the spin-flip term at $t = 0$, and at $\alpha_\rho(t) = 0$.

It has been suggested that, in addition to the zeros to kill ghosts, C and D should also both vanish when $\alpha(t) = -(2N)$, with N integer (Arbab & Chiu, 1966). This would produce further observable dips in the differential cross-section.

Other exchange reactions

There are a number of other exchange reactions that are accessible to experiment and involve the exchange of a ρ pole in the term that is believed to be dominant at high energy. For example, the charge exchange reactions

$$\pi^+ + p \to \pi^0 + N^{*++}(1238), \quad (9.2.24)$$

$$\pi^- + p \to \pi^0 + N^{*0}(1238), \quad (9.2.25)$$

both involve only the ρ trajectory amongst the five leading meson trajectories. Experimental results (Morrison, 1966) indicate that the differential cross-section for the reaction (9.2.24) is almost zero at $-t \sim 0.5\,(\text{Gev}/c)^2$. Whether this is a dip has yet to be ascertained. There are a number of complications associated with unequal mass difficulties (see § 5.6), that have yet to be solved before the theoretical interpretation is fully understood even when only single Regge pole exchange is allowed.

Another example of single pole exchange is given by the exchange reaction,

$$\pi^- + p \to \eta^0 + n. \tag{9.2.26}$$

The only known Regge pole that could fit the quantum numbers of the crossed channel,

$$\pi^+ + \eta^0 \to p + \bar{n}, \tag{9.2.27}$$

is that of the R trajectory. The signature is even. This trajectory is believed to go through the A_2 resonance, at mass $1\cdot3$ Gev, having spin and parity equal to 2^+. The present data is consistent with single pole exchange in this reaction (Phillips, 1966), but the polarisation has not yet been measured.

9.3   Total cross-sections using several Regge poles

In this and the following sections we will briefly describe further experimental tests of Regge theory. Examples of reactions are given to illustrate the main classes of experiment that can be usefully related to Regge theory. These examples are intended only as a selection from possible experiments, and references to other experimental tests are given in the reviews by Leader (1966), Phillips (1966), Van Hove (1966, 1967).

For nucleon-nucleon scattering, we have listed in (9.1.29–36) the contributions from the five Regge poles that are assumed to be dominant as $s \to \infty$. The values of $\alpha(t)$ at $t = 0$ can be found from experimental results on total cross-sections. They are approximately

$$\alpha_P = 1; \quad \alpha_{P'} = 0\cdot65; \quad \alpha_\rho = 0\cdot57; \quad \alpha_\omega = 0\cdot5; \quad \alpha_R = 0\cdot3. \tag{9.3.1}$$

These numerical values will have to be modified if other trajectories are found to contribute, and also if Regge branch cuts give significant contributions. For example, at $t = 0$ there is believed to be a branch cut along $l \leqslant 1$, which would give a contribution to σ(total) of magnitude,

$$P\,(\text{branch cut}) \sim \text{constant}\,(\log s)^{-\gamma}, \tag{9.3.2}$$

where $\gamma > 0$. This contribution from the Pomeron branch cut would be additive to the pole contributions for forward elastic scattering and for total cross-sections. Important differences between the effects of poles and branch cuts can arise for inelastic contributions (see 'parity exchange' in §9.7), because the Pomeron branch cut contributions can correspond to the exchange of odd parity (Gribov, 1966), but the Pomeron itself has even parity.

The Pomeron branch cut does not contribute to the differences between suitably chosen total cross-sections, but other branch cuts may be significant. If these branch cuts are neglected we will have

$$\sigma(\bar{p}, p) - \sigma(p, p) \sim 2(\rho + \omega), \tag{9.3.3}$$

$$\sigma(p, n) - \sigma(p, p) \sim 2(\rho - R). \tag{9.3.4}$$

The energy dependence of these differences can be used to determine the values of α at $t = 0$, quoted in (9.3.1), for ρ, ω and R.

For pion-nucleon scattering, we have noted in (9.2.13) that only ρ exchange contributes to the difference of $\pi^- p$ and $\pi^+ p$ total cross-sections,

$$\sigma(\pi^-, p) - \sigma(\pi^+, p) \sim 2\rho. \tag{9.3.5}$$

For kaon-proton scattering, one obtains (Phillips & Rarita, 1965)

$$\sigma(K^-, p) - \sigma(K^+, p) \sim 2(\omega + \rho). \tag{9.3.6}$$

In (9.3.5, 6), compared with (9.3.3, 4), there must be appropriate modification of the residue factors to allow for the different π and K coupling to ρ and to ρ, ω respectively. However, the energy variation of a ρ term is the same in each of (9.3.3, 4, 5 and 6).

It is interesting to compare (9.3.5) and (9.3.6) with the Johnson–Treiman relation derived from SU 6 invariance in the forward direction (Johnson & Treiman, 1965). This states that the differences of total cross-sections satisfy

$$\tfrac{1}{2}[\sigma(K^-, p) - \sigma(K^+, p)] \sim [\sigma(K^0, p) - \sigma(\bar{K}^0, p)] \sim [\sigma(\pi^-, p) - \sigma(\pi^+, p)] \tag{9.3.7}$$

as $s \to \infty$. From (9.3.5, 6) this implies not only that $\alpha_\rho(0) \sim \alpha_\varpi(0)$, which agrees with suggested values in (9.3.1), but it also implies relations between coupling constants. The experimental comparison of (9.3.7) at 20 Gev is reasonably good (Johnson & Treiman, 1965) but its implications must await better data and a better understanding of SU 6.

Assuming isospin invariance, the relation (9.3.7) can be rewritten in the form,

$$\begin{aligned} \tfrac{1}{2}[\sigma(K^-, p) - \sigma(K^+, p)] &\sim \sigma(K^-, n) - \sigma(K^+, n), \\ &\sim \sigma(\pi^-, p) - \sigma(\pi^+, p). \end{aligned} \tag{9.3.8}$$

Experimental data on these relations is given by Lindenbaum (1965, 1967) and Foley *et al.* (1967). Due to the mass difference of π and K, it is not clear at what energies the comparison should be made (but see § 10.5).

9.4   Exchange cross-sections and backward scattering

In this section we consider (*a*) exchange cross-sections that involve contributions from more than one Regge pole, in the near forward direction, (*b*) near backward scattering, in which the differential cross-section may be dominated by a single Regge pole, and (*c*) a model for the energy dependence of backward scattering that includes resonances (Regge trajectories) in both the direct channel and in the relevant exchange channel.

(*a*)   *Exchange cross-sections*

In kaon-nucleon charge exchange scattering there can be contributions from exchange of both the ρ and the $R(A_2)$ Regge poles. These give (Phillips & Rarita, 1965),

$$F(K^-p \to \bar{K}^0 n) \sim 2F_\rho + 2F_R, \tag{9.4.1}$$

$$F(K^+n \to K^0 p) \sim -2F_\rho + 2F_R. \tag{9.4.2}$$

The corresponding elastic scattering amplitudes near the forward direction will satisfy

$$F(K^-p) - F(K^-n) \sim 2F_\rho + 2F_R, \tag{9.4.3}$$

$$F(K^+p) - F(K^+n) \sim -2F_\rho + 2F_R. \tag{9.4.4}$$

The optical theorem and the measured total cross-sections give the differences of the imaginary parts of these amplitudes. The experimental data (Galbraith *et al.* 1965) near 10 Gev indicates that

$$\mathrm{Im}\, F_\rho \sim \mathrm{Im}\, F_R \quad \text{at} \quad t = 0. \tag{9.4.5}$$

The only contributions to the forward amplitudes, and the total cross-sections, come from the non-spin flip terms. The phases of F_ρ and F_R can be found from (9.1.17 and 23), using $\alpha_\rho(0) = 0.57$ and $\alpha_R(0) = 0.3$. They show that the real parts of F_ρ and F_R are nearly equal in magnitude but of opposite signs, namely

$$\mathrm{Re}\, F_\rho \sim \mathrm{Im}\, F_\rho \tan\left[\tfrac{1}{2}\pi\alpha_\rho(0)\right], \tag{9.4.6}$$

$$\mathrm{Re}\, F_R \sim -\mathrm{Im}\, F_R \cot\left[\tfrac{1}{2}\pi\alpha_R(0)\right]. \tag{9.4.7}$$

When (9.4.5, 6 and 7) are substituted into (9.4.1 and 2) we see that in the forward direction $F(K^-p \to \bar{K}^0 n)$ should be dominantly pure imaginary, but $F(K^+n \to K^0 p)$ should be dominantly real This conclusion is supported by the experimental evidence (Phillips & Rarita

1965). The imaginary parts of the amplitudes (9.4.1 and 2) are obtained from experimental values of the total cross-sections, using (9.4.3 and 4). The real parts are then obtained by extrapolating the differential cross-sections to the forward direction.

(b) Scattering near the backward direction (baryon exchange)

Other exchange cross-sections can be studied in which non-zero baryon number, or non-zero strangeness, must be exchanged. For example, consider backward π^-p elastic scattering,

$$s\text{ channel:} \quad \pi^- + p \to \pi^- + p. \tag{9.4.8}$$

This process can be assumed to be dominated at high energy by exchange of a Regge particle in the u channel,

$$u\text{ channel:} \quad \pi^+ + p \to \pi^+ + p. \tag{9.4.9}$$

This channel involves isospin $\frac{3}{2}$ but it does not involve isospin $\frac{1}{2}$. Consequently near backward π^-p scattering may be quite different from near backward π^+p scattering, for which we have,

$$s\text{ channel:} \quad \pi^+ + p \to \pi^+ + p, \tag{9.4.10}$$

$$u\text{ channel:} \quad \pi^- + p \to \pi^- + p. \tag{9.4.11}$$

The process (9.4.11) can involve a neutron intermediate state. The possible dominance of near backward π^+p scattering by neutron exchange produces effects analogous to those discussed in §9.2 for ρ exchange.

The neutron exchange contribution to π^+p backward scattering was given in §5.5 (eq. (5.5.27)). The derivatives of the Legendre polynomials in that expression lead to a factor

$$[\alpha_N(W_u) + \tfrac{1}{2}], \tag{9.4.12}$$

where W_u is the neutron 'energy'. This factor occurs in both the spin-flip and the non-spin-flip terms and is zero when the neutron trajectory goes through $-\frac{1}{2}$,

$$l = \alpha_N(W_u) = -\tfrac{1}{2}. \tag{9.4.13}$$

A dip is observed in the experimental π^+p cross-section near the backward direction, when $u = W_u^2$ satisfies

$$u \approx -0 \cdot 2\,(\text{Gev/c})^2. \tag{9.4.14}$$

The point $\alpha_N = -\frac{1}{2}$, $u = -0 \cdot 2$ lies on the continuation of a Regge trajectory along a straight line through the physical neutron ($\alpha_N = \frac{1}{2}$, $u = 0 \cdot 88$) and the physical N^* ($\alpha_N = \frac{5}{2}$, $u = 2 \cdot 88$). The detailed Regge analysis of Stack (1966) on backward π^+p scattering confirms this qualitative agreement, although one would like to see more precise experimental results in the neighbourhood of the dip and to have more information about its energy variation.

For K^-p backward elastic scattering the relevant crossed channel is K^+p. No baryon of strangeness plus one has been observed. The very small magnitude of K^-p backward scattering suggests that no such baryon exists (at least, not with charge 2). This magnitude (at 5 Gev) is less than one quarter of the magnitude for π^-p backward scattering —for which the only known trajectory in the crossed channel is Δ_δ (see Fig. 9.4.1 (b)). This provides some evidence also that the background integral gives only a small contribution in this instance.

(c) *Backward scattering, resonances in the direct and exchange channels*

There are some remarkable interference phenomena observed in the dependence of the differential cross-section for backward scattering on the energy (see § 2.5). It is not surprising that interference is observed, but its effects are unexpectedly strong. In considering the backward direction, we note that non-resonant partial wave amplitudes can be assumed to vary smoothly with l and with energy. The smooth l dependence leads to partial cancellation when $\cos\theta = -1$, due to the factor $(-1)^l$ from the Legendre polynomial in the partial wave series. The smooth energy dependence leads to a slow variation of both the phase and the amplitude for the sum of these non-resonant terms. We will call the sum of the non-resonant terms, the background term $B(s)$ of $F(s,t)$ at $\theta = \pi$.

If there is one partial wave f_l that depends strongly on s near a resonance energy s_0, we will have

$$F(s, \theta = \pi) \sim [B(s) + (-1)^l \eta_l(s) \exp(i\delta_l) \sin\delta_l]. \qquad (9.4.15)$$

In any region of real s, where $B(s)$ and $\eta_l(s)$ are of comparable magnitude, the backward differential cross-section will have a sharp dip (negative relative parity), or a maximum (positive relative parity), at a resonance.

Backward scattering at high energy corresponds to $u = 0$, and for nearly backward scattering one might reasonably expect dominant effects from Regge poles exchanged in the u channel. As with nearly forward scattering, this will be most marked when only a limited number (preferably one) of exchange poles in the u channel are allowed by conservation laws.

An interesting approximation for π^-p backward scattering, which involves combining resonance poles and exchange poles, has been studied by Barger & Cline (1966). Their work is undoubtedly a prelude

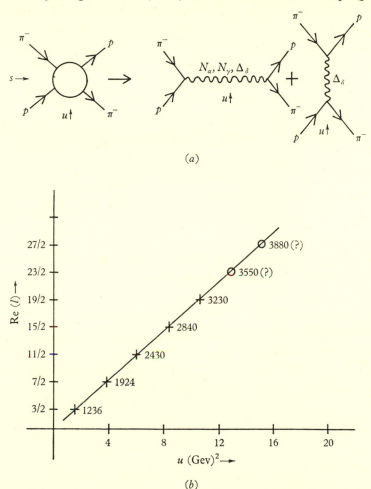

(a)

(b)

Fig. 9.4.1. (a) The approximation of Barger & Cline (1966) for backward π^-p scattering expressed as direct channel resonances plus exchange (u channel) Regge poles; (b) one of the Regge trajectories used in the Barger & Cline model. Resonances marked + have been observed, those marked O are predicted by the model (i.e. they are required so that it agrees with experiment).

to more detailed studies of this type. Their basic approximation is shown in Fig. 9.4.1 (a). In the s channel they include all observed (and some extrapolated) resonances that are allowed by conservation laws, and they do the same in the u channel. They extrapolate to possible new resonances along the Regge trajectory through observed resonances (whose angular momentum is assumed in some cases). Enough new resonances are inserted to fit the observed data. These are their 'predicted' resonances. One of the relevant Regge trajectories is shown in Fig. 9.4.1(b); this is the Δ_δ trajectory, where δ denotes ($P = +$, $\tau = -$) and $I = \frac{3}{2}$, $Y = 1$. The other two trajectories that are considered cannot be exchanged, and are relevant only in the direct channel, where they appear as resonances. They are denoted by N_α and N_γ; both these trajectories have $I = \frac{1}{2}$, $Y = 1$, and α denotes ($P = +$, $\tau = +$), while γ denotes ($P = -$, $\tau = -$).

The equations of these trajectories (Barger & Cline, 1966; see also Murphy, 1966), are given by

$$\Delta_\delta: \quad \mathrm{Re}\,\alpha(W_u) = (0\cdot 15) + (0\cdot 90)\,u, \tag{9.4.16}$$

$$N_\alpha: \quad \mathrm{Re}\,\alpha(W_u) = -(0\cdot 39) + (1\cdot 01)\,u, \tag{9.4.17}$$

$$N_\gamma: \quad \mathrm{Re}\,\alpha(W_u) = -(0.90) + (0\cdot 92)\,u, \tag{9.4.18}$$

with $u = W_u^2$ measured in (Gev)^2. Although this must be regarded as a very approximate model, it should indicate the orders of magnitude to be expected in a more detailed study.

9.5   Differential cross-sections, polarisation and spin correlations

(a) Differential cross-sections

In general, differential cross-sections will involve interference between different trajectories. The effects of the relative phases for $t < 0$ have been studied by Leader & Slansky (1966), for $\bar{p}p$ and pp scattering. In analogy with (9.1.21), the cross-section for nucleon-nucleon scattering (Wagner, 1963) is given by

$$\frac{d\sigma(NN \to NN)}{dt} \sim \frac{1}{4\pi} \sum_{ij} \mathrm{Re}\,[\zeta_i^* \zeta_j] \left(\frac{s}{s_{0i}}\right)^{\alpha_i - 1} \left(\frac{s}{s_{0j}}\right)^{\alpha_j - 1} [\eta_N^i \eta_N^j + \phi_N^i \phi_N^j]^2. \tag{9.5.1}$$

Let l denote the even signature poles P, P' and R. Then noting that ρ and ω have odd signature, we can write,

$$\frac{d\sigma(\bar{p}p \to \bar{p}p)}{dt} - \frac{d\sigma(pp \to pp)}{dt}$$

$$\sim \sum_l \mathrm{Re}\,[\zeta_l^* \zeta_\rho] \left(\frac{s}{s_{0l}}\right)^{\alpha_l - 1} \left(\frac{s}{s_{0\rho}}\right)^{\alpha_\rho - 1} [\eta_N^l \eta_N^\rho + \phi_N^l \phi_N^\rho]^2$$

$$+ \text{(a term with } \rho \leftrightarrow \omega). \tag{9.5.2}$$

Since ρ has odd signature, and each particle denoted by l has even signature,

$$\text{Re}\,[\zeta_l^* \zeta_\rho] = \tfrac{1}{4}[1 - \cot \tfrac{1}{2}\pi\alpha_l \tan \tfrac{1}{2}\pi\alpha_\rho]. \tag{9.5.3}$$

A similar formula holds for terms involving ω. From the values of $\alpha(t)$ and their slopes at $t = 0$, one can estimate whether or when (9.5.2) changes sign on this model. Experimentally the left-hand side changes sign at $t = -0.15$. There is doubt (Leader & Slansky, 1966) whether this can be achieved theoretically with previously determined values of the $\alpha(t)$. It may be that here also is some evidence of the effects of branch cuts, or it may cast doubt on the factorisation hypothesis.

(b) Polarisation and spin correlations

The opposite signatures of the ρ and R trajectories, give terms of opposite phase for the charge exchange reactions, (9.4.1) and (9.4.2), when

$$\alpha_\rho(t) \approx \alpha_R(t). \tag{9.5.4}$$

This approximate equality happens near $t = -0.5$ (Gev)². This should lead to relatively large polarisation (Phillips, 1966), which suggests an important but difficult class of experiments on systems for which only these two trajectories are allowed. These experiments include the exchange reactions,

$$K^- + p \to \bar{K}^0 + n, \tag{9.5.5}$$

$$K^+ + n \to K^0 + p. \tag{9.5.6}$$

In the interesting neighbourhood of $t = -0.5$, the differential cross-sections unfortunately decrease rather fast with increasing energy.

The importance of measurements of polarisation and other spin parameters has also been emphasised by Leader (1966) and by Leader & Slansky (1966). The latter study critical tests for Regge theory in nucleon-nucleon scattering, with special references to tests of the factorisation theorem. They find that the spin correlation parameter C_{NN} in nucleon-nucleon scattering contains terms like

$$\sum_{n,\,l} |r_{11}^n r_{22}^l - r_{12}^n r_{12}^l|^2 \left(\frac{s}{s_0}\right)^{\alpha_n + \alpha_l - 2}. \tag{9.5.7}$$

The leading power of s comes from the Pomeron, giving for $t = 0$,

$$C_{NN} \sim \text{constant}\,|r_{11}^P r_{22}^P - (r_{12}^P)^2|^2. \tag{9.5.8}$$

If the factorisation theorem holds, this term will be zero, and instead one finds a dominant term,

$$C_{NN} \sim \text{constant}\,|r_{11}^P r_{22}^\omega - r_{12}^P r_{12}^\omega|^2 \left(\frac{s}{s_0}\right)^{\alpha_\omega - 1}, \tag{9.5.9}$$

where $(\alpha_\omega - 1) \approx -\tfrac{1}{2}$ at $t = 0$.

This clear distinction between the behaviour with and without the factorisation hypothesis will unfortunately be somewhat obscured if contributions from branch cuts are important.

9.6   Sum rules

Suppose we have a dispersion relation of the form

$$f(E) = \frac{1}{\pi} \int_m^\infty \frac{dx \, \mathrm{Im} f(x)}{x - E}. \tag{9.6.1}$$

Suppose also that $f(E)$ is subject to the asymptotic bound

$$|f(E)| < E^\beta \quad \text{for} \quad E \to \infty. \tag{9.6.2}$$

Then, if this bound is strong enough that

$$\beta < -1, \tag{9.6.3}$$

we will have $\qquad Ef(E) \to 0 \quad \text{as} \quad E \to \infty. \tag{9.6.4}$

Substituting from the dispersion relation (9.6.1) into (9.6.4), we obtain

$$\int_m^\infty \mathrm{Im} f(x) \, dx = 0. \tag{9.6.5}$$

This is described as a 'superconvergence sum rule', and was first observed to hold for certain amplitudes by Alfaro, Fubini, Rossetti & Furlan (1966). Similar results had been derived earlier by Fubini (1966) using current algebra. They are related to the Adler–Weissburger sum rule (Adler, 1965; Weissburger, 1965)

When one or more of the colliding particles has non-zero spin, invariant amplitudes are introduced that have simple analytic properties. In the expressions for the scattering amplitude, the invariant amplitudes are multiplied by factors that depend on the energy (see (3.6.13) for πN scattering). Using Regge pole assumptions, each of these invariant amplitudes (times its energy factor) cannot have a faster rate of increase than the full amplitude. The discussion in § 6.7 suggests that this is a quite general result.

We will outline the argument of Alfaro *et al.* (1966) for $\rho\pi$ scattering, using the formalism of Frampton & Taylor (1966). The $\rho\pi$ scattering amplitude can be written $\qquad \sum_{\mu\nu} M_{\mu\nu} e^\mu e'^\nu, \tag{9.6.6}$

where e and e' denote the polarisations of the initial and final ρ mesons.

In terms of invariant amplitudes $A(s,t)$, B, C, D,

$$M_{\mu\nu} = AP_\mu P_\nu + B(P_\mu Q_\nu + P_\nu Q_\mu) + CQ_\mu Q_\nu + Dg_{\mu\nu}; \qquad (9.6.7)$$

$$P_\mu = \tfrac{1}{2}(P_\pi + P'_\pi)_\mu; \quad Q_\mu = \tfrac{1}{2}(P_\rho + P'_\rho)_\mu. \qquad (9.6.8)$$

Each of the invariant amplitudes will have a component that arises from the exchange of a particular isospin. Write A_I for the component of A coming from isospin I in the crossed channel. Assuming that no Regge poles exist with $I = 2$ and with $\alpha(0) > 0$, and assuming values for $\alpha(0)$ as in (9.3.1), one obtains the asymptotic behaviour,

$$s^2 A_2(s,t) \to 0; \quad s A_1(s,t) \to 0; \quad s B_2(s,t) \to 0, \qquad (9.6.9)$$

as $s \to \infty$, for $t \leqslant 0$.

This behaviour gives the following superconvergence relations, for $t \leqslant 0$,

$$\int_{s_0}^\infty ds\, \nu\, \mathrm{Im}\, A_2(s,t) = 0, \qquad (9.6.10)$$

$$\int_{s_0}^\infty ds\, \mathrm{Im}\, A_1(s,t) = 0, \qquad (9.6.11)$$

$$\int_{s_0}^\infty ds\, \mathrm{Im}\, B_2(s,t) = 0, \qquad (9.6.12)$$

where

$$2\nu = s - m_\rho^2 - m_\pi^2. \qquad (9.6.13)$$

From these relations, one can derive six sum rules (Low, 1966),

$$\int ds\, \mathrm{Im}\, A_1(s,0) = 0, \qquad (9.6.14)$$

$$\int ds\, \nu\, \mathrm{Im}\, A_2(s,0) = 0, \qquad (9.6.15)$$

$$\int ds\, \mathrm{Im}\, B_2(s,0) = 0 \qquad (9.6.16)$$

and

$$\int ds\, \mathrm{Im}\left[\frac{d}{dt} A_1(s,t)\right]_{t=0} = 0, \qquad (9.6.17)$$

$$\int ds\, \nu\, \mathrm{Im}\left[\frac{d}{dt} A_2(s,t)\right]_{t=0} = 0, \qquad (9.6.18)$$

$$\int ds\, \mathrm{Im}\left[\frac{d}{dt} B_2(s,t)\right]_{t=0} = 0. \qquad (9.6.19)$$

Equations (9.6.14) and (9.6.16) were tested by Alfaro *et al.* (1966), keeping only π, ω, and ϕ contributions to the intermediate states in $\mathrm{Im}\, A \approx A^*A$. They obtain

$$[g_{\omega\rho\pi}^2 + g_{\phi\rho\pi}^2]m_\rho^2 - 4g_{\rho\pi\pi}^2 = 0, \qquad (9.6.20)$$

$$(\nu_\omega + m_\rho^2)\, g_{\omega\rho\pi}^2 + (\nu_\phi + m_\rho^2)\, g_{\phi\rho\pi}^2 - 4g_{\rho\pi\pi}^2 = 0, \qquad (9.6.21)$$

with
$$\nu_{\omega,\phi} = \tfrac{1}{2}(m_{\omega,\phi}^2 - m_\rho^2 - m_\pi^2). \tag{9.6.22}$$

Subtracting (9.6.20) from (9.6.21)

$$\nu_\phi g_{\phi\rho\pi}^2 + \nu_\omega g_{\omega\rho\pi}^2 = 0. \tag{9.6.23}$$

Since ν_ϕ is very small, this equation shows that the ratio

$$g_{\phi\rho\pi}/g_{\omega\rho\pi} \tag{9.6.24}$$

is small; this is in good agreement with experiment. Similarly, the relation obtained between $g_{\rho\pi\pi}$ and $g_{\rho\omega\pi}$ is reasonable.

The complete set of six sum rules (9.6.14) to (9.6.19) has been tested by Frampton & Taylor (1966) assuming that the integrals are saturated by the ρ, ω, A_1 and A_2 (assuming also that $g_{\phi\rho\pi}$ is zero). The agreement is good for the sum rules (9.6.14–16), but not good for (9.6.17–19). Since we know that multiparticle inelastic processes (not quasi two-body) are experimentally important (§ 2.7), it seems likely that they would have to be included for any accurate approximation to these sum rules. Conversely, one might be able to use sum rules to obtain information about inelastic processes.

9.7 Further questions in Regge theory

It is appropriate to emphasize again the developing nature of Regge theory by noting a number of open questions on which further theoretical and experimental study is desirable.

(a) Parity exchange

An interesting possibility for the experimental investigation of Regge branch cuts has been suggested by Gribov (1966, see also Van Hove, 1966, whose account is used here). The branch cut generated by the exchange of several Pomerons (see § 5.6) is attached to a fixed branch point at $l = 1$ when $t \leqslant 0$. This is likely to be the most significant branch cut at high energy.

Gribov considers the reaction, in the s channel,

$$a+b \to c+d, \tag{9.7.1}$$

with the t channel taken to be

$$a+\bar{c} \to \bar{b}+d. \tag{9.7.2}$$

One can define an 'intrinsic' parity P_r in the t channel by

$$P_r = (-1)^J P, \tag{9.7.3}$$

where J is the angular momentum and P is the intrinsic parity of the $a + \bar{c}$ state (assuming meson exchange). Gribov remarks that P_r is $+1$ for Pomeron exchange but it can be either $+1$ or -1 for the Pomeron branch cut. This has significant experimental consequences:

If b and d each denote a proton, but a and c in (9.7.1) denote 0^- and 0^+ mesons respectively, there will be no contribution to the reaction from Pomeron exchange. However the P cut *can* contribute, so one would expect a cross-section with a slow logarithmic decrease. But if there is no P cut, the cross-section will decrease rapidly with s, since in this case ρ exchange would dominate. There will be similar but smaller differences of behaviour (due to the existence or non-existence of branch cuts) in the reaction

$$\pi + p \to \rho + p, \tag{9.7.4}$$

and in proton isobar production.

(b) Parametrisation of branch cut contributions

If it is found that branch cuts make significant contributions at high energy, it will be necessary to find ways of representing them parametrically. The success of the one-pole and several-pole approximations suggests either that the branch cut terms are small, or that they are readily included in pole terms for some processes. This emphasises the importance of seeking quantities, like those involving parity exchange, where the effects of the leading branch cut and the leading pole are different.

(c) Background terms

The background integral that is neglected in Regge pole models may become dominant in some reactions. For example, the process

$$K^- + p \to K^- + p, \tag{9.7.5}$$

near the backward direction, cannot proceed by exchange of any known Regge particle. This process and others having a similar property require detailed study if the nature of the background integral is to be better understood.

(d) Other experimental measurements

The following measurements would be of particular value in relation to Regge theory:

(i) Polarisation and spin correlation parameters, especially their energy dependence.

(ii) A systematic two-dimensional plot of differential cross-sections as functions of both s and t. Generally it is more useful to have the value of $(d\sigma/dt)$ for fixed t and varying s rather than fixed s and varying t.

(iii) Multiple production for suitably chosen ranges of the energy variables (this should show evidence of Regge behaviour).

(iv) Further detailed measurements of quasi two-body reactions are required, particularly angular correlations when high spin resonances are involved in the final state.

(e) Other theoretical questions

(i) The connection between Regge theory and the peripheral model is essential for an understanding of collisions as the energy varies from the peripheral range (a few Gev) up to the Regge range (a few tens of Gev). We will consider this further in Chapter 10.

(ii) Further theoretical and experimental study is required of the cancellation phenomena discussed in §5.6, which arise for collisions with general (unequal) masses and for collisions of particles having non-zero spin.

(iii) Large angle scattering ($\theta \approx 90°$) seems to be remote from the Regge model. The behaviour

$$\left.\frac{d\sigma}{dt}\right|_{90°} \sim \frac{C}{s^2}\exp[-C's^{\frac{1}{2}}], \tag{9.7.6}$$

that is observed experimentally (see §§2.3 and 10.4) suggests that there is a complicated cancellation between different Regge pole contributions. It may be that Regge theory is not suitable here. In that case, one must study the way in which the Regge model for small angles joins on to a 'large angle model' in the intermediate region.

(f) SU3 symmetry

Since π, η, K and $\bar{K}$ belong to the same SU3 octet, the meson vertices ($\pi\pi\rho$, $K\bar{K}\rho$, $\pi\eta R$, $K\bar{K}R$) are related ($R \equiv A_2$). Thus

$$\begin{aligned} g(\rho, K^-, K^0) &= \tfrac{1}{2}g(\rho, \pi^-, \pi^0), \\ g(A_2, K^-, K^0) &= (\tfrac{3}{2})^{\frac{1}{2}} g(A_2, \pi^-, \eta^0). \end{aligned} \left.\right\} \tag{9.7.7}$$

Since the reactions,

$$\pi^- + p \to \pi^0 + n; \quad \pi^- + p \to \eta^0 + n, \tag{9.7.8}$$

involve only ρ and A_2 exchange, they are directly related through the coupling constant relations (9.7.7) to the reactions

$$K^- + p \to K^0 + n; \quad K^- + p \to \eta^0 + n. \tag{9.7.9}$$

The agreement between these prediction and experimental results is good (Phillips, 1966; Phillips & Rarita, 1965).

SU 3 indicates that hypercharge exchange reactions are analogous to the charge exchange reactions discussed earlier. Examples that have been considered include (Arnold, 1966),

$$
\left.
\begin{aligned}
&\pi^- + p \to K^0 + \Lambda^0; \quad K^- + p \to \pi^0 + \Lambda^0, \\
&\pi^- + p \to K^0 + \Sigma^0; \quad K^- + p \to \pi^0 + \Sigma^0, \\
&\hspace{3.2cm} K^- + p \to \pi^- + \Sigma^+.
\end{aligned}
\right\} \tag{9.7.10}
$$

The mesons π, K, belong to the same octet; similarly the baryons Λ, Σ, belong to the nucleon octet. The exchanged particles in the t channel involve the trajectory that goes through the vector resonance K^* (892) (analogous to ρ in charge exchange) and the trajectory through the spin 2 resonance K^* (1411) (analogous to the R trajectory through the A_2). Further processes of this type include those involving meson or baryon resonances in the final states of any of the above reactions.

CHAPTER 10

SPECIAL MODELS FOR HADRON COLLISIONS

In previous chapters we have been concerned primarily with collisions at very high energies for which assumptions about asymptotic behaviour should be valid. These assumptions were used not only in Chapters 6 to 8, where we discussed results that become exact in the asymptotic limit, but also in Chapter 9 for our discussion of applications of Regge theory. In practice, Regge theory is remarkably successful even at quite low energies in the region of 10 Gev or less. However, as experiment and theory become more refined, it is to be expected that discrepancies will be found that are due to effects from the Regge background integral or from non-leading poles. It is therefore important to consider other models, even if they are primarily designed for lower energies, to see whether any of their features can lead to useful modifications of Regge theory. In regions where that theory does not apply, for example in large angle scattering or multiple production, it is necessary to develop alternative models.

In this chapter we consider the peripheral model and its various modifications of which the most successful is the absorption model. After this we consider large angle scattering, and the quark model for high energy scattering.

10.1 Peripheral processes

We will consider collision processes at energies above the resonance region, in the range 2 to 10 Gev/c. Their cross-sections are strongly peaked in the near forward direction, which suggests dominance by a peripheral interaction in which many partial waves are involved. One can also obtain a forward peak by absorption of the lower partial waves so that the collision is dominated by diffraction or absorption effects. Both approaches have defects when considered alone, but when combined they lead to the peripheral model with absorption which is successful for certain types of reaction.

Reviews of peripheral processes have been given by Jackson (1965) and Drell & Hearn (1966) who give further references to the extensive literature in this field.

The interaction of longest range, which might be expected to dominate the scattering amplitude $F(s, t)$ near the forward direction, is assumed to arise from the singularity in the momentum transfer variable t that is nearest to the physical region. If single particle exchange is allowed this singularity will be a pole, at $t = m^2$ say, giving a differential cross-section

$$\frac{d\sigma}{dt} = \frac{G(t)}{sk^2(t - m^2)^2},$$
(10.1.1)

where k is the centre of mass momentum and $G(t)$ is not singular at the pole. This one particle exchange model was suggested by Drell (1960, 1961), Salzman & Salzman (1960) and Ferrari & Selleri (1962).

It is not at all obvious, except by the most naive reasoning, that the nearest singularity has the longest effective range. In some circumstances, for example, with the form factor for a nucleus, the effects of the nearest singularity are in fact dominated by the effects of an oscillatory spectral function associated with more distant singularities (Eden & Goldstone, 1963). It is probable that the spectral function of $F(s, t)$ in $t > 4m^2$ will have oscillations, and for large s these may dominate over the effects from the pole near to $t = 0$ in the physical region. This possibility is supported by Regge theory in the one-pole (or several poles) approximation. However, at lower energies than those required for Regge theory, it is possible that such complications from the spectral function are less important, or that they can be included by some absorption modification (see below).

In nucleon-nucleon scattering with exchange of a pion, Fig. 10.1.1 (a), the numerator $G(t)$ in (10.1.1) is inconveniently zero at $t = 0$ which is the boundary of the physical region nearest to the pion pole. The differential cross-section from the pole alone is

$$\frac{d\sigma}{dt} = \frac{4\pi f_{\pi NN}^4 t^2 |F(t)|^2}{sk^2(t^2 - m^2)^2},$$
(10.1.2)

where $F(0)$ is approximately equal to $F(m^2) = 1$. The situation in the complex t-plane is shown in Fig. 10.1.1 (b). The zero in the numerator of (10.1.2) reduces the effect of the pole near $t = 0$, and for experimental comparison other more distant singularities must be included (Hamilton & Woolcock, 1963).

However, for the reaction

$$\pi^+ + p \to \rho^+ + p,$$
(10.1.3)

the differential cross-section given by the one-pion exchange diagram Fig. 10.1.1 (c) is (Jackson, 1965)

$$\frac{d\sigma}{dt} = \frac{(-\pi t)}{4m_\rho^2 m_N^2 p^2} \left(\frac{g^2}{4\pi}\right)\left(\frac{G^2}{4\pi}\right)\frac{[(m_\rho - m_\pi)^2 - t][(m_\rho + m_\pi)^2 - t]}{(t - m_\pi^2)^2}, \quad (10.1.4)$$

where g denotes the coupling $g(\rho\pi\pi)$ and G denotes the coupling $G(\pi NN)$; p is the laboratory momentum of the incident pion.

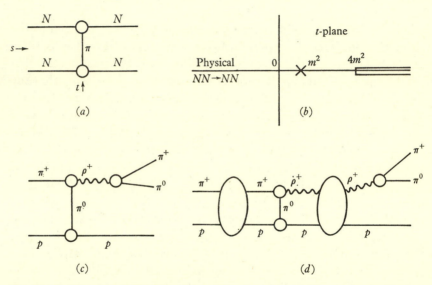

Fig. 10.1.1. Diagrams illustrating the peripheral model: (a) Pion exchange diagram for NN scattering; (b) the location of the pion pole and $\pi\pi$ branch cut relative to the physical region for NN scattering. Note that, in this case, the pion pole contribution is zero at $t = 0$; (c) the one-pion exchange diagram for ρ production and decay; (d) diagram illustrating absorption before and after OPE.

The comparison with experiments at 2·75 Gev/c, of the approximation (10.1.4) for the reaction (10.1.3), is shown in Fig. 10.1.2. The unmodified one-pion exchange term (upper curve (a)) disagrees with unitarity for small partial waves, and it disagrees with experiment both in its magnitude and in its variation with t. Also shown are the modified theoretical curves (b) including absorption and (c) including form factor variation. Both of these are consistent with unitarity.

The modifications due to absorption will be considered in § 10.3 after we have discussed the simplest absorption model—that of diffraction scattering. Qualitatively, the absorption allows for modification of the initial state (preceding the one-particle exchange (OPE)) and the final state (following OPE), and it is illustrated by the diagram (*d*) in Fig. 10.1.1.

The form-factor modifications allow for the fact that the exchanged pion is off its energy shell so that the propagator should be adjusted, and the coupling constants g and G in (10.1.4) should be replaced by the appropriate vertex functions $gF_{\rho\pi\pi}(t)$ and $GF_{\pi NN}(t)$. The use of

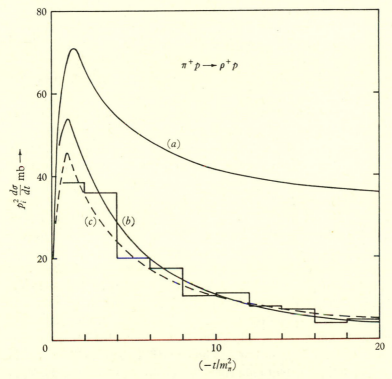

Fig. 10.1.2. Comparison of theory and experiment for the reaction $\pi^+ p \to \rho^+ p$ at 2·75 Gev/c. Curve (*a*) corresponds to the unmodified OPE model; curve (*b*) to the peripheral model with absorption (using OPE); curve (*c*) to the peripheral model with Amaldi–Selleri form factors; (references to the data are given by Jackson, 1965).

form factors in this way has been developed extensively by Ferrari & Selleri (1962), who also review earlier work (see also Selleri (1964)). The pion-nucleon form factor satisfies a dispersion relation,

$$F_{\pi NN}(t) = 1 + \frac{(t - m^2)}{\pi} \int_{9m^2}^{\infty} \frac{dx\, f(x)}{(x - m^2)\,(x - t)}. \qquad (10.1.5)$$

The modification of $F_{\rho\pi\pi}(t)$ is more complicated since the ρ meson is unstable and the singularities in the variable t are differently located.

However, in practice, little use is made of the details of these singularities and a simple phenomenological form is assumed for the product of the vertex parts. In this the term gG in (10.1.4) is replaced by

$$G(t) = A + \frac{Bm^2}{t - Cm^2}. \tag{10.1.6}$$

The parameters A, B, C are then determined from experiment. The same function has been found to fit a variety of experiments (involving only pion exchange) in the region of a few Gev. However at higher energies, a more complicated parametrisation is necessary in order to reduce the predicted differential cross-section at large values of t, (see (10.1.4)).

It appears that modifications of the OPE model that depend only on minus the exchanged momentum squared (t) are not adequate except at quite low energies, where for pion exchange the form-factor modifications give reasonable agreement with experiment. However for processes that involve ρ exchange, the t dependence of the differential cross-section requires a parametric form (10.1.6) having singularities nearer to $t = 0$ than the exchanged pole (Daudin *et al.* 1963). Aside from this unsatisfactory feature, the energy dependence with ρ exchange does not agree with experiment.

Treiman & Yang (1962) have proposed a test for the validity of the OPE model which we consider in relation to pion exchange in the reaction (10.1.3). If diagram (c) in Fig. 10.1.1 dominates this reaction, the possibility of decay correlations can be ascertained. The exchanged pion has zero spin, so it can only carry information about the direction of the momentum transfer but not information about helicity. In the rest system of the incident pion there can be no correlation of the plane of the two decay pions (Fig. 10.1.1 (c)) and the scattering plane of the initial and final nucleons. Let θ be the angle between the incident pion and one of the final pions in the centre of mass system of the two final pions. Then the Treiman–Yang test shows that the decay amplitude depends only on the angle θ (and s and t).

If we impose the further condition that the decay pions come entirely from the ρ meson (of spin one), there can be no component of spin (in the rest frame of the ρ meson) parallel to the direction of the incident pion. Hence the production probability in this frame must vanish when $\theta = 90°$, giving a probability distribution,

$$W(\theta, \phi) = C \cos^2 \theta. \tag{10.1.7}$$

No higher powers of $\cos \theta$ are possible since the ρ meson has spin one. Analogous results can be obtained when the exchanged particle in the OPE model has non-zero spin. The simple prediction (10.1.7) is not in agreement with experimental results (Hagopian *et al.* 1966). This result cannot be modified except by introducing vector exchange or by some departure from the OPE picture involving both s and t dependence, so that the ρ meson can have a different polarisation from that deduced above.

The OPE model is intuitively unsatisfactory because low partial waves do not correspond to a peripheral interaction. Since these low partial waves involve short distance strong interactions it is reasonable to assume that they lead to multiple production. If such inelastic processes are important it will be necessary to include absorption effects in the low partial waves at least. In the next section we consider the diffraction model as a preliminary to including absorption in the peripheral model.

10.2   The diffraction model

We begin by considering non-relativistic potential scattering of spinless particles by an absorbing target. The partial wave expansion is

$$F(s, \cos \theta) = \frac{1}{k} \sum_{0}^{\infty} (2l + 1) f_l(s) P_l(\cos \theta), \qquad (10.2.1)$$

where we have made a change in normalisation $(8\pi W)^{-1}$ compared with §3.5, and

$$f_l(s) = \frac{1}{2i} [\eta_l \exp (2i\delta_l) - 1]. \qquad (10.2.2)$$

The diffraction model assumes that all the scattering arises from the inelasticity $\eta_l(s)$. For absorption by a black disc of radius R, $\delta_l = 0$ and,

$$\begin{aligned} \eta_l &= 0 \quad (l \leqslant kR), \\ \eta_l &= 1 \quad (l > kR). \end{aligned} \qquad (10.2.3)$$

This gives a scattering amplitude,

$$F(s, \cos \theta) = -\frac{1}{2ik} \sum_{0}^{kR} (2l + 1) P_l(\cos \theta). \qquad (10.2.4)$$

For small angles, $\qquad P_l(\cos \theta) \approx J_0(2l \sin \tfrac{1}{2}\theta); \qquad (10.2.5)$

$$\begin{aligned} F(s, \cos \theta) &\approx -\frac{1}{ik} \int_0^{kR} l \, dl \, J_0(2l \sin \tfrac{1}{2}\theta) \\ &\approx \frac{iRJ_1(2kR \sin \tfrac{1}{2}\theta)}{2 \sin \tfrac{1}{2}\theta}. \end{aligned} \qquad (10.2.6)$$

This has sharp diffraction minima which are in obvious disagreement with experiment. The minima can be modified by avoiding the sharp cut off in the partial waves at $l = kR$. Instead one could take for example
$$\eta_l = \exp[-l^2/k^2R^2].$$

More generally one can use the approximation (10.2.5) in the partial wave series (10.2.1). Assuming that many partial waves contribute, the sum can be replaced by an integral, giving an expression of the form
$$F(s, \cos\theta) \approx ik \int_0^\infty b\, db\, J_0(2kb \sin\tfrac{1}{2}\theta)\, M(b, k), \qquad (10.2.7)$$

where $b = l/k$ is the impact parameter. This formula is closely related to the eikonal approximation which gives a method for calculating $M(b, k)$ from a given potential (Glauber, 1958).

This method was used by Blankenbecler & Goldberger (1962) to derive a form of Regge theory. In the more restricted sense of a diffraction model, this method was used by Serber (1963), Van Hove (1964) and Byers & Yang (1966). Another approach based on the impact parameter form, has been developed by Predazzi (1966) using, instead of (10.2.5),
$$P_l(\cos\theta) = \int_0^\infty dx\, J_0(x \sin\tfrac{1}{2}\theta)\, J_{2l+1}(x). \qquad (10.2.8)$$
This gives from (10.2.1)
$$F(s, \cos\theta) = 2 \int_0^\infty db\, J_0(2bk \sin\tfrac{1}{2}\theta) f(b, k), \qquad (10.2.9)$$

$$f(b, k) = \sum_0^\infty (2l+1) f_l(k)\, J_{2l+1}(2bk). \qquad (10.2.10)$$

The problem, in both the approximation (10.2.7) and in (10.2.9), is how to determine the unknown functions $M(b, k)$ and $f(b, k)$ respectively. The authors quoted above have considered various parametric forms for these functions that are based on models or intuitive arguments. There are two types of difficulty in producing a convincing derivation, one is that the derivation may be so refined that it cannot be evaluated (or unknown correction terms must be discarded, as in Regge theory at medium energies). The other difficulty is that the model itself may involve a drastic simplification whose physical meaning is not very clear. However, it can be argued that the impact parameter method is based rather directly on a physical picture, and one should therefore continue to refine it until a better understanding is obtained of the unknown weight functions. One possible approach is to make analyticity assumptions that are analogous to those used for partial wave amplitudes in Regge theory (Dean, 1967). Other developments have made use of the model proposed by Bugg (1963), in which it is assumed that low partial waves should be subtracted from the single particle exchange amplitudes. This model has been used in a generalised form based on the impact parameter method by Dar & Tobocman (1964) and by Dar (1964). It is related to the peripheral model with absorption which we consider in the following section.

10.3   The peripheral model with absorption

(a) Absorption corrections

In order to take account of absorption in a peripheral process Sopkovich (1962) assumed that the initial and final wave functions are distorted by an absorbing potential. This idea was developed by Durand & Chiu (1964, 1965), by Gottfried & Jackson (1964) and Jackson *et al.* (1965), (see also Drell & Hearn 1966).

Denote the partial wave reaction amplitude between initial and final states by $M^{fi}(s)$, and the corresponding one-particle exchange amplitude by $B^{fi}(s)$. This is evaluated approximately in potential scattering, when B represents the Born approximation. Then the reaction amplitude, in the distorted wave Born approximation for potential V, is

$$M^{fi} = \langle f|M|i \rangle = \langle \psi_f(\text{out})\,|\,V\,|\,\psi_i(\text{in}) \rangle, \qquad (10.3.1)$$

where $\psi_i(\text{in})$ denotes an incoming state with an optical potential U_i between the incident particle and the scattering centre. Similarly $\psi_f(\text{out})$ denotes an outgoing state for the final particle using an optical potential U_f.

The above amplitude has been calculated by Gottfried & Jackson (1964) using the Glauber (1958) form for the wave functions, which should be valid at high energy and small angles. This is

$$\psi_i(\mathbf{b}, z) = \psi_i(\text{in}) \approx \exp\left(i\mathbf{q}_i \cdot \mathbf{r}\right) \exp\left[-\frac{im}{q_i} \int_{-\infty}^{z} dx\, U_i(\mathbf{b} + \mathbf{K}_0 x)\right], \; (10.3.2)$$

where $\mathbf{q}_i$ denotes the initial three-momentum; $\mathbf{K}_0$ is a unit vector along $\mathbf{K} = \mathbf{q}_i + \mathbf{q}_f$; the impact parameter vector $\mathbf{b}$ is perpendicular to $\mathbf{K}$; and $\mathbf{r} = \mathbf{b} + \mathbf{K}_0 z$.

When (10.3.2) and an analogous form for $\psi_f(\text{out})$ are substituted into (10.3.1), an expression is obtained for M^{fi}, which can be evaluated if it is assumed that the initial and final states have the same absorbing potentials $U_i = U_f = U$, and that $q_i = q_f = k$. This gives

$$M^{fi} = 2\pi \int_0^\infty b\, db\, J_0(2kb \sin \tfrac{1}{2}\theta) \exp\left[2i\,\delta(b)\right] B(b), \qquad (10.3.3)$$

$$2\delta(b) = -\frac{m}{k} \int_{-\infty}^\infty dz\, U(\mathbf{b} + \mathbf{K}_0 z), \qquad (10.3.4)$$

$$B(b) = \int_{-\infty}^\infty dz\, V(\mathbf{b} + \mathbf{K}_0 z). \qquad (10.3.5)$$

In deriving the above equations we have used the relation,

$$J_0(z) = \frac{1}{2\pi} \int_0^{2\pi} d\phi \exp\left(iz \cos \phi\right).$$

Thus $\delta(b)$ is the complex phase shift of a wave packet passing through a potential U with impact parameter $\mathbf{b}$, and $B(b)$ is the Born approximation at this impact parameter.

The expression (10.3.3) can be compared with (10.2.7). Allowing for the different normalisation for F and M one gets

$$M(b, k) = B(b) \exp\left[2i\,\delta(b)\right]. \qquad (10.3.6)$$

This is the basic result on absorption that is included in the peripheral model with absorption. There still remains the problem of what to use for the complex phase shift $\delta(b)$.

The absorption model assumes that the phase $\delta(b)$ is pure imaginary,

$$\exp\left[2i\delta(b)\right] = 1 - C \exp\left[-(b-b_0)^2/2A\right], \qquad (10.3.7)$$

with $0 < C < 1$. This is analogous to $\delta_l = 0, 0 < \eta_l < 1$, in the diffraction model § 10.2. Although the formula (10.3.6) was derived under the assumption that initial and final states have the same absorption, for most applications it is replaced by

$$M(b, k) \approx \exp\left[i\delta_f\right] B(b) \exp\left[i\delta_i\right], \qquad (10.3.8)$$

with $$\exp\left[2i\delta_f(b)\right] = 1 - C_f \exp\left[-(b-b_0)^2/2A_f\right], \qquad (10.3.9)$$

and a similar expression for $\exp\left[2i\delta_i(b)\right]$, with parameters C_i and A_i. When one or more of the colliding particles has non-zero spin, the helicity formalism should be used. Instead of (10.3.8) one obtains (Durand & Chiu, 1964; Gottfried & Jackson, 1964),

$$\langle \lambda_c \lambda_d | M(b,k) | \lambda_a \lambda_b \rangle \approx \exp\left[i\delta_{cd}\right] \langle \lambda_c \lambda_d | B | \lambda_a \lambda_b \rangle \exp\left[i\delta_{ab}\right], \quad (10.3.10)$$

where the matrix element of M denotes the helicity amplitude (see § 3.6) and B is the corresponding Born approximation.

If $\delta_i(b)$ is interpreted literally as the phase shift for elastic scattering of the incident particle i with angular momentum $l = bk$, the optical theorem relates the constants A and C to the total cross-section giving

$$C_i = \frac{\sigma_i(\text{total})}{4\pi A_i}. \tag{10.3.11}$$

For experimental comparison, b_0 is taken to be the lowest allowed value. For example, for $j = \frac{1}{2}$ in $\pi N \to \rho N$, a typical choice for the parameters gives

$$C_f = 1; \quad b_0 = l_0/k = 1/(2k); \quad A_f = \tfrac{4}{3} A_i. \tag{10.3.12}$$

The constants C_i and A_i are taken from elastic scattering data for the incident particle; note that this procedure imposes an energy dependence on the reaction cross-section that is related to the energy dependence of the elastic cross-section (in addition to the OPE energy dependence).

The modification of the partial waves given by (10.3.8, 9 and 12) gives strong damping of the low partial waves, which becomes smaller with increasing $b = l/k$. This is a refinement of the proposal of Bugg (1963) whose method was equivalent to taking $\exp[i\delta(b)]$ to be zero for $b < R$, and to be unity for $b > R$ where it is reasonable to assume that the peripheral interaction given by OPE will be a suitable approximation. In terms of minus the square of the momentum transfer (t), one should expect the peripheral model to be a valid approximation only for small t. Then the experimental values are relatively close to the pole singularity due to the exchanged particle. At large values of t, the relative importance of this pole is not so great compared with other singularities. In terms of distance of approach, large angle scattering depends on small values of the impact parameter b. This is determined by (10.3.8, 9 and 12) only in a very qualitative way with no detailed theoretical basis. It is therefore not to be expected that the absorption model will be valid at large angles in general.

(b) Angular correlations

Experimental tests of the absorption model involve angular correlations. These are of a more complicated nature than those from the OPE model considered in §10.1. The general formalism for studying correlations in the absorption model is given by Gottfried & Jackson (1964). For the reaction

$$\pi^+ + p \to \rho^+ + p \atop \quad\;\; \hookrightarrow \pi^+ + \pi^0 \tag{10.3.13}$$

the angular distribution of the pions is found to be

$$W(\theta, \phi) = \frac{3}{4\pi} \sum_{m, m'} \rho_{mm'} \exp\left[i(m - m')\phi\right] d^1_{m0}(\theta) d^1_{m'0}(\theta). \quad (10.3.14)$$

Parity conservation implies

$$\rho_{-m, -m'} = (-1)^{m-m'} \rho_{mm'}. \quad (10.3.15)$$

Hermiticity implies $\rho_{mm'} = (\rho_{m'm})^*. \quad (10.3.16)$

Since the ρ meson has spin 1, the spin density matrix $\rho_{mm'}$ will have the form

$$\rho_{mm'} = \begin{pmatrix} \rho_{11}, & \rho_{10}, & \rho_{1,-1} \\ \rho_{10}^*, & \rho_{00}, & -\rho_{10}^* \\ \rho_{1,-1}, & -\rho_{10}, & \rho_{11} \end{pmatrix} \quad (10.3.17)$$

with all matrix elements real except ρ_{10}; and $\rho_{00} = 1 - 2\rho_{11}$. This gives an angular distribution for the final state (decay) pions (see Fig. 10.1.1 (d)),

$$W(\theta, \phi) = \frac{3}{4\pi} [\rho_{00} \cos^2\theta + \rho_{11} \sin^2\theta - \rho_{1,-1} \sin^2\theta \cos 2\phi$$

$$- 2^{\frac{1}{2}} \operatorname{Re}(\rho_{10}) \sin 2\theta \cos\phi]. \quad (10.3.18)$$

The angle θ is the scattering angle between the incident pion and one of the decay pions in the rest frame of the ρ meson. The angle ϕ is the angle between the plane of the final pions and the scattering plane of the initial and final proton, also in the ρ meson rest frame. The experimental comparison by Hagopian et al. (1966) shows that ρ_{10} differs significantly from zero. For pure pion exchange only ρ_{00} is non-zero, and for pure ω exchange only ρ_{11} and $\rho_{1,-1}$ are non-zero. This confirms that there are essential correlations not contained in the OPE model.

(c) Experimental tests of the absorption model

The peripheral model with absorption has been remarkably successful in its comparison with experiments on reactions that involve only exchanged particles having zero spin. The model successfully describes the reaction (10.3.13) for incident momenta from 2 to 8 Gev/c. The differential cross-section, in both shape and absolute value, is well described on the basis of pion exchange with absorption, for example see Fig. 10.1.2 and Fig. 10.3.1. It also gives good agreement with angular distributions of the decay pions for this reaction.

A wide variety of other reactions have been studied by Jackson *et al.* (1965) using the absorption model. Their conclusions are as follows:

(i) Angular correlations of the decay products give the most decisive distinction between the absorption model and the form-factor model (see §10.1). For the $\pi N \to \rho N$ reaction with π exchange, the latter

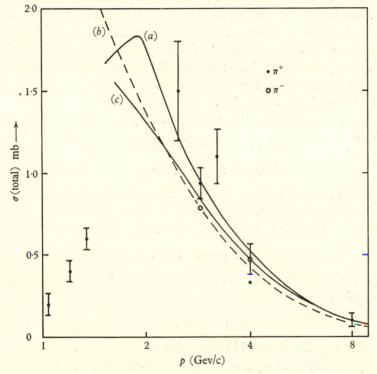

Fig. 10.3.1. The total cross-sections for the reactions $\pi^{\pm}p \to \rho^{\pm}p$ as functions of the incident pion momentum, assuming only pion exchange. The curve (*a*) is for the π^+ reaction, and curve (*c*) is for the π^- reaction, the difference being due to different absorption effects. The dashed curve (*b*) is given by the one-pion exchange model with an empirical form factor (reference Jackson *et al.* 1965).

model gives $\rho_{00} = 1$, and other spin density matrix elements are zero. If it is modified to include some vector exchange so as to fit the decay correlations then it fails to fit the energy variation of the differential cross-section. The failure of the peripheral model with form factors is even more decisive in the reaction $\pi p \to \omega N$, where with ρ exchange it predicts $\rho_{00} = 0$ in contrast with the experimental value ≈ 0.6 at 3.75 Gev/c. In the following we therefore consider only the model with absorption; as noted above, this fits the angular correlations of the $\pi N \to \rho N$ reactions in the range 2 to 8 Gev/c. For example at $2 \cdot 75$ Gev/c, $\rho_{00} \approx 0 \cdot 7$; Re (ρ_{10}) decreases with t to about $-0 \cdot 2$; and $\rho_{1,-1}$ increases slowly from zero to about $0 \cdot 1$.

(ii) For the reaction
$$K + N \to K^* + N, \tag{10.3.19}$$

it is possible to obtain agreement at 2 to 3 Gev/c using the absorption model. Vector meson exchange is necessary in addition to pion exchange in order to fit correlations. However, after determining the parameters at 3 Gev/c, the model fails to agree with data at 5 Gev/c. The forward peak is much too broad, and it is clear that the absorption is not adequate to reduce the effect of vector exchange, which by itself gives an increasing cross-section.

(iii) Results similar to those for (10.3.19) are obtained for the reactions
$$K^+ + p \to K^0 + N^{*++}(1238), \tag{10.3.20}$$

$$K^- + p \to \pi^- + Y^*(1385). \tag{10.3.21}$$

The data at low energies can be fitted using ρ exchange for (10.3.20) and K^* exchange for (10.3.21), but again the energy variation is wrong.

(iv) The model fails to account for the very sharp forward peak in the reaction
$$\pi^+ + p \to \pi^0 + N^*, \tag{10.3.22}$$

nor does it agree with the observed rapid decrease in the total reaction cross-section.

(v) The reaction
$$\pi^+ + n \to \omega + p, \tag{10.3.23}$$

can be fitted by the model at low energies using ρ exchange, and gives correctly the large value $\rho_{00} \approx 0 \cdot 6$, (which would be predicted as zero without absorption). However, there is the same difficulty over energy variation; the model predicts too wide a peak at higher energies and too large a reaction cross-section.

(vi) For double resonance production such as
$$\pi p \to \rho N^*; \quad KN \to K^* N^*, \tag{10.3.24}$$

angular distributions and decay correlations can be correctly fitted at low energies (momentum 3 to 4 Gev/c), but the energy variation is wrong.

The general conclusion is that the absorption model is successful when it involves only exchange of spin-zero particles. It gives angular distributions and decay correlations reasonably well at low energies when spin-one exchange is involved, though the fit is less good than with spin-zero exchange processes. With spin-one exchange the variation with energy is quite wrong; at higher energies ($p \sim 8\,\text{Gev/c}$) the angular distribution gives much too wide a forward peak, and the differential cross-section can be wrong by orders of magnitude at $|t| \approx 0.5\,(\text{Gev})^2$.

It appears to be essential to incorporate some Reggeisation of the exchanged particles. The dependence on t of the angular momentum $\alpha(t) = l$, of an exchanged Regge particle could provide the damping of the energy variation that is required when spin one is exchanged in the absorption model. However, at the present time, no satisfactory consistent method has been developed for combining Regge theory with the absorption (peripheral) model.

10.4   Large angle scattering

It has been noted by Orear (1964) that the differential cross-section for proton-proton scattering at high momentum transfer (outside the diffraction peak) can be represented by the formula

$$\frac{d\sigma}{d\Omega} = \frac{A}{s}\left[\exp\left(-\frac{p\sin\theta}{B}\right)\right], \tag{10.4.1}$$

where $\qquad A = 595\,(\text{Gev})^2\,\text{mb/ster}, \quad B = 0.158\,\text{Gev/c};$ $\qquad$ (10.4.2)

p denotes the centre of mass momentum in Gev/c, and s the centre of mass energy squared. The agreement with experiment at fixed angle (90°) was illustrated in Chapter 2 (Fig. 2.3.2), but the agreement is not confined to the energy variation alone. It has been noted also that π^+p and π^-p scattering at large angles follows the Orear formula with the same constants, though the agreement is not so good as for pp scattering (Cocconi, 1966). Similar agreement has been noted for $\bar{p}p$ scattering (Bialas & Czyzewski, 1966) for incident momenta 3, 4 and 5·7 Gev/c, but it is not known whether the agreement remains at higher energies. They have also noted that the n, p data agree with the Orear formula below 90° scattering angle in centre of mass scattering, but the agreement is less satisfactory for scattering angles above 90°.

There is no satisfactory theory of large angle scattering at high energies. The statistical model (Hagedorn, 1965) gives the right order of magnitude for large angle scattering but if applied literally it would predict fluctuations with angle (Ericson & Mayer-Kuckuk, 1966) that are not observed (Van Hove, 1966). A model proposed by

Wu & Yang (1965) is based on incoherence of the interactions at high momentum transfers; their model leads to the interesting prediction that electromagnetic form factors at large momentum transfer are related by a power law to differential cross-sections at high momentum transfer. They predict that the electromagnetic form factor $G(q^2)$ is proportional to the one-fourth power of the p-p differential cross-section. Using the Orear formula (10.4.1) this gives for the form factor

$$G(q^2) = \exp[-q/4B], \quad q > 1 \,\text{Gev/c}, \tag{10.4.3}$$

where $4B \approx 0.63 \,\text{Gev/c}$. This has been compared with experiment by Drell (1966) and gives good agreement as q varies from 1 to 3 Gev/c.

Another approach to large angle scattering has been proposed by Van Hove (1964). This approach assumes that elastic scattering is pure shadow scattering corresponding to absorption into multiparticle inelastic channels. It gives a reasonable fit to πp and pp data (Cottingham & Peierls, 1965). However, it is necessary in the Van Hove model to make specific assumptions to obtain the t dependence. He suggests that the most common inelastic states may have only weak correlations between them due to the many degrees of freedom excited at high energy. This 'uncorrelated jet' model leads to an exponential decrease of the differential cross-section with t rather than with $t^{\frac{1}{2}}$ as suggested by the Orear formula. It is not clear whether the assumption of weak correlations between inelastic states is justified at the energies available on present high energy accelerators.

10.5 The quark model

The quark model is based on the suggestion (Gell-Mann, 1964 and Zweig, 1964) that hadrons are compound states of quarks. The quantum numbers of the quarks are those of a spin-half particle and the triplet representation of SU 3 (Gell-Mann, 1962 a; Ne'eman, 1961), each with baryon number one third. Denoting the quark states by Q_α ($\alpha = 1, 2, 3$) their quantum numbers are:

	J	B	I	I_3	Charge	S
Q_1	$\frac{1}{2}$	$\frac{1}{3}$	$\frac{1}{2}$	$\frac{1}{2}$	$\frac{2}{3}$	0
Q_2	$\frac{1}{2}$	$\frac{1}{3}$	$\frac{1}{2}$	$-\frac{1}{2}$	$-\frac{1}{3}$	0
Q_3	$\frac{1}{2}$	$\frac{1}{3}$	0	0	$-\frac{1}{3}$	-1

$$\tag{10.5.1}$$

Baryons are regarded as being made up of three quarks in this model, and mesons as being made up of a quark and an anti-quark. The meson octets correspond to a formal combination (Dalitz 1965)

$$M_\beta^\alpha = (Q^\alpha Q_\beta - \tfrac{1}{3} Q^\gamma Q_\gamma \, \delta_{\alpha\beta}), \qquad (10.5.2)$$

where $Q^\alpha = \bar{Q}_\alpha$ denotes an anti-quark. The baryon octet cannot be written using only $SU\,3$ indices, but must include spin directions. The proton contains $Q_1 Q_1 Q_2$ and can be written (Weisskopf, 1966),

$$(\text{proton}) = \frac{1}{3\sqrt{2}} \left[2 \begin{pmatrix} \uparrow\uparrow\downarrow & \uparrow\downarrow\uparrow & \downarrow\uparrow\uparrow \\ 1\,1\,2 + 1\,2\,1 + 2\,1\,1 \end{pmatrix} - \begin{pmatrix} \uparrow\downarrow\uparrow & \uparrow\uparrow\downarrow & \uparrow\uparrow\downarrow \\ 1\,1\,2 + 1\,2\,1 + 2\,1\,1 \end{pmatrix} \right.$$

$$\left. - \begin{pmatrix} \downarrow\uparrow\uparrow & \downarrow\uparrow\uparrow & \uparrow\downarrow\uparrow \\ 1\,1\,2 + 1\,2\,1 + 2\,1\,1 \end{pmatrix} \right]. \qquad (10.5.3)$$

From (10.5.1, 2 and 3) we see that a π^+ meson contains Q_1 and $\bar{Q}_2 = Q^2$; K^+ contains Q_1 and $\bar{Q}_3 = Q^3$; p contains $Q_1 Q_1 Q_2$, etc.

Mesons may also occur in singlet states represented by $S = Q^\alpha Q_\alpha$. All observed meson states are consistent with singlet or octet multiplets, that can be characterised by quantum numbers s, l, j, (denoting total spin s, with $s = 0$ or 1; orbital angular momentum l, and total angular momentum j).

The simplest form of the quark model for a meson assumes that a quark Q and anti-quark $\bar{Q}$ are bound essentially non-relativistically. Although the forces must be very strong it is possible to argue that, for a high quark mass M_Q, one can work non-relativistically if the range R of the Q-$\bar{Q}$ force is large compared with $(\hbar/M_Q c)$. Assuming that the force is due to a neutral vector field coupled to the baryon current, the $Q\bar{Q}$ system forms a series of nonets; each state in a nonet having approximately the same mass value. Symmetry breaking can be introduced by assuming that the quark Q_3 is heavier than Q_1 and Q_2.

In this model baryons are composed of three quarks. Since the symmetries give an $SU\,3$ decomposition,

$$(3) \times (3) \times (3) = (1) + (8)_1 + (8)_2 + (10), \qquad (10.5.4)$$

only singlet, octet and decuplet states can be formed for baryons in this way.

Other multiplets for mesons and baryons can exist only if combinations of $Q\bar{Q}$ with (10.5.2) or (10.5.3) exist and are stable. Thus an excited state of a meson might be composed $Q\bar{Q}Q\bar{Q}$. It is an important hypothesis in the quark model that states of this type are rather highly excited compared with those from the simpler combinations (10.5.2).

The quark model has been remarkably successful in fitting known mesons and meson resonances. A survey is given by Dalitz (1966) and Squires (1967). Here we will consider only those aspects that are relevant to high energy collisions.

The quark model provides a representation of SU 6. We have already noted that SU 3 symmetry leads to predictions about high energy cross-sections that come from a direct extension of the formalism of isospin symmetry SU 2, which was discussed in Chapter 8. Apart from the difficulties due to symmetry breaking, the SU 3 symmetries seem to be well established. The difficulties from symmetry breaking are not trivial. For example, it is found that the following triangle inequality predicted by SU 3 is not in agreement with high energy experiments (Van Hove, 1967); at $t = 0$ unbroken SU 3 symmetry predicts (Harari & Lipkin, 1964),

$$\left[\frac{d\sigma}{dt}(K^-p \to \pi^-\Sigma^+)\right]^{\frac{1}{2}} \geqslant \left[\frac{d\sigma}{dt}(\pi^-p, \text{elastic})\right]^{\frac{1}{2}} - \left[\frac{d\sigma}{dt}(K^-p, \text{elastic})\right]^{\frac{1}{2}}.$$

(10.5.5)

Experimentally the left-hand side is found to decrease rapidly whereas the right-hand side stays nearly constant (and positive).

If SU 6 symmetry is assumed, further relations are obtained between cross-sections. Amongst these is the Johnson–Treiman relation (1965) for total cross-sections,

$$\tfrac{1}{2}[\sigma_t(K^+p) - \sigma_t(K^-p)] = \sigma_t(\pi^+p) - \sigma_t(\pi^-p) = \sigma_t(K^+n) - \sigma_t(K^-n),$$

(10.5.6)

which gives reasonable agreement with experiment (Lipkin, 1966).

For its use in high energy scattering, the basic assumption of the quark model is that the interactions are additive (a similar assumption is made in its application to electromagnetic and weak interactions but for these it is easier to understand its validity). Although the interactions between the quarks are strong, the assumption that they are additive, with their sum equal to the hadron interactions, does lead to certain results in agreement with experiment (Levin & Frankfurt, 1965; Lipkin & Scheck, 1966; Kokkedee & Van Hove, 1966; Van Hove, 1967). We will consider its implications for the scattering process

$$A + B \to A' + B',$$

(10.5.7)

where A and A' are assumed to be both mesons (or both baryons), and similarly for B and B'. Thus A and A' have the same quark structure in the model (also B and B').

For this scattering process the additivity assumption is represented schematically in Fig. 10.5.1. It is assumed that the scattering amplitude for meson on baryon can be represented by the sum of the individual quark-scattering amplitudes F_{ij} modified only by form factors $f_i(\Delta p)$,

$$\langle A'B'|\,F\,|AB\rangle = \sum_{ij} \langle Q_i^{A'}Q_j^{B'}|\,F_{ij}\,|Q_i^A Q_j^B\rangle\, f_i^{AA'}(\Delta p)f_j^{BB'}(-\Delta p), \quad (10.5.8)$$

where $\Delta p = p^{A'} - p^A = p^B - p^{B'}$. The matrix element of F_{ij} in (10.5.8) describes the scattering of two quarks from their initial states within

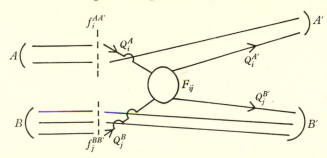

Fig. 10.5.1. Schematic illustration of the additivity hypothesis in the quark model for meson-baryon scattering $A + B \to A' + B'$ (Van Hove, 1966).

A and B to final states in A' and B' respectively. The momentum transfer variable t_{ij} for the quark scattering will be

$$t_{ij} = (\Delta p)^2 = t, \quad (10.5.9)$$

since momentum is exchanged between A and B only via the pair of interacting quarks. The form factor $f_i^{AA'}$ corresponds to the transition distribution of the quark Q_i between the states A and A' (Rivers, 1966). When the quark undergoes no change of state, as with forward elastic scattering we have $A = A'$ and $f_i^{AA}(0) = 1$.

The energy dependence of the factors in (10.5.8) is more difficult to justify in the model. It is assumed that, at least for certain scattering processes, the quarks effectively share the total momentum equally between themselves. Thus, in satisfying

$$p^A = \sum_i p_i^A; \quad p^B = \sum_j p_j^B,$$

it is assumed that

$$p_i^A = C_A p^A; \quad p_j^B = C_B p^B, \quad (10.5.10)$$

where $C_A = \tfrac{1}{2}$ for a meson, and $C_B = \tfrac{1}{3}$ for a nucleon. This assumption implies that the relative motion of a quark within a meson or a baryon is small. This will be a reasonable approximation only if the effective quark mass within the meson or baryon is given by

$$m_Q \text{ (effective)} \approx \tfrac{1}{2}m_A \quad \text{for a meson } A, \qquad (10.5.11a)$$

$$m_Q \text{ (effective)} \approx \tfrac{1}{3}m_B \quad \text{for a baryon } B. \qquad (10.5.11b)$$

The effective mass differs essentially from the mass of a 'free' quark, for which

$$m_Q \gg m_B. \qquad (10.5.12)$$

No satisfactory explanation has been put forward for the origin of this suggested low effective mass and low relative motion for quarks bound in hadrons. However, let us consider its consequences. At high energies, the meson-baryon mass difference is small compared with $s^{\frac{1}{2}}$, and

$$s_{ij} \approx C_A C_B s_{AB}. \qquad (10.5.13)$$

This gives the same quark-quark energy for meson-meson, meson-baryon, and baryon-baryon collisions if they have their energies squared in ratio $4:6:9$, respectively.

The experimental consequences of the additivity assumption are listed by the authors quoted above, and (taking spin into account) by Itzykson & Jacob (1966). We will give only a few here for illustration. For forward elastic scattering $A' = A$, $B' = B$ and $\Delta p = 0$. The elastic-scattering amplitudes can be obtained simply by counting quarks, but reaction amplitudes require more care with the symmetries of the wave functions or states. We will denote the elastic-scattering amplitudes by,

$$(AB) = \langle AB| F |AB\rangle, \qquad (10.5.14)$$

$$(Q_i Q_j) = \langle Q_i Q_j| F_{ij} |Q_i Q_j\rangle. \qquad (10.5.15)$$

Then
$$(\pi^+ p) = 2(Q_1 Q_1) + (Q_1 Q_2) + 2(\bar{Q}_2 Q_1) + (\bar{Q}_2 Q_2), \qquad (10.5.16)$$

$$(\pi^- p) = 2(\bar{Q}_1 Q_1) + (\bar{Q}_1 Q_2) + 2(Q_2 Q_1) + (Q_2 Q_2), \qquad (10.5.17)$$

$$(pp) = 4(Q_1 Q_1) + 4(Q_1 Q_2) + (Q_2 Q_2), \qquad (10.5.18)$$

$$(np) = 2(Q_1 Q_1) + 5(Q_1 Q_2) + 2(Q_2 Q_2), \qquad (10.5.19)$$

$$(\bar{p}p) = 4(\bar{Q}_1 Q_1) + 2(\bar{Q}_1 Q_2) + 2(\bar{Q}_2 Q_1) + (\bar{Q}_2 Q_2). \qquad (10.5.20)$$

$$(\bar{n}p) = 2(\bar{Q}_1 Q_1) + 4(\bar{Q}_2 Q_1) + (\bar{Q}_1 Q_2) + 2(\bar{Q}_2 Q_2). \qquad (10.5.21)$$

From these relations between forward amplitudes it follows that

$$(pp) + (np) + (\bar{p}p) + (\bar{n}p) = 3[(\pi^+ p) + (\pi^- p)]. \qquad (10.5.22)$$

From the optical theorem this gives a relation between total cross-sections,

$$\frac{2[\sigma(\pi^+ p) + \sigma(\pi^- p)]}{\sigma(pp) + \sigma(np) + \sigma(\bar{p}p) + \sigma(\bar{n}p)} = \frac{2}{3}. \tag{10.5.23}$$

For baryon-baryon scattering at 20 Gev, if (10.5.13) is used so that meson-baryon scattering involves the same quark-quark energy, the meson-baryon scattering must be evaluated at 13 Gev. Then the experimental ratio (10.5.23) is (0.59 ± 0.05) (Van Hove, 1967).

If we consider the asymptotic limit of high energy scattering, the pion-nucleon cross-sections tend to the limit $\sigma(\pi N)$, and both the nucleon-nucleon and anti-nucleon-nucleon cross-sections tend to $\sigma(NN)$ (see Chapters 2 and 8). Then (10.5.23) becomes

$$\frac{\sigma(\pi N)}{\sigma(NN)} = \frac{2}{3}. \tag{10.5.24}$$

The experimental high energy limit $\sigma(\pi N)$, obtained by polynomial extrapolation, is approximately 24 mb, and $\sigma(NN)$ is approximately 38 mb, which gives a ratio in reasonable agreement with this prediction of the quark model.

Further consequences of the quark model are listed by the authors quoted above. These include many relations between amplitudes for elastic scattering. For the latter, two extreme approximations have been particularly emphasised. In the first of these the dependence of the forward peak in any hadron-hadron collision is assumed to be determined mainly by the form factors $f_i^{AA'}(\Delta p)$ and $f_j^{BB'}(-\Delta p)$ in (10.5.8). In this picture the quark-quark scattering amplitudes will change slowly with $t_{ij} = t$, compared with the change in the form factors within the forward peak. This rapid variation in the form factors can be interpreted in terms of the relative size of a hadron and its component quarks. It implies that quarks are small compared with the hadrons, and if this was so it could be regarded as an intuitive justification for the additivity assumption of the quark model.

The second extreme approximation attributes the rapid variation of F_{AB} in the forward peak to a rapid change in the quark-quark scattering amplitude F_{ij}, whilst the form factors f are taken to change slowly with the momentum transfer variable t. This gives a natural explanation of the similarity of the forward elastic peaks for meson-baryon and for baryon-baryon collisions when no quantum numbers are exchanged. Then each is proportional to the basic quark-quark differential cross-section. The slow variation of the form factors compared with the rapid change in F_{ij} can be interpreted as implying that the quarks are large and comparable in size with the hadrons (by 'size' is meant the extent of the wave function of the quark). However, from (10.5.18), the quark-quark total cross-section at high energy should satisfy

$$\sigma(QQ) = \tfrac{1}{9}\sigma(NN). \tag{10.5.25}$$

Thus in this extreme picture, the quarks are very transparent compared with nucleons and once again the additivity assumption appears to be reasonable. The results for these two extreme pictures suggest that the additivity assumption may also be intuitively justified in the physical situation, which presumably lies between the two extremes.

REFERENCES

Aachen, Berlin, Cern (1965). *Phys. Lett.* **19**, 608.

Adler, S. L. (1965). *Phys. Rev.* **140**, B 736.

Ahmadzadeh, A., Burke, P. G. & Tate, C. (1963). *Phys. Rev.* **131**, 1315.

Ahmadzadeh, A. & Leader, E. (1964). *Phys. Rev.* **134**, B 1058.

Alfaro, V. de, Fubini, S., Rossetti, C. & Furlan, G. (1966). *Phys. Rev. Lett.* **21**, 576.

Alfaro, V. & Regge, T. (1963). *Non-relativistic Potential Scattering* (Benjamin, New York).

Amati, D. *et al.* (1963). *Nuovo Cimento*, **28**, 639.

Amati, D., Foldy, L. L., Stanghellini, A. & Van Hove, L. (1964). *Nuovo Cimento*, **32**, 1685.

Amati, D., Fubini, S. & Stanghellini, A. (1962). *Nuovo Cimento*, **26**, 896.

Anderson, E. W. *et al.* (1966). *Phys. Rev. Lett.* **16**, 855.

Andrews, M. & Gunson, J. (1964). *J. Math. Phys.* **5**, 1391.

Aramaki, S. (1963). *Progr. Th. Phys.* **30**, 265.

Arbab, F. & Chiu, C. B. (1966). Berkeley, U.C. Report.

Arnold, R. C. (1966). Argonne Laboratory Report.

Ashkin, J. (1959). *Nuovo Cimento Suppl.* **14**, 221.

Atkinson, D. A. & Barger, V. (1965). *Nuovo Cimento*, **38**, 634.

Balázs, L. A. P. (1963). *Phys. Rev.* **128**, 1939; **129**, 872.

Baldin, A. M., Gol'danski, V. I. & Rosental, I. L. (1961). *Kinematics of Nuclear Reactions* (Pergamon Press).

Barashenkov, V. S. (1966). Berkeley Conference (reported by Van Hove, 1966).

Bardardin-Otwinowska, M. *et al.* (1966). *Phys. Lett.* **21**, 351.

Barger, V. & Cline, D. (1966). Berkeley Conference (reported by Murphy, 1966).

Bargmann, V. (1949). *Rev. Mod. Phys.* **21**, 488.

Bateman, H. (1953). See Erdelyi (1953).

Belyakov, V. A. *et al.* (1966). Berkeley Conference (reported by Van Hove, 1966).

Bellettini, G. *et al.* (1965). *Phys. Lett.* **14**, 164.

Bessis, J. D. (1966). CERN preprint TH. 653.

Bessis, J. D. & Kinoshita, T. (1966). CERN preprint 66/1482/5-TH. 729.

Bethe, H. A. (1958). *Ann. Phys.* **3**, 190.

Białas, A. & Czyzewski, O. (1966). CERN report 66/1206/5 TH. 710.

Białas, A. & Kotanski, A. (1966). Cracow preprint TPJU-5/66.

Bjorken, J. D. (1959). Doctoral dissertation, Stanford University.

Bjorken, J. D. & Drell, S. D. (1965). *Relativistic Quantum Fields* (McGraw-Hill).

Blankenbecler, R. & Goldberger, M. L. (1961). Proceedings of La Jolla Conference (unpublished).

Blankenbecler, R. & Goldberger, M. L. (1962). *Phys. Rev.* **126**, 766.

Blankenbecler, R., Goldberger, M. L., Khuri, N. N. & Treiman, S. B. (1960). *Ann. Phys.* **10**, 62.

Boas, R. P. (1954). *Entire Functions* (Academic Press, New York).

Bogoliubov, N. N. (1958). (Unpublished proof of dispersion relation.)

Borgeaud, P. *et al.* (1966). Stony Brook Conference (see also Wetherell, 1966).

Bottino, A. & Longoni, A. M. (1962). *Nuovo Cimento*, **24**, 353.

Bottino, A., Longoni, A. M. & Regge, T. (1962). *Nuovo Cimento*, **23**, 954.

Bremermann, H. J., Oehme, R. & Taylor, J. G. (1958). *Phys. Rev.* **109**, 2178.

Bros, J., Epstein, H. & Glaser, V. (1964). *Nuovo Cimento*, **31**, 1265.

Bros, J., Epstein, H. & Glaser, V. (1965). *Com. Math. Phys.* **1**, 240.

Bugg, D. V. (1963). *Phys. Lett.* **7**, 365.

Byers, N. & Yang, C. N. (1966). *Phys. Rev.* **142**, 976.

Calogero, F., Charap, J. M. & Squires, E. J. (1963). *Annals of Physics*, **25**, 325.

Carruthers, P. A. (1966). *Introduction to Unitary Symmetry* (Interscience, New York).

Carruthers, P. A. & Krisch, J. P. (1965). *Annals of Physics*, **33**, 1.

Carter, A. A. (1966). Rutherford Laboratory preprint RPP/G/14.

Castillejo, L., Dalitz, R. & Dyson, F. (1956). *Phys. Rev.* **124**, 1258.

Cerulus, F. & Martin, A. (1964). *Phys. Lett.* **8**, 80.

Challifour, J. & Eden, R. J. (1963 a). *Phys. Rev.* **129**, 2349 (see also *Nuovo Cimento*, **27**, 1104).

Challifour, J. & Eden, R. J. (1963 b). *J. Math. Phys.* **4**, 359.

Chandler, C. & Stapp, H. P. (1966). Berkeley Lectures by H. P. Stapp.

Chew, G. F. (1961). *S-Matrix Theory of Strong Interactions* (Benjamin, New York).

Chew, G. F. (1962). *Rev. Mod. Phys.* **34**, 394.

Chew, G. F. (1966). *The Analytic S Matrix* (Benjamin, New York).

Chew, G. F. & Frautschi, S. C. (1961). *Phys. Rev. Lett.* **7**, 394.

Chew, G. F. & Frautschi, S. C. (1962). *Phys. Rev. Lett.* **8**, 41.

Chew, G. F., Goldberger, M. L., Low, F. E. & Nambu, Y. (1957). *Phys. Rev.* **106**, 1337.

Chew, G. F. & Jacob, M. (1963). *Strong Interaction Physics* (Benjamin, New York).

Chew, G. F. & Mandelstam, S. (1960). *Phys. Rev.* **119**, 467.

Chisholm, R. (1952). *Proc. Camb. Phil. Soc.* **48**, 300.

Cocconi, G. (1966). Stony Brook Conference Report.

Condon, E. U. & Shortley, G. H. (1957). *The Theory of Atomic Spectra* (Cambridge University Press).

Cornille, H. (1964). *Nuovo Cimento*, **31**, 1101.

Cottingham, W. N. & Peierls, R. F. (1965). *Phys. Rev.* **137**, B 147.

Czyzewski, O. (1966). Berkeley Conference Report and CERN/TC/Physics 66–28.

Dalitz, R. H. (1953). *Phil. Mag.* **44**, 1068.

Dalitz, R. H. (1965). Oxford Conference Report.

Dalitz, R. H. (1966). Berkeley Conference Report.

Dar, A. (1964). *Phys. Rev. Lett.* **13**, 91.
Dar, A. & Tobocman, W. (1964). *Phys. Rev. Lett.* **12**, 511.
Daudin, A. *et al.* (1963). *Phys. Lett.* **7**, 125.
Dean, N. W. (1967). Cavendish Laboratory, Cambridge (preprint).
De Swart, J. J. (1963). *Rev. Mod. Phys.* **35**, 916.
De Swart, J. J. (1964). *Nuovo Cimento*, **31**, 420.
Dirac, P. A. M. (1958). *The Principles of Quantum Mechanics*, 4th Edition. (Oxford University Press).
Donnachie, A. (1966). CERN Lectures, CERN 66/1042/5-TH. 590.
Donnachie, A. & Hamilton, J. (1965). *Ann. Phys.* **31**, 410.
Drell, S. D. (1960). *Phys. Rev. Lett.* **5**, 342.
Drell, S. D. (1961). *Rev. Mod. Phys.* **33**, 458.
Drell, S. D. (1966). Berkeley Conference Report.
Drell, S. D. & Hearn, A. C. (1966). Chapter 9 of *High Energy Physics* (Academic Press, New York).
Durand, L. & Chiu, Y. T. (1964). *Lectures in Theoretical Physics* (University of Colorado Press).
Durand, L. & Chiu, Y. T. (1965). *Phys. Rev.* **137**, B 1530 and **139**, B 640.
Dyson, F. J. (1949). *Phys. Rev.* **75**, 486, 1736.
Dyson, F. J. (1951). *Phys. Rev.* **82**, 428.
Eden, R. J. (1952). *Proc. Roy. Soc.* A **210**, 388.
Eden, R. J. (1960). *Proc. 1960 Rochester Conference on High Energy Physics* (Interscience, New York).
Eden, R. J. (1961). *Phys. Rev.* **121**, 1567 (see also 1960 *Phys. Rev.* **119**, 1763, **120**, 1514).
Eden, R. J. (1966*a*). *Phys. Lett.* **19**, 695.
Eden, R. J. (1966*b*). *Phys. Rev. Lett.* **16**, 39.
Eden, R. J. (1966*c*). Lectures at the Scottish Universities Summer School (see Preist & Vick, 1967).
Eden, R. J. (1967). *J. Math. Phys.* February.
Eden, R. J & Goldstone, J. (1963). *Nucl. Phys.* **49**, 33.
Eden, R. J. & Landshoff, P. V. (1965). *Ann. Phys.* **31**, 370.
Eden, R. J., Landshoff, P. V., Polkinghorne, J. C. & Taylor, J. C. (1961). *J. Math. Phys.* **2**, 656.
Eden, R. J., Landshoff, P. V., Olive, D. I. & Polkinghorne, J. C. (1966). *The Analytic S-Matrix* (Cambridge University Press).
Eden, R. J. & Łukaszuk, L. (1967). *Nuovo Cimento*, **47**, 817.
ELOP (1966). See Eden, R. J., Landshoff, P. V., Olive, D. I. & Polkinghorne, J. C. (1966).
Erdelyi, A. (1953) (Ed.). *Higher Transcendental Functions*, (Bateman manuscript project), (McGraw-Hill, New York).
Ericson, T. E. O. (1964). CERN Report TH. 406.
Ericson, T. E. O. & Mayer-Kuckuk, T. (1966). *Ann. Rev. Nucl. Sci.* **16**.
Federbush, P. G. & Grisaru, M. T. (1963). *Ann. Phys.* **22**, 263, 299.
Ferrari, E. & Selleri, F. (1962). *Nuovo Cimento Suppl.* **24**, 453.
Ferro-Luzzi, M. (1966). Berkeley Conference Report.
Feynman, R. P. (1949). *Phys. Rev.* **76**, 749, 769.

Foldy, L. F. & Peierls, R. F. (1963). *Phys. Rev.* **130**, 1585.

Foley, K. J. *et al.* (1965*a*) *Phys. Rev. Lett.* **14**, 862.

Foley, K. J. *et al.* (1965*b*). *Phys. Rev. Lett.* **15**, 45.

Foley, K. J. *et al.* (1967). *Phys. Rev.* (and see Lindenbaum, 1967).

Fotiadi, D., Froissart, M., Lascoux, J. & Pham, F. (1964). (Unpublished reprint) see articles reprinted in Hwa & Teplitz (1966).

Fox, G. C. (1966). Cambridge (preprint). *Phys. Rev.* 1967.

Frampton, P. H. & Taylor, J. C. (1966). Clarendon Laboratory, Oxford (preprint).

Frautschi, S. C. (1963). *Regge Poles and S-Matrix Theory* (Benjamin, New York).

Frautschi, S. C., Gell-Mann, M. & Zachariasen, F. (1962). *Phys. Rev.* **126**, 2204.

Frazer, W. & Fulco, J. (1960). *Phys. Rev.* **117**, 1603; *ibid.* **119**, 1420.

Freedman, D. Z. & Wang, J. M. (1966). Berkeley preprint (see Low, 1966).

Froissart, M. (1961*a*). Proceedings of La Jolla Conference (unpublished).

Froissart, M. (1961*b*). *Phys. Rev.* **123**, 1053.

Froissart, M. (1963). *Trieste Lectures on Theoretical Physics* (Ed. A. Salam, published by International Atomic Energy Agency, Vienna, 1963).

Froissart, M. (1966). Berkeley Conference Report.

Frye, G. & Warnock, R. L. (1962). *Phys. Rev.* **130**, 478.

Fubini, S. (1966). *Nuovo Cimento*, **43** A, 475.

Galbraith, W. *et al.* (1965). *Phys. Rev.* **138**, B 913.

Gell-Mann, M. (1962*a*). *Phys. Rev.* **125**, 1067.

Gell-Mann, M. (1962*b*). *CERN High Energy Conference Proceedings*, p. 533.

Gell-Mann, M. (1962*c*). *Phys. Rev. Lett.* **8**, 263.

Gell-Mann, M. (1964). *Phys. Lett.* **8**, 214.

Gell-Mann, M., Goldberger, M. L., Low, F., Marx, E. & Zachariasen, F. (1963). *Phys. Rev.* **133**, B 145, *ibid.* **133**, B 161.

Gell-Mann, M., Goldberger, M. L. & Thirring, W. (1954). *Phys. Rev.* **95**, 1612.

Gell-Mann, M. and Leader, E. (1966). Berkeley Conference Report, see Low (1966) and Leader. Cambridge preprint (1967).

Gell-Mann, M. & Ne'eman, Y. (1964). *The Eightfold Way* (Benjamin, New York).

Glauber, R. (1958). *Lectures in Theoretical Physics* (Ed. W. E. Brittin & L. G. Dunham) (Interscience, New York).

Goldberger, M. L. (1955). *Phys. Rev.* **99**, 979.

Goldberger, M. L. (1960). *Dispersion Relations and Elementary Particles* (Ed. C. De Witt & Omnes, R.) (Wiley, New York).

Goldberger, M. L., Grisaru, M. T., Macdowell, S. W. & Wong, D. Y. (1960). *Phys. Rev.* **120**, 2250.

Goldberger, M. L. & Jones, E. (1966). Princeton preprint (see Low 1966), also *Phys. Rev. Lett.* **17**, 105.

Goldhaber, G. (1966). Berkeley Conference Report.

Good, M. L. & Walker, W. D. (1960). *Phys. Rev.* **120**, 1957.

Gottfried, K. & Jackson, J. D. (1964). *Nuovo Cimento*, **33**, 309, *ibid.* **34**, 735.

Greenberg, O. W. (1964). *Lectures at the Istanbul Summer School* (Ed. F. Gursey) (Gordon and Breach).

Greenberg, O. W. & Low, F. E. (1961). *Phys. Rev.* **124**, 2047.

Gribov, V. N. (1962a). *Soviet Physics JETP*, **14**, 1395.

Gribov, V. N. (1962b). *Soviet Physics JETP*, **15**, 873.

Gribov, V. N. (1963). *Soviet Physics JETP*, **16**, 1080.

Gribov, V. N. (1966). CERN Conference (reported by Van Hove, 1966).

Gribov, V. N. & Pomeranchuk, I. Ya. (1962a). *Phys. Rev. Lett.* **8**, 343.

Gribov, V. N. & Pomeranchuk, I. Ya. (1962b). *Phys. Rev. Lett.* **8**, 412.

Gribov, V. N. & Pomeranchuk, I. Ya. (1962c). *Nucl. Phys.* **33**, 516.

Gribov, V. N. & Pomeranchuk, I. Ya. (1962d). *Phys. Rev. Lett.* **9**, 238.

Gribov, V. N., Pomeranchuk, I. Ya. & Ter Martirosyan, K. A. (1965). *Phys. Rev.* **139**, B 184.

Gunson, J. (1963 preprint published in 1965). *J. Math. Phys.* **6**, 827, 845, 852.

Hagedorn, R. (1965). *Nuovo Cimento*, **35**, 216 and other papers quoted there.

Hagopian, V. W. *et al.* (1966). *Phys. Rev.* **145**, 1128.

Hamilton, J. (1959). *Reports on Progress in Nuclear Physics*, **7**.

Hamilton, J. (1964). *Strong Interactions and High Energy Physics* (Ed. R. G. Moorhouse) (Oliver and Boyd).

Hamilton, J. (1966). *Acta Physica Austriaca Suppl.* **3**, 229.

Hamilton, J. & Woolcock, W. S. (1963). *Rev. Mod. Phys.* **35**, 737.

Hara, Y. (1964). *Phys. Rev.* **136**, B 507.

Harari, H. & Lipkin, H. J. (1964). *Phys. Rev. Lett.* **13**, 208.

Hartle, J. & Jones, E. C. (1965). *Phys. Rev.* **140**, B 90.

Heisenberg, W. (1943). *Z. Phys.* **120**, 513, 673.

Heisenberg, W. (1944). *Z. Phys.* **123**, 93.

Hepp, K. (1963). *Helv. Phys. Acta*, **36**, 355.

Hepp, K. (1964). *Helv. Phys. Acta*, **37**, 639.

Hofstadter, R. (1963). *Nuclear and Nucleon Structure* (Benjamin, New York).

Höhler, G. *et al.* (1966). *Phys. Lett.* **20**, 79, *ibid.* **21**, 223.

Hwa, R. C. & Teplitz, V. L. (1966). *Homology and Feynman Integrals* (Benjamin, New York).

Itzykson, C. & Jacob, M. (1966). Saclay preprint.

Jackson, J. D. (1965). *Rev. Mod. Phys.* **37**, 484.

Jackson, J. D. (1966). Stony Brook Conference Report.

Jackson, J. D., Donahue, J. T., Gottfried, K., Keyser, R. & Svensson, B. E. Y. (1965). *Phys. Rev.* **139**, B 428.

Jacob, M. (1963). See Chew & Jacob (1963).

Jacob, M. & Wick, G. C. (1959). *Annals of Physics*, **7**, 404.

Jaffe, A. (1966). Berkeley Conference (see report by Froissart 1966).

Jin, Y. S. (1960). *Nucl. Phys.* **15**, 102.

Jin, Y. S. (1966). *Phys. Rev.* **143**, 975.

Jin, Y. S. & Macdowell, S. W. (1965). *Phys. Rev.* **138**, B 1279.

Jin, Y. S. & Martin, A. (1964). *Phys. Rev.* **135**, B 1369 and B 1375.

Johnson, K. & Treiman, S. B. (1965). *Phys. Rev. Lett.* **14**, 189.

Jost, R. (1947). *Helv. Phys. Acta*, **20**, 256.

Källen, G. (1964). *Elementary Particle Physics* (Addison Wesley, New York).

Källen, G. & Wightman, A. (1958). *Dan. Vid. Selsk. Mat-fys. Skr.* **1**, no. 6.

Karplus, R., Sommerfield, C. M. & Wichmann, E. H. (1958). *Phys. Rev.* **111**, 1187, see also *Phys. Rev.* **114**, 376.

Khuri, N. N. (1957). *Phys. Rev.* **107**, 1148.

Khuri, N. N. & Kinoshita, T. (1965*a*). *Phys. Rev. Lett.* **14**, 84.

Khuri, N. N. & Kinoshita, T. (1965*b*). *Phys. Rev.* **137**, B 720.

Khuri, N. N. & Kinoshita, T. (1965*c*). *Phys. Rev.* **140**, B 706.

Kinoshita, T. (1964*a*). *Phys. Rev. Lett.* **12**, 256.

Kinoshita, T. (1964*b*). *Lectures in Theoretical Physics* (University of Colorado Press).

Kinoshita, T. (1966). CERN preprint 66/1207/5-TH. 713 (see also Cornell preprints).

Kinoshita, T., Loeffel, J. J. & Martin, A. (1964). *Phys. Rev.* **135**, B 1464.

Kirrilova, L. *et al.* (1965). *Soviet J. of Nucl. Phys.* **1**, 379.

Kokkedee, J. J. J. & Van Hove, L. (1966). *Nuovo Cimento*, **42**, 1827.

Kormanyos, S. W. *et al.* (1966). *Phys. Rev. Lett.* **16**, 709 (see also Murphy 1966).

Landau, L. D. (1959). *Nucl. Phys.* **13**, 181.

Landshoff, P. V. (1965). Oxford Conference Report.

Landshoff, P. V., Polkinghorne, J. C. & Taylor, J. C. (1961). *Nuovo Cimento*, **19**, 939.

Lautrup, B. & Olesen, P. (1965). *Phys. Lett.* **17**, 62.

Leader, E. (1963). *Phys. Lett.* **5**, 75.

Leader, E. (1966). *Rev. Mod. Phys.* **38**, 476.

Leader, E. & Omnes, R. (1966). Strasbourg preprint.

Leader, E. & Slansky, R. (1966). *Phys. Rev.* **148**, B 1491.

Lee, B. E. & Sawyer, R. F. (1962). *Phys. Rev.* **127**, 2266.

Lehmann, H. (1958). *Nuovo Cimento*, **10**, 579.

Lehmann, H. (1964). Varenna Lecture Notes, *Nuovo Cimento Supplement*.

Lehmann, H., Symanzik, K. & Zimmerman, W. (1955). *Nuovo Cimento*, **1**, 1425.

Levin, E. M. & Frankfurt, L. L. (1965). *JETP Lett.* **2**, 65.

Levinson, N. (1949). *Kgl. Danske Vid. Selskab. Mat-fys. Medd.* **25**, no. 9.

Levintov, I. I. & Adelson-Velsky, G. M. (1964). *Phys. Lett.* **13**, 185.

Lindenbaum, S. J. (1965). Oxford Conference Report.

Lindenbaum, S. J. (1967). Coral Gables Conference Report.

Lipkin, H. J. (1965). *Lie Groups for Pedestrians* (Interscience, New York).

Lipkin, H. J. (1966). *Phys. Rev. Lett.* **16**, 1015.

Lipkin, H. J. & Schek, F. (1966). *Phys. Rev. Lett.* **16**, 71.

Logunov, A. A., Nguyen van Hien, Todorov, I. T. & Krustalev, O. A. (1963). *Phys. Lett.* **7**, 69.

Lohrman, H. *et al.* (1964). *Phys. Lett.* **13**, 78.

Lovelace, C. & Masson, D. (1963). *Nuovo Cimento*, **26**, 472.

Low, F. (1966) Berkeley Conference Report.

Łukaszuk, L. & Martin, A. (1967). Cavendish Laboratory, Cambridge (preprint).

MacDowell, S. W. (1960). *Phys. Rev.* **116**, 774.

MacDowell, S. W. & Martin, A. (1964). *Phys. Rev.* **135**, B 960.

Mandelstam, S. (1958). *Phys. Rev.* **112**, 1344 (see also Mandelstam, 1959).

Mandelstam, S. (1959). *Phys. Rev.* **115**, 1741 and 1752.

Mandelstam, S. (1960). *Nuovo Cimento*, **15**, 658.

Mandelstam, S. (1963). *Nuovo Cimento*, **30**, 1127 and 1148.

Martin, A. (1963*a*). *Phys. Rev.* **129**, 1432.

Martin, A. (1963*b*). Lectures at Scottish Universities Summer School (see Moorhouse (1963)).

Martin, A. (1963*c*). *Nuovo Cimento*, **29**, 993.

Martin, A. (1965*a*). *Nuovo Cimento*, **37**, 671.

Martin, A. (1965*b*). CERN preprint 65/1699/5-Th. 630 (N.C. 1966).

Martin, A. (1966*a*). *Nuovo Cimento*, **42**, 930, *ibid.* **44**, 1219.

Martin, A. (1966*b*). CERN report TH. 702.

Mathews, J. (1959). *Phys. Rev.* **113**, 381.

Meiman, N. N. (1962). *JETP* (USSR) **43**, 227 (see also (translation) *Soviet Physics JETP*, **16**, 1609 (1963)).

Meshkov, S., Levinson, C. A. & Lipkin, H. J. (1963). *Phys. Rev. Lett.* **10**, 361.

Meshkov, S., Snow, G. A. & Yodh, G. B. (1964). *Phys. Rev. Lett.* **12**, 87.

Møller, C. (1932). *Ann. Phys.* **14**, 531.

Møller, C. (1945). *K. Danske Vid. Selsk.* **23**, no. 1.

Møller, C. (1946). *K. Danske Vid. Selsk.* **22**, no. 19.

Moorhouse, R. G. (1963). *Strong Interactions and High Energy Physics, Proceedings of Scottish Universities Summer School* (Oliver and Boyd).

Morrison, D. R. O. (1966). Stony Brook Conference Report.

Murphy, P. G. (1966). Berkeley Conference Report.

Nakanishi, N. (1959). *Progr. Theor. Phys.* **21**, 135.

Ne'eman, Y. (1961). *Nucl. Phys.* **26**, 222.

Newton, R. G. (1960). *J. Math. Phys.* **1**, 319.

Newton, R. G. (1964). *The Complex j-Plane; Complex Angular Momentum in Non-Relativistic Quantum Theory* (Benjamin, New York).

Oehme, R. (1964). *Strong Interactions and High Energy Physics* (Ed. R. G. Moorhouse, Oliver and Boyd).

Okun, L. B. & Pomeranchuk, I. Ya. (1956). *Soviet Physics JETP*, **3**, 307.

Oleson, P. (1965). *Phys. Lett.* **14**, 66.

Olive, D. I. (1964). *Phys. Rev.* **135**, B 745.

Omnes, R. (1958). *Nuovo Cimento*, **8**, 316.

Omnes, R. (1961). *Nuovo Cimento*, **21**, 524.

Omnes, R. (1966). *Phys. Rev.* **146**, 1123.

Omnes, R. & Froissart, M. (1963). *Mandelstam Theory and Regge Poles* (Benjamin, New York).

Omnes, R. & Leader, E. (1966). Strasbourg preprint.

Orear, J. (1964). *Phys. Rev. Lett.* **13**, 190.

Phillips, R. J. N. (1966). Erice Summer School Lecture Notes (to be published).

Phillips, R. J. N. & Rarita, W. (1965). *Phys. Rev.* **138**, B 723 and **139**, B 1336.

Polkinghorne, J. C. (1963 a). *Phys. Lett.* **4**, 24.

Polkinghorne, J. C. (1963 b). *J. Math. Phys.* **4**, 503.

Polkinghorne, J. C. (1963 c). *J. Math. Phys.* **4**, 1393.

Polkinghorne, J. C. (1963 d). *J. Math. Phys.* **4**, 1396.

Polkinghorne, J. C. & Screaton, G. R. (1960). *Nuovo Cimento*, **15**, 289, 925.

Pomeranchuk, I. Ya. (1956). *Soviet Physics JETP*, **3**, 306.

Pomeranchuk, I. Ya. (1958). *Soviet Physics JETP*, **7**, 499.

Pomeranchuk, I. Ya. & Feinberg, E. L. (1956). *Suppl. Nuovo Cimento*, **3**, 652.

Predazzi, E. (1966). *Annals of Phys.* **36**, 228, 250.

Preist, T. & Vick, L. (1967). (Editors) *Particle Interactions at High Energy, Proceedings of the Scottish Universities Summer School* 1966 (Oliver and Boyd).

Regge, T. (1959). *Nuovo Cimento*, **14**, 951 and **18**, 947.

Regge, T. (1960). *Nuovo Cimento*, **18**, 947.

Regge, T. (1963). Trieste lectures (see Salam 1963).

Rivers, R. J. (1966). *Phys. Lett.* **22**, 514.

Rose, M. E. (1957). *Elementary theory of Angular Momentum* (Wiley, New York).

Rosenfeld, A. H. (1966). Annual Review of Particles and Resonances issued by the Lawrence Radiation Laboratory, Berkeley. See also (1967) *Rev. Mod. Phys.* **39**, 1.

Salam, A. (1963). (Editor) *Theoretical Physics* (International Atomic Energy Agency).

Salam, A. (1966). Trieste Institute Seminar Proceedings.

Salzman, F. & Salzman, G. (1960). *Phys. Rev. Lett.* **5**, 377.

Schweber, S. S. (1961). *Introduction to Relativistic Quantum Field Theory* (Harper & Row, New York).

Selleri, F. (1964). *Boulder Summer School Lectures* (University of Colorado Press).

Serber, R. (1963). *Phys. Rev. Lett.* **10**, 357.

Shohat, J. A. & Tamarkin, J. D. (1943). *The Problem of Moments* (American Math. Soc., New York).

Singh, V. (1962). *Phys. Rev.* **127**, 632.

Singh, V. (1963). *Phys. Rev.* **129**, 1889.

Soding, P. (1964). *Phys. Lett.* **8**, 285.

Solov'ev, L. D. (1966). *Soviet Physics JETP*, **22**, 205.

Sommer, G. (1966). CERN preprint 65/1591/5—TH. 731.

Sommerfeld, A. (1949). *Partial Differential Equations in Physics* (Academic Press, New York).

Sonderegger, P. *et al.* (1966). *Phys. Lett.* **20**, 75.

Sopkovich, N. J. (1962). *Nuovo Cimento*, **26**, 186.

Squires, E. J. (1962). *Nuovo Cimento*, **25**, 242.

Squires, E. J. (1963). *Complex Angular Momenta and Particle Physics* (Benjamin, New York).

Squires, E. J. (1967). Scottish Universities Summer School Lecture Notes (see Preist & Vick, 1967).

Stack, J. D. (1966). *Phys. Rev. Lett.* **16**, 286.

Stapp, H. P. (1962*a*). *Phys. Rev.* **125**, 2139 and Berkeley preprint UCRL-10289.

Stapp, H. P. (1962*b*). *Rev. Mod. Phys.* **34**, 390.

Stapp, H. P. (1966). Berkeley Lectures on S-Matrix Theory (unpublished).

Stevenson, M. L. *et al.* (1962). *Phys. Rev.* **125**, 687.

Sugawara, M. (1965). *Phys. Rev. Lett.* **14**, 336 (see also Sugawara, M., *Lectures on Dispersion Theory*, Helsinki Research Institute for Theoretical Physics, Finland, 1965).

Sugawara, M. & Kanazawa, A. (1961). *Phys. Rev.* **123**, 1895.

Sugawara, M. & Kanazawa, A. (1962). *Phys. Rev.* **126**, 2251.

Sugawara, M. & Tubis, A. (1963). *Phys. Rev.* **130**, 2127.

Symanzik, K. (1957). *Phys. Rev.* **100**, 743.

Symanzik, K. (1960). *J. Math. Phys.* **1**, 249, Appendix B.

Szego, G. (1948). *Orthogonal Polynomials* (Edwards Brothers, Ann Arbor, Michigan).

Taylor, A. E. *et al.* (1965). *Phys. Lett.* **14**, 54.

Taylor, J. R. (1966). *J. Math. Phys.* **7**, 181.

Ter Martirosyan, K. A. (1966). Berkeley Conference Report (see Van Hove, 1966).

Titchmarsh, E. C. (1939). *Theory of Functions* (Oxford University Press).

Toll, J. (1952). *Princeton thesis*.

Toll, J. (1956). *Phys. Rev.* **104**, 1760.

Treiman, S. B. & Yang, C. N. (1962). *Phys. Rev. Lett.* **8**, 140.

Trueman, T. L. & Wick, G. C. (1964). *Annals of Physics*, **26**, 322.

Udgaonkar, B. M. (1962). *Phys. Rev. Lett.* **8**, 142.

Van Hove, L. (1963). *Nuovo Cimento*, **28**, 798.

Van Hove, L. (1964). *Rev. Mod. Phys.* **36**, 655.

Van Hove, L. (1965). CERN Lecture Notes CERN 65–22.

Van Hove, L. (1966). Berkeley Conference Report (and other papers quoted by Van Hove).

Van Hove, L. (1967). Lectures at Scottish Universities Summer School 1966 (see Preist & Vick, 1967).

Vaughn, M. T., Aaron, R. & Amado, R. D. (1961). *Phys. Rev.* **124**, 1258.

Volkov, D. V. & Gribov, V. N. (1963). *Soviet Physics JETP*, **17**, 720.

Wagner, W. G. (1963). *Phys. Rev. Lett.* **10**, 202.

Wagner, W. G. & Sharp, D. H. (1962). *Phys. Rev.* **128**, 2899.

Wang, L. (1966). *Phys. Rev.* **142**, 1187 and Berkeley preprint, UCRL-17053.

Warburton, A. E. A. (1964). *Nuovo Cimento*, **32**, 122.

Watson, G. N. (1918). *Proc. Roy. Soc.* **95**, 83.

Weissburger, W. I. (1965). *Phys. Rev. Lett.* **14**, 1047.

Weisskopf, V. F. (1966). CERN Report 66–19.

Wetherell, A. M. (1966). Berkeley Conference Report (see also Van Hove, 1966).

Wheeler, J. A. (1937). *Phys. Rev.* **52**, 1107.

Wightman, A. S. (1960). *Dispersion Relations and Elementary Particles* (Editors, C. de Witt and R. Omnes) (Wiley, New York).

Williams, D. N. (1963). Lawrence Radiation Laboratory Report UCRL-1113.

Wilson, R. (1967). Lectures at Scottish Universities Summer School 1966 (see Preist & Vick, 1967).

Wit, R. (1965*a*). *Acta Physica Polonica*, **28**, 295.

Wit, R. (1965*b*). *Phys. Lett.* **15**, 350.

Wolfenstein, L. (1954). *Phys. Rev.* **96**, 1654.

Wu, T. T. & Yang, C. N. (1965). *Phys. Rev.* **137**, B 708.

Yamamoto, K. (1963). *Nuovo Cimento*, **27**, 1277.

Yang, C. N. (1963). *J. Math. Phys.* **4**, 52.

Zachariasen, F. (1964). *Strong Interactions and High Energy Physics* (Ed. R. G. Moorhouse) (Oliver and Boyd, Edinburgh).

Zweig, G. (1964). CERN preprint 8419/TH. 412.

INDEX

Where a topic is discussed on several consecutive pages, reference is made
only to the first of these pages

A